The Practical Approach in Chemistry Series

SERIES EDITORS

L. M. Harwood
Department of Chemistry
University of Reading

C. J. Moody
Department of Chemistry
University of Exeter

The Practical Approach in Chemistry Series

Organocopper reagents
Edited by Richard J. K. Taylor

Macrocycle synthesis
Edited by David Parker

Preparation of alkenes
Edited by Jonathan M. J. Williams

High-pressure techniques in chemistry and physics
Edited by Wilfried B. Holzapfel and Neil S. Isaacs

Transition metals in organic synthesis
Edited by Susan E. Gibson (née Thomas)

High-pressure Techniques in Chemistry and Physics

A Practical Approach

Edited by

WILFRIED B. HOLZAPFEL

Department of Physics,
University of Paderborn, Germany

and

NEIL S. ISAACS

Department of Chemistry,
University of Reading, UK

OXFORD NEW YORK TOKYO
OXFORD UNIVERSITY PRESS
1997

Oxford University Press, Great Clarendon Street, Oxford OX2 6DP

Oxford New York
Athens Auckland Bangkok Bogota Bombay Buenos Aires
Calcutta Cape Town Dar es Salaam Delhi Florence Hong Kong
Istanbul Karachi Kuala Lumpur Madras Madrid Melbourne
Mexico City Nairobi Paris Singapore Taipei Tokyo Toronto

and associated companies in
Berlin Ibadan

Oxford is a trade mark of Oxford University Press

Published in the United States
by Oxford University Press Inc., New York

A catalogue record for this book is available from the British Library

Library of Congress Cataloging-in-Publication Data
Data available

ISBN 0 19 855811 2 (Hbk)

Typeset by Footnote Graphics, Warminster, Wilts
Printed in Great Britain by
Bookcraft (Bath) Ltd
Midsomer Norton, Avon

Contents

Contributors

D. C. BASSETT
Department of Physics, University of Reading, Reading RG6 6AD, UK.

J. M. BESSON
Physique des Milieux Condensés, Université Pierre et Marie Curie, T13-E4, Place Jussieu, F-75252 Paris, France.

M. BUBACK
Institut für Physikalische Chemie, Universität Göttingen, D-34 Göttingen, Germany.

CHRISTIAN CROS
Laboratoire de Chimie du Solide du CNRS, 351 cours de la Libération, 33405 Talence Cedex, France.

GÉRARD DEMAZEAU
Laboratoire de Chimie du Solide du CNRS, 351 cours de la Libération, 33405 Talence Cedex, France.

RUDI VAN ELDIK
Institut für Anorganische Chemie, der Universität Erlangen–Nürnberg, Egerlandstrasse 1, 91058 Erlangen, Germany.

LOTHAR HELM
Institut de Chimie Minérale et Analytique, Université de Lausanne, CH-1015 Lausanne, Switzerland.

C. HINTON
Formerly Institut Für Physikalische Chemie, Universität Göttingen, D-34 Göttingen, Germany.
Current address: Institute of Environmental Science and Research Ltd, P.O. Box 29-181, 27 Crayke Road, Christchurch, New Zealand.

W. B. HOLZAPFEL
Fachbereich Physik, Universität G-H Paderborn, Warberger Strasse 100, 33098 Paderborn, Germany.

NEIL S. ISAACS
Department of Chemistry, University of Reading, Reading RG6 6AD, UK.

H. E. KING JR
Exxon Research and Engineering Company, Corporate Research Science Laboratories, Annandale, NJ 08801, USA.

Contributors

H.-D. LÜDEMANN
Institut für Biophysik, Universität Regensberg, 93040 Regensberg, Germany.

DOUGLAS MAGDE
Department of Chemistry, University of California, La Jolla, CA 92093-0314, US.

PATRICK MASSON
Centre de Recherches du Service de Santé des Armées, Unité de Biochemie, 24 Avenue du Maquis du Grésivaudan, BP 87-38702, La Tronche Cédex, France.

ANDRÉ E. MERBACH
Département de Chimie, Université de Lausanne, Place du Château 3, 1005 Lausanne, Switzerland.

HUGH POWELL
Département de Chimie, Université de Lausanne, Place du Château 3, 1005, Lausanne, Switzerland.

W. E. PRICE
Department of Chemistry, University of Wollongong, Wollongong, NSW, Australia.

ALAN SHARP
President, UK High-Pressure Technology Group, 11 Eymore Close, Selly Oak, Birmingham B29 4LB, UK.

THOMAS W. SWADDLE
Department of Chemistry, University of Calgary, 2500 University Drive N.W., Calgary, Alberta, Canada T2N 1N4.

1

Pressure generation

J. M. BESSON

The instruments and methods which have been used for the generation of high pressures are probably more diverse than those in any other field of instrumentation. Since the high-pressure field has long been considered as a closed domain, devoted to the study of matter under extreme or exotic conditions, it has frequently progressed under the impulse of highly dedicated individuals, or schools of thought, who drove to its ultimate limits one given technology or one given concept and more or less ignored alternative solutions. Thus, high-pressure technology seems to have evolved in the present century in a completely irrational way, in sharp contrast to low-temperature technology, which, since Kamerlingh Onnes at least, has followed a logical pattern of improvement. A few examples will illustrate this. About half a century ago, P. W. Bridgman devised and operated an intricate two-stage system to compress large volumes of gas (fluid) to 10 GPa. At the present time, no laboratory could repeat this feat. By contrast, the diamond-anvil cell (DAC), which is now the dominant device for high-pressure studies, was perfected in its complete form, that is, as a gasketed device, with a ruby-fluorescence scale only in the 1970s, following work by S. Block and G. J. Piermarini. Nevertheless, all the necessary know-how and materials existed a century ago, and it is strange that no one thought of it before.

For these reasons, this chapter deals only with the pressure-generation methods that are in practical use at the time of writing. Instruments which were widely used but which are no longer in general use, such as the Hall tetrahedral press, will not be dealt with. In the same way, no description will be made of devices which *could* bring about technological improvements but which have not yet been fully validated. This means that even logical and well-supported predictions for future pressure-generation methods have been avoided.

Another voluntary omission concerns dynamic (shock wave) methods. These methods have provided precious knowledge of the behaviour of high-density matter: indeed they form the basis of our understanding of equations of state at high pressures and temperatures, and the high-pressure scale is founded upon such measurements. But their use requires too much highly specialized know-how and expertise to be readily instrumented in most academic institutions, and therefore such methods fall outside the scope of this book.

Pressure-generation methods can be divided into three categories corresponding to existing domains of high-pressure methods.

The first category (Section 1) deals with hydraulic techniques for the compression of fluids, whether the fluid is the sample itself or a pressure-transmitting medium for the compression of a solid sample. This concerns, as a rule, pressures up to 1.4 GPa, or at most 1.8 GPa, which are applied to large volumes of fluid (> 1 mm^3). Higher pressures can be applied but, in practice, DAC methods are currently used at higher pressures.

The second category (Section 2) deals with the compression of large-volume solid samples in the range from 2 to 20–30 GPa. This is the domain of high-temperature high-pressure experiments in materials science and geophysical studies, and quite different instruments and methods must be used.

The third category (Section 3) deals with DACs. Although, in principle, these cells are simple, Bridgman opposed-anvil systems which should belong to the second category, in practice they have such original characteristics and widespread applications that it is more convenient to treat them separately.

1. Hydraulic techniques for the compression of fluids

1.1 Introduction

In practice, the compression of fluids in large-volume systems is performed at pressures up to 1.4 GPa, or 1.8 GPa at the very most. The reason is that standard construction steels and alloys for the various components of a high-pressure set-up have maximum yield strengths of 2 GPa. The design and operation of a 1.4 GPa set-up is now comparatively simple, but a 2 GPa instrument requires specialized expertise. In this range of pressure, two separate regions must be distinguished: pressures up to ~600 MPa, and pressures up to 1.4 GPa.

The first region is that of standard high-pressure equipment for which commercial high-pressure components can be purchased off the shelf. This applies to valves, fittings, connectors, and Bourdon gauges for use up to 700 MPa, and flexible tubing up to 500 MPa. Simple plunger pistons and reciprocating manual or pneumatic pumps are available from various suppliers for this pressure range.

Between 0.7 and 1.4 GPa, one has to use more specialized set-ups. Very few high-pressure accessories are commercially rated for routine operation in this range of pressure. For instance although valves do exist for pressures up to a nominal rating of 1.4 GPa, it is best to dispense with them, since their practical lifetime is limited to a few operations only.

Thus, the first factor to consider when selecting a pressure-generation unit is the maximum desired pressure. One should have extremely good reasons for operating at a pressure of, say, 850 MPa rather than being satisfied with 650 MPa. In the latter case, all the components will be readily available and certified, and the pump or compressor will be a simple self-contained unit. In the former case, by contrast, a multistage intensifier unit of the 1.4 GPa class will be necessary, which will be ten times the mass and the cost.

The nature of the fluid to be compressed is the second parameter to consider when selecting a pressure-generating system. Obviously light 'gases' such as helium are more compressible and thus more difficult to compress than 'liquids'. Also, the compression of corrosive liquids, such as water, involves more care than the compression of noncorrosive fluids, such as hydraulic oil or pentane. When compressing large quantities of gas, in industrial processes, multistage compression and interstage cooling must be used. We shall not deal with this aspect here, since we will assume that the mass or volume of the compressed fluid is small enough to cause only transient heating of the set-up. The most convenient liquids for use up to pressures of 0.7 GPa are industrial hydraulic mixtures of the Plexol type or commercial brake fluids, which are inert and retain low viscosities up to 1 GPa (see, for example, the review by Sherman and Stadtmuller,[1] pp. 136–142). For pressures up to 1.4 GPa and above, pentane or a pentane–isopentane 1:1 mixture are practical. These should be preferred to ethanol–methanol mixtures, which are polar and much more active chemically and have an unwanted ability to absorb water from the air. In numerous cases the pressure-transmitting fluid has to be a neutral gas, especially for work at low temperatures. The choice of the gas depends on the temperature range: argon can be used down to 200 K at 600 MPa, but this is not much better than isopentane, which solidifies at 210 K at this pressure. Helium solidifies at $\sim$ 1.4 GPa at liquid nitrogen temperature, and it is the best pressure medium from this viewpoint. Unfortunately it is also the most difficult gas to pressurize because of its high compressibility, and it is the most difficult to contain since its leak rate is the highest of all gases.

A number of factors other than the maximum pressure and the nature of the fluid to be compressed should also be considered before selecting a high-pressure generating system. Here we shall not discuss rotary pumps and compressors of the centrifugal or turbine type, which are used for low-pressure, high-flow processes. Only reciprocating devices are described. They are called *pumps* for pressurizing 'liquids' (that is, comparatively incompressible fluids) or they are called *compressors* for more compressible 'gases', and they have different characteristics which should be weighed by the user such as the following three factors.

1. *Mode of operation.* There is a vast difference between a set-up that will be used continuously (for example, for high-pressure synthesis) and one which will be used intermittently for a small number of cycles. High-pressure

systems have a limited cyclic lifetime; this is generally calculated by the manufacturer and should be known to the user.

2. *Size and weight.* Although this seems an obvious criterion, it is seldom properly taken into account. Evidently, in a laboratory, the smaller a set-up is the better. The size can be minimized by using a pump or compressor which has just sufficient capacity for the experiment. This minimizes, among other factors, the amount of compressed fluid, and thus the stored free energy, which is important for safety reasons. In most cases, it is possible to reduce the experimental compressed volume, to make it fit with the compressor's capacity, by using filler pieces.

3. *Manual or automatic operation.* Pumps and compressors often exist in manual, electrically operated, or pneumatically operated versions. Whenever practical, manual instruments should be preferred for laboratory applications, especially when a high-pressure set-up is used close to its limit. This forces experimenters to closely control the system by requiring their constant attention and vigilance.

1.2 Pumps and compressors for use up to 600 MPa

1.2.1 Single-stroke pumps

One of the simplest devices for compressing fluids is a rotating-crankshaft compressor. In this instrument, a plunger piston is used to compress a volume of a few cubic centimetres of fluid. The thrust on the piston is generated by a hand screw with a capstan wheel. A typical pressure × volume value for these devices is

$$P\,V \approx 2\ \text{GPa cm}^3, \tag{1.1.1}$$

so that 5 cm^3, say, of fluid can be compressed to 400 MPa, which takes about 40 rotations of the crankshaft. Isolation valves allow these devices to compress gases through several compression cycles, although this is a time-consuming process.

1.2.2 Reciprocating pumps and compressors

Reciprocating pumps are units in which fluid is forced in and out of the pressure chamber through one-way check valves. The force applied to the piston can be generated manually (Fig. 1.1) by a handle. The compression

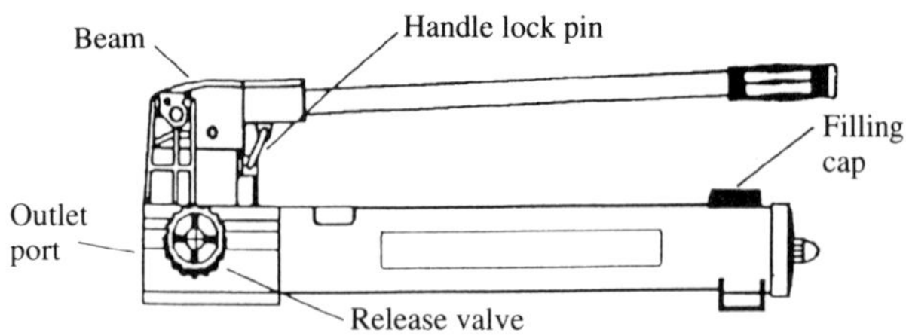

Fig. 1.1 A 280 MPa hand pump (Enerpac).

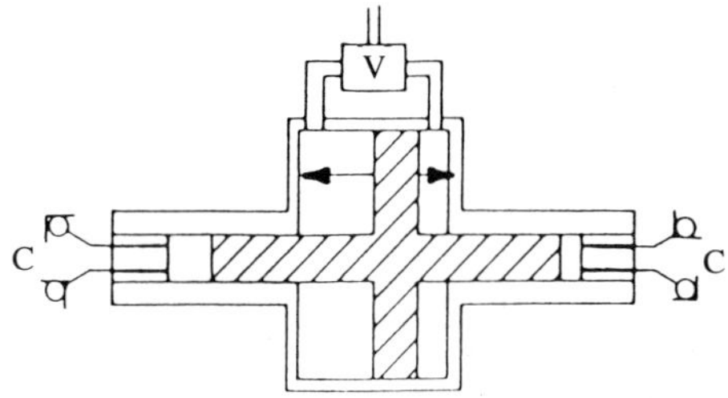

Fig. 1.2 The principle of an air-driven reciprocating pump: (V) a compressed-air distributor valve, and (C) one-way check valves.[2] (Reprinted with permission from *High Pres. Tech.* (1977).)

fluid may be either an industrial hydraulic fluid or any other liquid (water, pentane, etc.) for which adequate construction materials have been used for the moving parts and seals. For the compression of gases, hand operation would be quite tedious, and such compressors are usually driven electrically or pneumatically (Fig. 1.2) by compressed air. Multistage units also exist for the compression of gases up to ~ 500 MPa. Pumps such as that shown in Fig. 1.1 are the cheapest and most reliable pressure-generating systems for the compression of liquids up to 400 MPa. They work best when used with a specialized hydraulic fluid, which ensures that the moving parts and seals have a long lifetime. There are versions which are compatible with other liquids, as noted above, but the lifetime and reliability of the seals is greatly diminished. To compress such liquids, one should prefer standard pumps, and compress the experimental liquid through a separator of the type shown in Fig. 1.3.

1.2.3 Diaphragm compressors

One drawback of the designs described thus far is that the pressurized fluid is in contact with the moving seals. This will cause contamination of the pres-

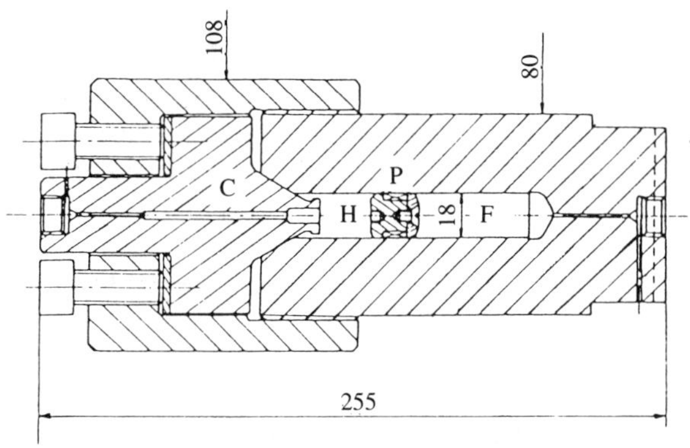

Fig. 1.3 A 700 MPa separator chamber (TOP Industrie): (C) cone fitting, (H) hydraulic fluid, (P) separator piston, and (F) pressurized fluid.

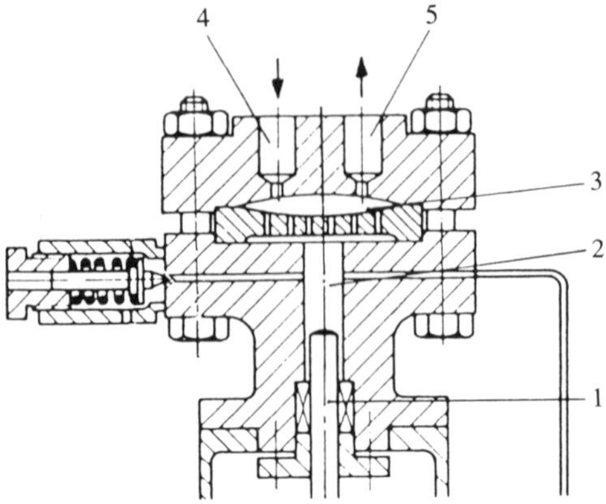

Fig. 1.4 A cross section of a diaphragm compressor. (1) plunger piston, (2) oil cylinder, (3) diaphragm (4) and (5) in and out feedthroughs, respectively, for compressed gas. (The check valves are not shown.) (From reference 2 with modifications.)

surized medium. In addition, the lifetime and reliability of the seals is lower than with hydraulic fluids, especially when compressing gases. These problems are solved by use of Corblin-type diaphragm compressors (Fig. 1.4). In these instruments, a piston (1) compresses a hydraulic fluid (2) which acts on a thin diaphragm (3) made of a stainless-steel or beryllium–copper sheet. This compresses a lens-like cavity containing the gas, which is let in and out by one-way check valves (4 and 5). Thus the gas is never in contact with the hydraulic fluid and it should be entirely free from contamination. Such units are usually pneumatically (but sometimes electrically) driven. Two-stage units allow compression of most gases up to 350 MPa.

1.3 Intensifier systems

1.3.1 Introduction

As noted in Section 1.1, the compression of fluids up to 0.6 GPa can readily be performed with industrial set-ups which are precisely rated and have been tested to maximum pressures 1.5 to 2 times higher than their nominal rating, depending on the nature of the system. For higher pressures, it is possible to purchase, for instance, a 1.8 GPa compressor unit from a manufacturer. But it should be borne in mind that such a unit will *not* have been tested to $1.5 \times 1.8 = 2.7$ GPa, which would exceed the yield strength of its components! There is no real guarantee on apparatus for fluid pressures above 1 GPa. At most, a manufacturer will, for instance, replace a binding ring which shatters under load after a few cycles only. Currently, 1.4 GPa is a kind of standard limit for the rating of such apparatus, since it provides a reasonable safety margin in the design, because of the characteristics of existing steels and alloys; but the use of such apparatus is at the user's risk. Thus some degree of technical expertise is expected from the users of such systems. This involves an elementary knowledge of standard safety precautions, especially with gas compressors, and some mechanical know-how to assemble, disassemble, and service the various parts. In this subsection, we shall deal with the design of an intensifier system for gas-pressure generation around 1.4 GPa and briefly describe the various high-pressure components which are now commonly used.

1.3.2 General layout

A typical intensifier system is shown diagrammatically in Fig. 1.5. As in all such systems, it consists of a low-pressure part which precompresses the gas to 200 ~ 400 MPa, and a high-pressure press and ram to achieve the final compression in an intensifier (Fig. 1.6). The first stage, shown in Fig. 1.5 consists of a separator (S) and a first-stage intensifier (I) with a ratio of piston areas of 1:4. A rotary oil compressor (C) provides a hydraulic oil pressure of 70 MPa, which is an industrial standard. Thus in the present example the first stage delivers a maximum gas pressure of 4×70 MPa = 280 MPa through valve V_3. This stage can be replaced by a two-stage diaphragm compressor, as mentioned previously. The low-pressure fluid which is fed into the intensifier system U through a lateral hole is then further compressed by the high-pressure piston through a hydraulic ram (P). Here, since the ratio of areas is 1:25,

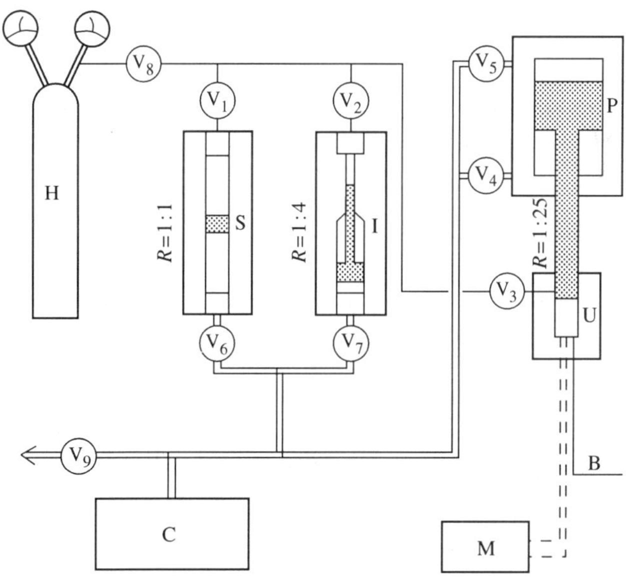

Fig. 1.5 A schematic layout of an intensifier system: (B) to experimental cell, (C) low-pressure compressor (700 MPa), (I) intensifier cylinder, (M) resistance manometer, (P) main oil ram and press, (S) separator cylinder, (U) compression cylinder, and (V) valves. Here, *R* is the multiplication ratio of the cylinders.

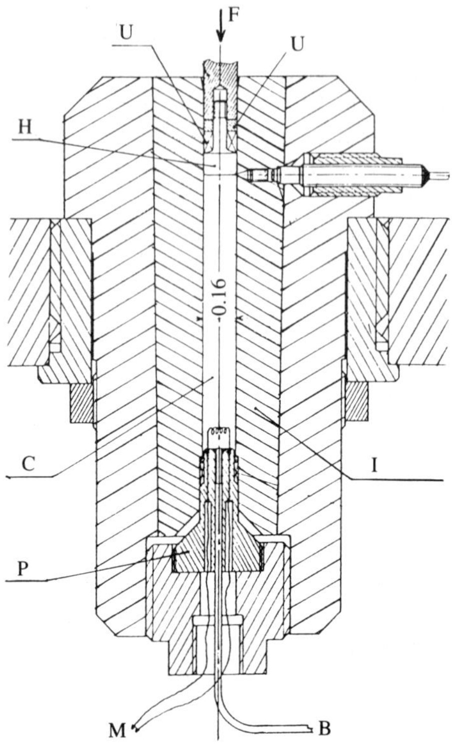

Fig. 1.6 A cross section of a 1.6 GPa cylinder: (B) to experimental cell, (C) compression chamber (diameter, 16 mm), (H) piston head, (I) inner cylinder under radial compression, (M) to resistance manometer, (P) high-pressure outlet plug with electrical leads, (U) unsupported-area, Bridgman-type, sliding-piston arrangement. The force applied to the piston by the oil ram for a pressure of 1.6 GPa is $F = 350$ kN.

the theoretical maximum pressure which could be attained would be 25 × 70 MPa = 1750 MPa, or perhaps in the region of 1600 MPa, to allow for friction losses. In the present set-up, no high pressure valve is used; V_3 in Fig. 1.5 is a standard 400 or 700 MPa system. When the piston has been pushed past the inlet hole the pressure in V_3 should drop to the ambient pressure. With the bore dimensions shown in Fig. 1.6 a thrust of 400 kN is necessary, and thus a hydraulic press (ram plus frame or yoke) of some 100 tonne (1 MN) capacity would be used. If a larger compressed volume is required, this cannot be provided by indefinitely increasing the stroke of the ram. Presses with long displacements tend to be large and heavy. In practice, for the same engineering design and the same performance, the mass—and cost—of a press increases as the square of the displacement, and as the cube of the bore diameter of the compression chamber. In the present case, to increase the compressed volume by a factor of 2 the stroke must be doubled, which increases the mass by

$2^2 = 4$, or the bore area may be doubled, which increases the mass by only $(\sqrt{2})^3 = 2.8$. A ratio of 1:6 between the diameter and the length of the pistons is usually adopted as a practical limit. In the intensifier system shown in Fig. 1.6, which can be used to 1.8 GPa, no high-pressure valve is needed, but if the gasket assembly is pushed once past the lateral hole it will suffer some damage and must then be replaced. Thus, for pressures of around 1 GPa, low-pressure gas may be fed in by tubing on the bottom plug and isolated during the compression by a high-pressure valve, which will have to support the highest pressure in the closed position on the needle only, and not on the gland seal.

1.3.3 Piston design

The piston is by far the most sensitive part in an intensifier system. Two classes of piston designs are currently used; they differ in their sealing arrangement. In the so-called piston seal, a dynamic seal is carried by the piston head along the bore. In a gland seal, the packing is stationary and the piston slides through it. Both types are fully discussed by Sherman and Stadtmuller,[1] and only a brief outline will be given here. It must be mentioned that a common characteristic of both these types of pistons is that they are brittle, since they are built from high-compressive-strength piston steel or tungsten carbide. Thus care should be taken to ensure that, when the maximum force is applied, the piston body is almost entirely inside the cylinder. This will ensure good alignment and so prevent failure. Otherwise, brittle fracture of a piston will give rise to high-velocity shrapnel made of hard material; shielding is, therefore, essential.

The drawback with piston seals is that they require: (i) a well-honed and/or polished bore in the high-pressure cylinder, and (ii) a constant clearance under load. The first condition is quite easy to meet for a new system but is harder to maintain under repeated use: a single grain of abrasive, when carried along with the packing will ruin a cylinder and will require painstaking rehoning procedures to eliminate the scratch as well as requiring the rebuilding of the piston assembly to the new bore diameter. The second condition simply cannot be met exactly. Under pressure, the diameter of the bore increases and so does the clearance between the piston head and the bore; thus this causes the gasket material to flow past the piston body, which increases friction. Piston seals, in contrast, have two advantages over gland seals for operation in the high-pressure range: (i) they allow the operation shown in Fig. 1.6, that is, operation with a lateral feedthrough for the low-pressure gas, which does not require the use of a high-pressure valve; and (ii) at the highest pressures, the length of the pressurized portion of the chamber is a minimum. Under these conditions, the ultimate strength of a thick cylinder is much higher than in an infinite-cylinder approximation, that is, when the cylinder is pressurized along its total length. Thus, the piston-seal arrangement should be preferred for work for pressures between 1.2 and 1.8 GPa, or

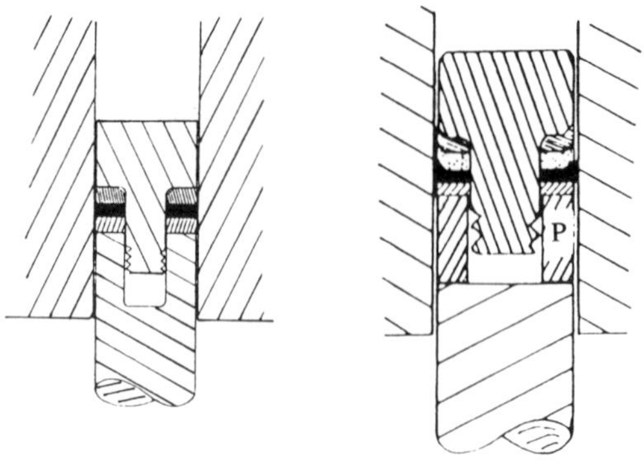

(a) (b)

Fig. 1.7 (a) A Bridgman mushroom-type piston seal with a hollow piston stem, and (b) the same, but with the push piece (P) separate from the stem. (From reference 1.)

more. The type of seal shown in Fig. 1.6 is a self-sealing Bridgman seal, which is also called a mushroom seal because of the shape of the piston head. It is remarkably leak-free since it works on the unsupported-area principle of P. W. Bridgman's design (Fig. 1.7(a)). In this configuration, the pressure in the gaskets is

$$P' = P\frac{D^2}{D^2 - d^2} \tag{1.1.2}$$

where P is the inner pressure, D is the diameter of the piston, d is the diameter of the stem. For $D = 2d$, $P' = 1.33P$.

Thus, for a pressure of 1.4 GPa, the overpressure in the gaskets is 470 MPa, which is enough to make the softer parts of the gasket packing flow and plug any leak. The drawback is that the bore of the cylinder and the piston-head stem undergo a compressive stress of $1.4 \times 1.33 \approx 1.9$ GPa, or more than 2 GPa if friction is included. This may cause a pinch-off of the stem and permanent deformation of the bore.

Another weak part of the configuration shown in Fig. 1.7(a) is the hollowed-out shape of the piston body, which may bring about failure at the highest loads. A much better stress distribution is obtained with the design shown in Fig. 1.7(b) where the annular push piece is free from the piston body. The main drawback of this second configuration is that the push-piece may get stuck in the bore, and thus the piston head and push-piece assembly cannot be extracted without disassembling the cell. Unsupported-area seals have been used in a variety of configurations. Such configurations are remarkably leak-free, but there is a large overpressure on some parts of the set-up, and this causes comparatively high friction values (10–20 per cent of the inner pressure). Wedge-type seals are different in this respect. Their basic

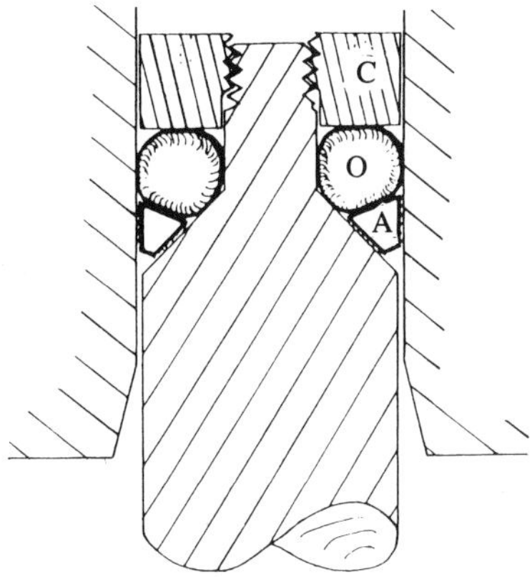

Fig. 1.8 A wedge-type piston seal: (A) antiextrusion ring (bronze), (C) retaining precompression cap, and (I) O-ring. (From reference 1.)

design is shown in Fig. 1.8. Due to the geometry of the metallic wedge ring, this type of seal is also self-sealing. It does not cause the same type of pinch-off as a Bridgman seal and has been used well above 2 GPa—in fact, up to 5 GPa—and thus it should be preferred for ultrahigh-pressure work.

An alternative to the moving-piston seal is the piston gland seal, which is fixed to the entrance of the cylinder. The piston is made to slide through the gland seal, so that only its cylindrical surface needs to be milled and polished (Fig. 1.9). The inner bore of the cylinder does not have to be precisely machined. Furthermore, the piston is a cylindrical rod, made of hard piston

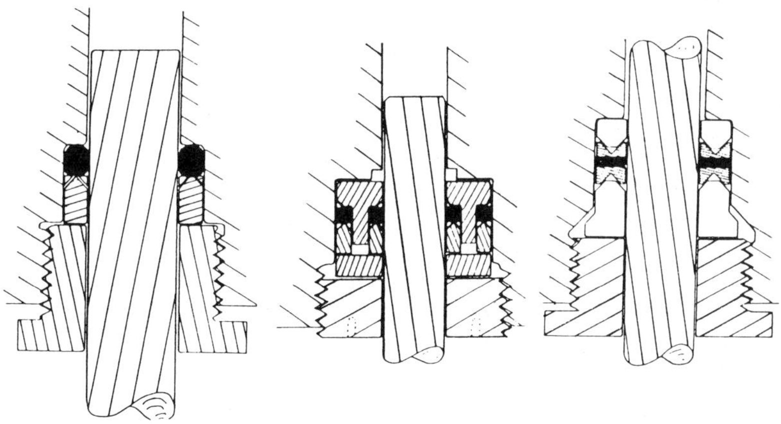

Fig. 1.9 (a) A piston-gland seal (an O-ring with an antiextrusion washer), (b) a Bridgman-type gland seal, and (c) a wedge-type gland seal. (From reference 1.)

steel or tungsten carbide, which is scratch-resistant and is easy to machine and to replace. Thus, from the viewpoint of maintenance, gland seals are preferable to piston seals. Their main drawback, as stated above, is that during their operation the whole length of the high-pressure cylinder is subjected to the maximum pressure. Gland seals may be used with O-rings up to 1 GPa or with a mushroom-type or wedge-type seals well above this value; both of these arrangements are illustrated in Fig. 1.9. Piston gland seals of the type shown in Fig. 1.9(a) have been used in reciprocating pumps up to 1 GPa, which bears testimony to their reliability for cyclic duty. An exhaustive review of piston and gland seals may be found in reference 1, and the references therein.

1.4 Experimental considerations

Caution! For safety procedures see Appendix 2. Compression of fluid media up to ~600 GPa can be readily achieved with commercially available apparatus with high reliability. Above this pressure, the performance and lifetime of existing set-ups rapidly decrease with the maximum pressure. Compression of large volumes of fluid between 1.2 and 1.8 GPa should be performed only by experienced staff, even when using a commercial apparatus.

2. Compression of solid media

Three types of apparatus are used for the compression of large-volume (>1 mm^3) samples above 2 GPa: Piston–cylinder devices are used up to 5 GPa, and opposed-anvil systems and multiple-anvil systems are used up to 20–30 GPa.

2.1 Piston–cylinder devices

Piston–cylinder devices, analogous to those used for fluids, can be used at high pressures for the compression of solids. In these systems, both the piston and the cylinder must be externally supported, if pressures of the order of 5 GPa must be reached.

The principle used in supporting a high-pressure piston is shown in Fig 1.10;

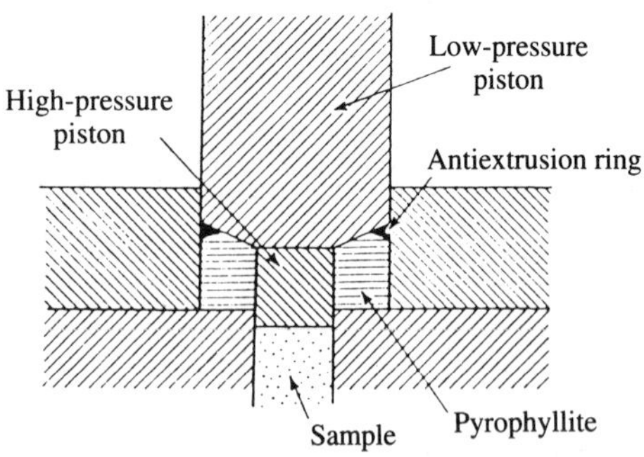

Fig. 1.10 A schematic of a supported-piston arrangement. (From reference 4.)

such arrangements have been used to generate pressures up to 8 GPa. It is essential also to support the cylinder, since it is subjected to tensile stresses and, in general, it will be the first part to fail. For pressures above 2 or 3 GPa, frettage procedures are used with compound cylinders, the inner liner being made of tungsten carbide or ultrahard steel, and the outer cylinders of softer material with a high tensile strength. These are shrink-fitted with positive interference at the interfaces. Thus, the hoop stress is compressive inside the cylinder, becoming more and more tensile at the outermost cylinders. Another way of supporting a cylinder is by winding a steel wire under tension around it. This allows a much lighter construction as well as safety of operation. Both methods are reviewed by Sherman and Stadtmuller,[1] and the principles of the calculation of the stress distribution are also discussed. To increase the yield strength of compound cylinders, different methods have been used to increase the external support gradually, as the internal pressure is increased. This can be achieved[2] by the use of a conical inner liner, which is allowed to slide into a matching outer jacket (Fig. 1.11). As the pressure is increased, the inner vessel itself is forced into the outer jacket by a larger-capacity ram to provide increased external support. With adequate programming of the thrust of both rams, pressures of 3 GPa, or more, have been achieved.[3]

Another way of providing external support to a cylinder is by exerting a longitudinal (axial) thrust. This decreases the inner diameter of the cylinder, and balances the effect of the inner pressure. This method is widely used for large-volume piston–cylinder devices used for high-pressure synthesis above 3 GPa. Such a system,[4] shown in Fig. 1.12, will generate pressures up to 5 GPa. Cylinders with variable support are very efficient devices, but some expertise

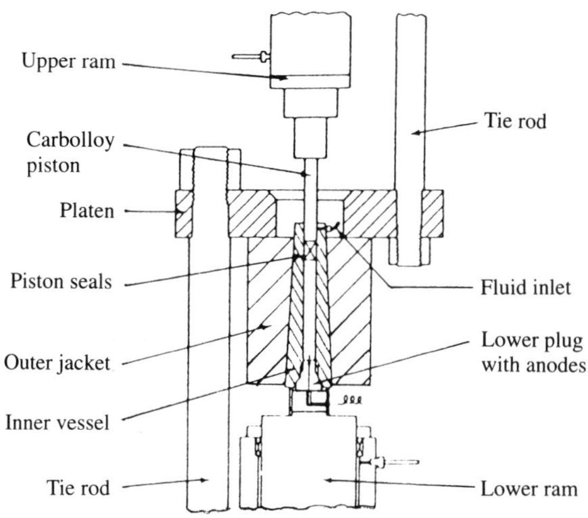

Fig. 1.11 A tapered-support apparatus, after a P.W. Bridgman design (Harwood Engineering). (Reference 2.)

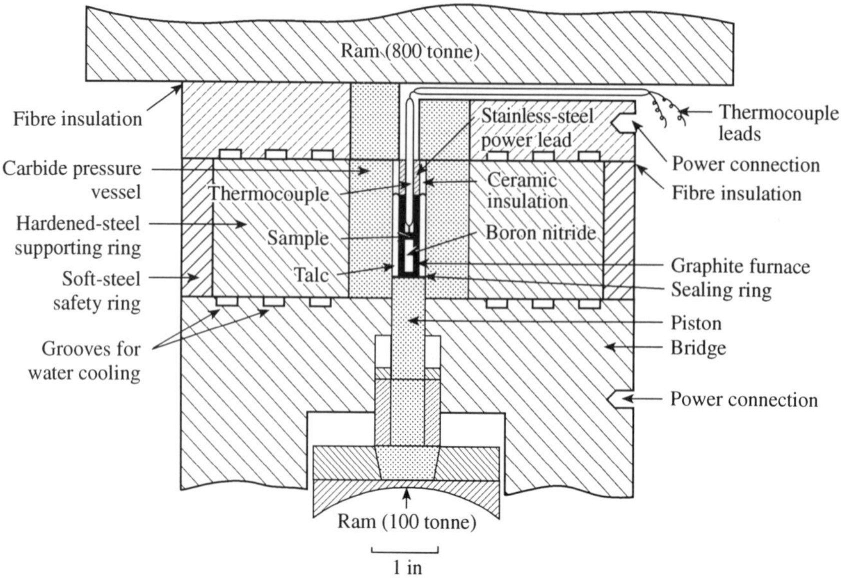

Fig. 1.12 A double-ram apparatus with an axially supported tungsten-carbide vessel. (From reference 4.)

is required in their use. When properly maintained and operated they are quite reliable. An instrument described by Bridgman[3] has been used by the present author thirty years after its construction, and it was still running quite smoothly. However, failures of moving parts (for example, the piston) can be catastrophically expensive, since massive, and thus expensive, parts can be ruined.

2.2 Opposed-anvil systems

Piston–cylinder devices rely on the compression of the sample itself to generate the pressure. In opposed-anvil systems, in contrast, the decrease in volume, and therefore the maximum pressure, is limited by the flow of deformable gaskets in various geometries. Thus, the starting geometry in such devices usually is critical; it depends on the maximum desired pressure and on the mechanical properties of the sample and the gasket materials.

2.2.1 Bridgman anvils

The principle of a Bridgman opposed-anvil apparatus is shown in Fig. 1.13. The sample is compressed between supported tungsten carbide (WC) anvils. It is contained by gaskets which may be pyrophyllite or metallic materials, depending on the measurement to be performed. This design makes use of the *massive-support mechanism*, in that the maximum pressure which may be

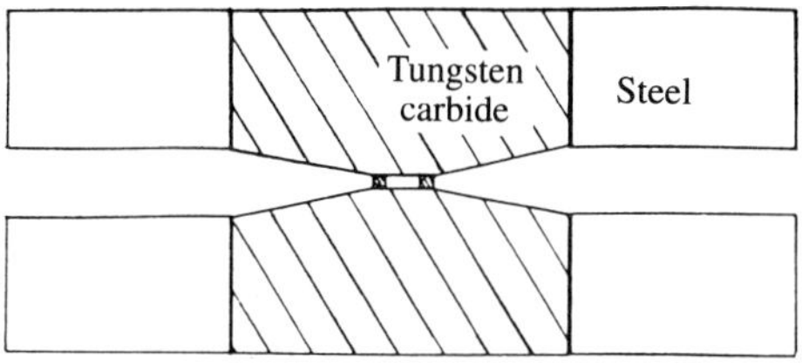

Fig. 1.13 The principle of a Bridgman opposed-anvil apparatus. (From reference 1.)

generated at the sample location vastly exceeds the compressive strength of the anvil material[5] at ambient conditions. Typically, for a cone angle $\alpha = 5°$, pressures of four times the compressive strength of tungsten carbide can be sustained by suitably supported anvils. Thus, sintered tungsten carbide with a compressive strength of 5 GPa can be used for anvils which will generate pressures of ~ 20 GPa in the Bridgman geometry. This strengthening factor[5] decreases as α increases and is only equal to 2 for $\alpha \approx 20°$. Clearly, this design implies the use of rather thin samples, the thickness of which is controlled only by the nature of the gaskets and by the friction between the experimental assembly and the anvil faces. To improve on this, a different set-up has been used by H. Drickamer to limit the extrusion of the gasket.

2.2.2 Drickamer anvils

In Drickamer supported-anvil devices (shown in Fig. 1.14), the gasket assembly is contained within a cylinder,[6] usually of a compound type. They are

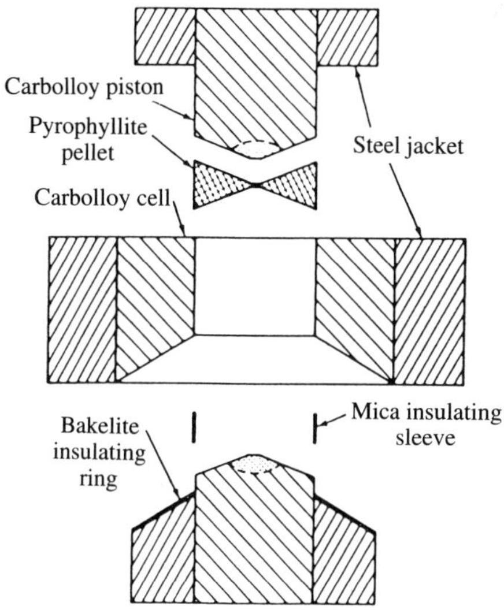

Fig. 1.14 The principle of Drickamer supported-anvil devices[6]. (Reprinted with permission from *Rev. Sci. Instr.* **32**, 308 (1961).)

thus intermediate between piston–cylinder devices and Bridgman opposed-anvil devices. Radial support is provided by the cylinder and massive support by the shape of the piston. Drickamer cells with tungsten-carbide pistons have been used at pressures up to 25 GPa. By inserting a sintered diamond tip into the piston, pressures of 35 GPa have been reported. In the configuration shown in Fig. 1.14, the device can only be used for electrical measurements. For other uses, the outer cylinder must be bored to accommodate a transparent material (for example, NaCl) to allow optical measurements,[7] at ambient or low[8] temperatures up to about 10 GPa. Three-window cells have been built and successfully used for Raman measurements up to 5 GPa at low temperatures, but DACs have now superseded Drickamer cells for such measurements.

2.2.3 Girdles and belts

It was mentioned in Section 2.1 (see Fig. 1.10) that, in order to apply pressures higher than 5 GPa to a cylindrical piston, some kind of radial support should be applied to it. Rather than applying such support in discrete stages along the piston's length, it is desirable to apply this support continuously, in a progressive fashion; this can be done by using tapered conical pistons. This continuous multistaging procedure is well described and discussed in reference 1 and the references therein. The resulting geometry, shown in Fig. 1.15(a), is usually called a girdle device. The tungsten-carbide piston and girdle both receive radial support by frettage into steel rings. Support for the girdle may be controlled by a separate press.[9]

A conical shape is by no means the optimum profile for the piston. The profile shown in Fig. 1.15(b) has been found to provide better support for the

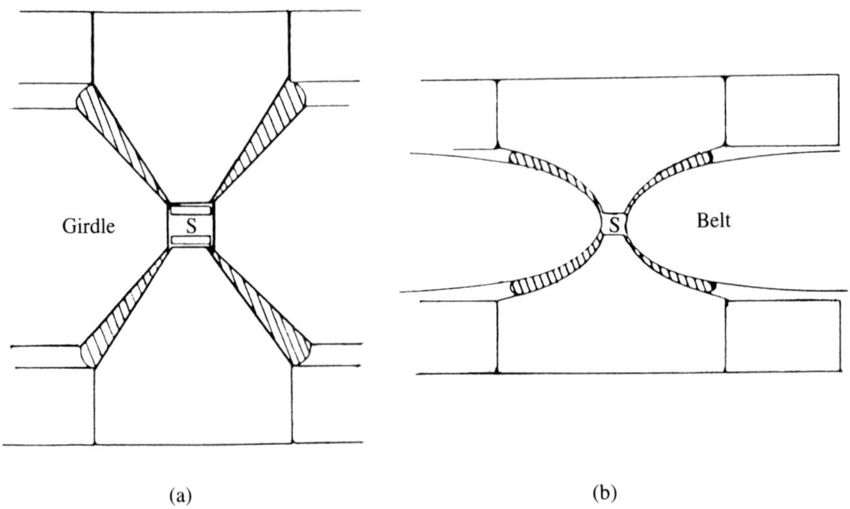

Fig. 1.15 (a) A girdle device with conical pistons and (b) a belt apparatus: (S) sample space. The shaded areas are gaskets. (From reference 1.)

cylinder in the low-pressure region, while retaining sufficient stroke length in the high-pressure region. This geometry is referred to as a *belt apparatus*. This is by far the most widely used system for materials synthesis, especially in industry. In these devices, staged binding rings are used to provide the optimum radial support (Fig. 1.16). The pistons themselves are insulated from the belt by the gasket assembly (usually pyrophyllite) and they are used as electrical current leads for the internal heater. Thermocouples or electrical probes can be run through the gasket. Belts are usually used up to 10 GPa, although pressures of up to 20 GPa have been reached on occasion. There is, in theory, no limit to the volume that can be compressed in this type of apparatus. Indeed a litre-class belt has been built with a bore of 120 mm. This

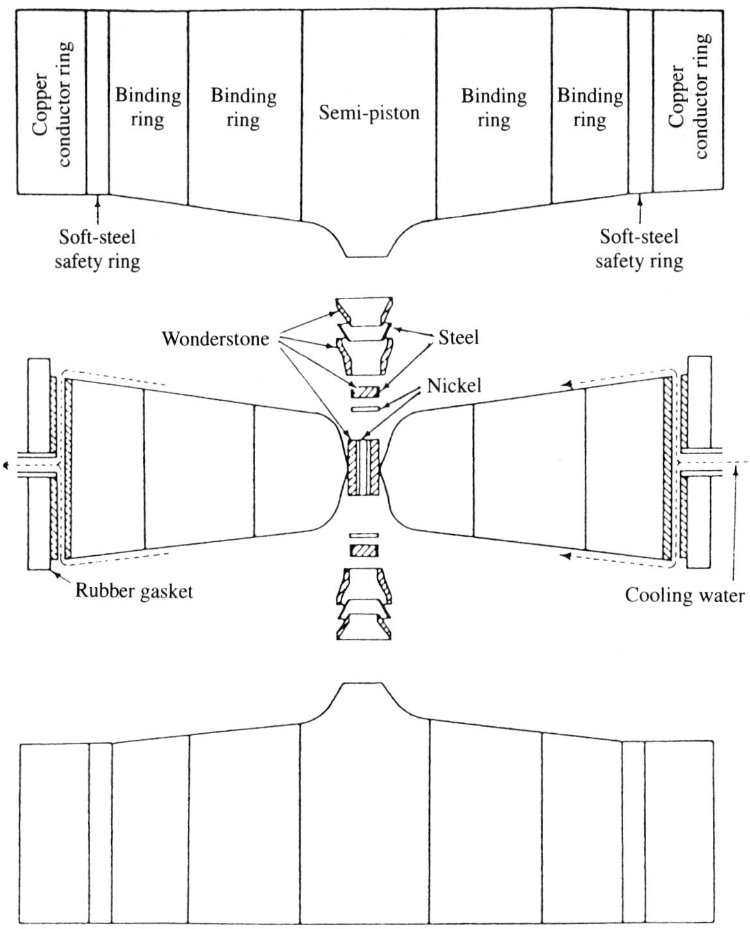

Fig. 1.16 A schematic cross section of a belt for high-pressure high-temperature synthesis. (Reprinted with permission from reference 4. Copyright Academic Press, 1963.)

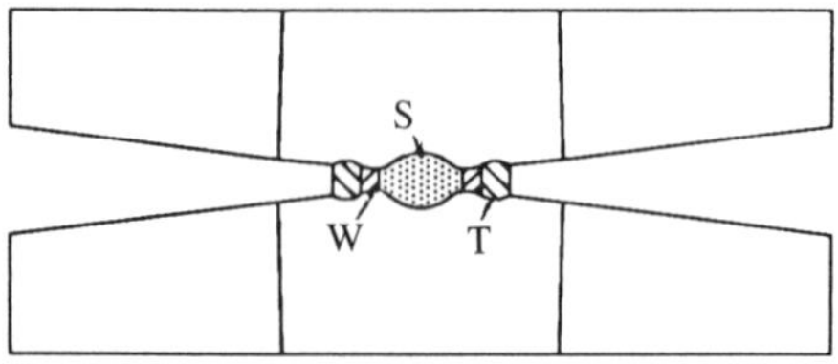

Fig. 1.17 Toroidal anvils used by Besson *et al.*[12] The sample space (S) has a volume of ~100 mm^3, a diameter of 6 mm: (W) a flat Cu–Be washer, (T) a toroidal hardened Cu–Be gasket.

was planned[10] for use up to 10 GPa, under a thrust of 300 MN. Unfortunately it could never be operated at this level of performance because of a lack of tungsten carbide pieces of suitable size. The sintering process for tungsten carbide involves precompression of the powder before heating. In large dies this precompression simply cannot be homogeneous due to the unavoidable size effect on the transmission of pressure, and therefore large (300 mm) tungsten carbide pieces with adequate mechanical properties simply cannot be manufactured at the present time.

2.2.4 Toroidal anvils

Drickamer anvils and girdle or belt devices are derived from Bridgman anvils by adding lateral support through some kind of cylindrical arrangement. In toroidal-anvil systems this lateral support is provided by a suitable gasket assembly[11] (Fig. 1.17). In addition, the sample space is hollowed out to provide a large experimental volume. This type of system has been successfully used in the former Soviet Union since 1970 for compressing large-volume samples to 6 GPa. It is still in use there for the industrial synthesis of hard materials. Nevertheless, it was rarely used in the West until 1991, because it was generally 'accepted' that it did not work. Uncharacteristically, this system is not even mentioned by Sherman and Stadtmuller in their otherwise excellent review.[1] In the geometry shown in Fig. 1.17 the sample space is a spheroid, some 100 mm^3 in volume. The flat washers may be made of pyrophyllite or beryllium copper. The toroidal gaskets are made of a harder metal such as heat-treated Cu–Be. With supported tungsten carbide anvils, pressures of 10 GPa can be achieved.[12] Tungsten carbide has not been used above 14 GPa in this geometry. To reach higher pressures, sintered diamond anvils can be used. These are machined to accommodate two staged toroids and flat washers. With this arrangement, pressures of 25 GPa could be generated[13] on a 30 mm^3 H_2O sample.

In the toroidal-anvil system, the pressure in the sample and gasket assembly is close to a parabolic function of the radial distance, r, to the centre.

$$P \approx P_{\max}\left(1 - \frac{r^2}{R^2}\right), \tag{1.2.1}$$

where P_{max} is the pressure at the centre and R is the outer radius of the gasket assembly after compression.

Thus the total thrust required for compression is

$$F = \int_0^R P_{max}\left(1 - \frac{r^2}{R^2}\right) 2\pi r \, dr,$$

$$F = \tfrac{1}{2}\pi R^2 P_{max}. \qquad (1.2.2)$$

Thus, for a gasket assembly with a typical diameter of 16 mm, a force of 2 MN is required to generate a pressure of 20 GPa in the sample. This can be done by use of small hydraulic rams such as those described in Chapter 3, Section 4.

For high-temperature operation such as high-pressure synthesis, insulating gaskets must be used so that the pistons can act as high-intensity current leads for the internal heater. The maximum pressures are then limited to 7–8 GPa. If pressures of only 5–6 GPa are sufficient, the set-up can be simplified to that shown in Fig. 1.18. The same geometry has been used for anvils made of sapphire for use in neutron-scattering experiments. In this case, the maximum pressure is severely limited by size effects on the sapphire material, whose mechanical quality decreases very rapidly with increasing size.[14]

2.3 Multianvil systems

2.3.1 Introduction

The piston–cylinder and opposed-anvil devices discussed in the previous subsections are most often used for the compression of solids. Nevertheless they can also be loaded with samples which are liquids or gases at ambient conditions, by cryogenic condensation methods. This is not the case with multianvil systems which require a massive environment for the experimental space and which are used exclusively for the compression of solid samples.

The principle of most multiple-anvil systems is to compress a sample volume, in the shape of a regular polyhedron, by a variable number of identically shaped pistons which advance towards each other. Thus, tetrahedral, cubic, and octahedral compression have successfully been achieved. In this type of compression, the thrust is strictly along the axis of the pistons. Other geometries which introduce nonaxial (shear) components, such as hexahedral geometries, have proven to be remarkably unsuccessful and have been abandoned. Sliding-

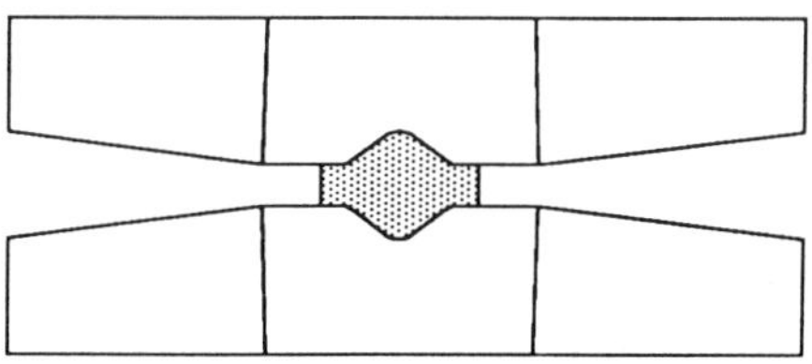

Fig. 1.18 A simplification of Fig. 1.17. (I. Makarenko, private communication.)

anvil cells, in which the experimental space is compressed not by advancing pistons, but by sliding components past each other, are capable of very high pressures[1] but they have not been used in practice and so they will not be discussed here. Tetrahedral presses were devised forty years ago; they were originally used in the industrial synthesis of hard materials but they are no longer in general use. A comprehensive review of such apparatus has been given by Lees.[5] In the same way, direct octahedral-compression systems and hydraulic split-sphere apparatus will not be considered here. The principle of, and literature on, such systems can be found in the review by Sherman and Stadtmuller.[1]

Two multiple-anvil devices are currently in practical use. They are cubic-anvil devices and the two-stage cubic–octahedral systems.

2.3.2 Cubic-anvil devices

In cubic devices, six pistons are made to advance simultaneously towards the barycentre[15] of the experimental volume. Different solutions have been used to solve this problem (Fig. 1.19). The original design with six separate rams (Fig. 1.19(a)) is no longer used except in large-capacity systems for industrial diamond synthesis. In Fig. 1.19(b) the compression is along the [1 1 1] direction of the cube, with three pistons being pushed by the upper and lower frames of a single hydraulic ram. In Fig. 1.19(c), the thrust axis is along the [1 0 0] directions of the cube. Two pistons are moved forward directly by the upper and lower platens, while the four lateral anvils are pushed in through 45° guide blocks. In Fig. 1.19(c), the sliding surfaces are replaced by 'beer bottle' knee links. The experimental space is filled with the sample, which is enclosed in an adequate pressure transmitter, which also acts as the gasket material and as an insulator for high-temperature operation. Pyrophyllite is classically used for this. For X-ray work, pyrophyllite is replaced by boron–epoxy mixtures, which are transparent to photons with energies greater than 2 keV. The anvil arrangement used for the geometry[16] of Fig. 1.19(b) is shown in Fig. 1.20. Used with tungsten-carbide pistons, this type of device is quite reliable up to some 10 GPa. Due to the geometry of the cube, access for the incident and scattered X-ray beams is provided between the anvils, and this device is now used for X-ray studies at synchrotron sources. Higher pressures can be achieved by use of sintered diamond tips for the tungsten-carbide anvils. Using this type of anvil, pressures up to 15–20 GPa can be generated. Other cubic compression devices exist, such as the so-called split-sphere system, which is now mostly used as the first stage of the two-stage compression devices described in the next subsection. In cubic devices, the compression efficiency is about 50 per cent. In the arrangement shown in Fig. 1.19(c), where the single ram thrust is three times the force on each anvil pair, the thrust which is required to compress anvils with 5×5 mm^2 faces to 10 GPa would thus be 1.5 MN. Wedge-type geometries such as that shown in Fig. 1.19(b) would require only $\sim$ 1 MN for the same performance.

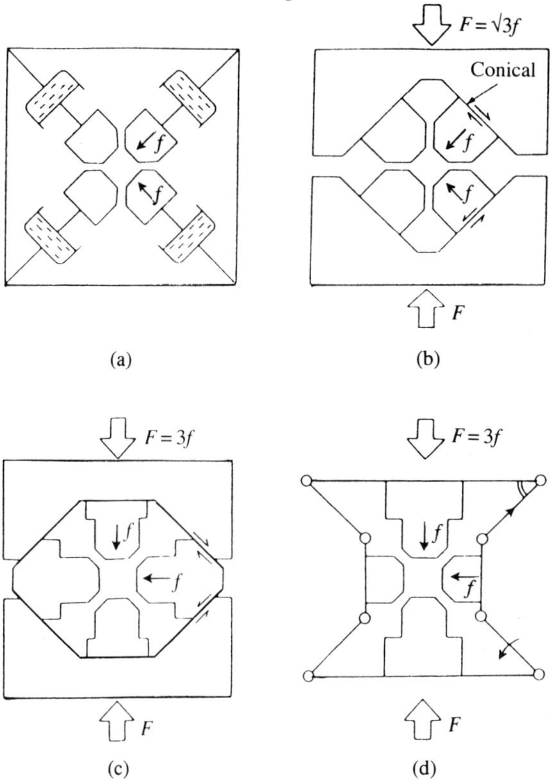

Fig. 1.19 Various types of cubic-anvil systems[15]. (Reprinted with permission from Kluwer Academic Publishers.)

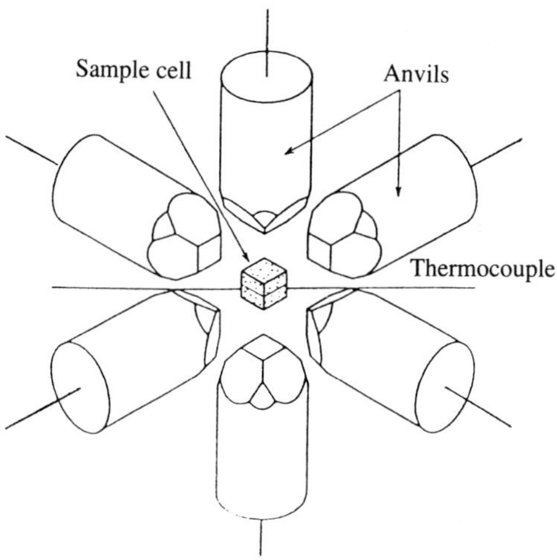

Fig. 1.20 The six pistons in a cubic-anvil system[16]. (Reprinted with permission of the publisher.)

2.3.3 Two-stage cubic–octahedral devices

The cubic space which is compressed in the multianvil devices described above may be used as the first stage of compression for a second anvil arrangement[17] which is sketched in Fig. 1.21. Eight tungsten carbide cubes are provided with a [1 1 1] truncation defining an octahedral space which is compressed by the thrust of the primary cubic device. A number of cubic-compression systems have been used in this two-stage device, such as the split-sphere system,[17,18] which can be simplified to a cylindrical arrangement which is easier to manufacture.[19] A cylindrical system can also be used within the cubic-anvil devices described in the previous subsection. These systems have been successfully used up to 30 GPa. However, at pressures above 20 GPa, permanent plastic deformation of the tungsten carbide cubes is to be expected,[17] especially at high temperatures. Another drawback of this type of set-up is its very poor efficiency. To compress an 8 mm^3 volume to 25 GPa, a force of 5 MN is required, so that rams with a 10 or 20 MN capacity are necessary in practice.[17]

2.4 Experimental considerations

Caution! For safety procedures see Appendix 2. Compression of large-volume solid samples in opposed-anvil or multianvil systems can be performed well above 10 GPa, at temperatures in the range of 1500–2000 K.

The experimental possibilities of such set-ups are critically dependent on the type of apparatus, and on its size. A given apparatus must be selected for a given type of experiment in a precise domain of pressure and temperature. In any case, the measurement of pressure in such set-ups is a problem in itself, especially at high temperatures. The precision on the pressure is two orders of magnitude poorer than in fluid systems.

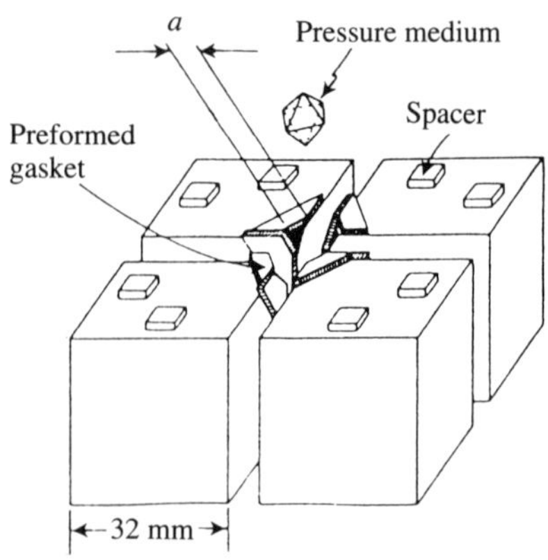

Fig. 1.21 The second stage of a cubic–octahedral system. Eight tungsten carbide cubic anvils with a truncation edge length of *a* compress an octahedral sample (reference 17).

3. Diamond-anvil cells (DACs)

The devices described in the previous subsections allow compression of large sample volumes ($\geqslant 1\ mm^3$). Diamond-anvil cells, in contrast, are used to pressurize samples which are several orders of magnitude smaller. Nevertheless, they are at present the most widely used devices because of their simplicity, safety, and performance. Diamond anvils appeared in the 1950s, but they became practical instruments only with the advent of gasketed DACs and the ruby-fluorescence scale[20] in the 1970s. Since then, DACs have revolutionized high-pressure experimentation and they are now by far the most widely used instruments above 1 GPa. Practically all physicochemical methods have been used in DACs except for measurements in the ultraviolet and soft X-ray regions (since diamond is opaque between 5 and 5000 eV), and in the very far infrared (where the gasket hole diffracts long-wavelength radiation). Comprehensive historical reviews of DAC methods[21] and techniques[22] exist, and a practical guide to this technology can be found in reference 23.

3.1 The working elements of DACs

The working principle of a DAC is shown in Fig. 1.22. Two brilliant-cut stones are bevelled at the tip (culet) with their large faces (table) resting on flat hard seats. Between the culets, a metal gasket encloses a sample which may be embedded in a suitable pressure medium. The pressure which is generated by a force acting on the anvils is measured by a pressure calibrant which may be the fluorescence of a ruby chip.

3.1.1 Diamond anvils

Anvils are usually cut with eight or sixteen facets; the latter geometry gives a

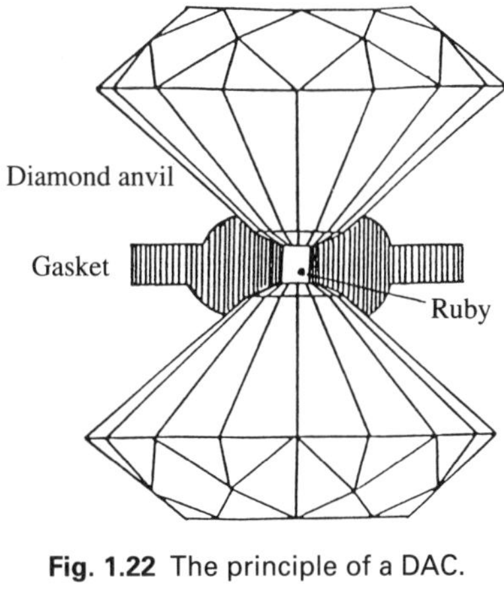

Fig. 1.22 The principle of a DAC.

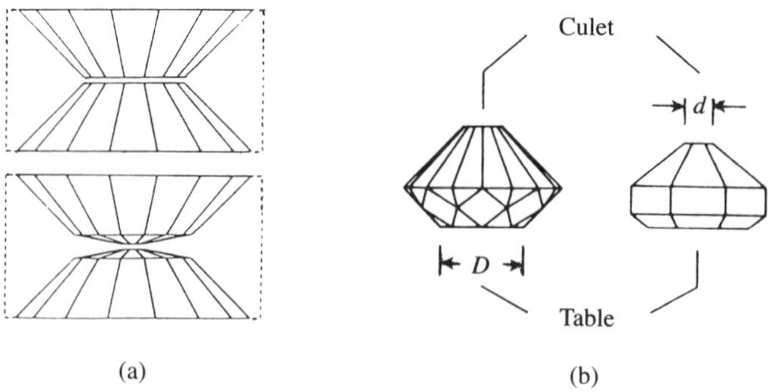

Fig. 1.23 (a) Diamond tips with flat (top) and bevelled (bottom) culets. (b) The usual shapes of diamond anvils: (*left*) brilliant cut, and (*right*) Drukker–Dubbledee standard[23]. (Reprinted with permission from *J. Phys. E.*, **22**, 913, (1989).)

closer approximation to a circle for the culet. The culet itself may be a single flat, or it may have a 5–10° bevel for the highest pressures (Fig. 1.23(a)). The brilliant cut shown in Fig. 1.22 is now often simplified to the so-called Drukker–Dubbledee design shown on the right of Fig. 1.23(b). Typical sizes for the culet would be 700 μm up to 20 GPa and 400 μm up to 50 GPa. A useful rule of thumb for the maximum safe pressure to expect from a culet with a diameter d is

$$P_{\mathrm{max}} = 10/d \text{ GPa mm}^{-1}. \tag{1.3.1}$$

This is valid for stones with a mass of 60 mg (0.3 carats) and a table ~3 mm in diameter—the most commonly used size. Larger sizes are expensive and prone to failure, as discussed in Chapter 3, Section 5.4. For studies at very high pressures, where bevelled culets are used, eqn (1.3.1) does not hold, and the exact shape depends on the pressure region to be explored.[24] In any case, it is worth noting that in the ultrahigh-pressure range, diamond failure must be expected. Experiments above 100 GPa are often one way only, and above 150 GPa they are almost always one way. The optical quality is a prime factor in selecting diamonds. Type I stones (with nitrogen impurities) have strong absorption at wavenumbers between 1100 and 1300 cm^{-1} (Fig. 1.24). Type II stones (nitrogen-free) which are much more expensive, only have the intrinsic two-phonon absorption around 2000 cm^{-1}. The luminescence of the anvil material is a most important problem[25] for Raman and Brillouin measurements, since it varies from sample to sample, and it depends on the exciting laser wavelength. This has to be specified when placing an order. Most vendors now provide *low-luminescence anvils*. The ultimate strength of an anvil, when properly mounted, will depend on its being free from cracks or faults. This is the reason why large stones are more fragile than small ones: the fault density increases with the size of the crystal. Diamond dealers will provide

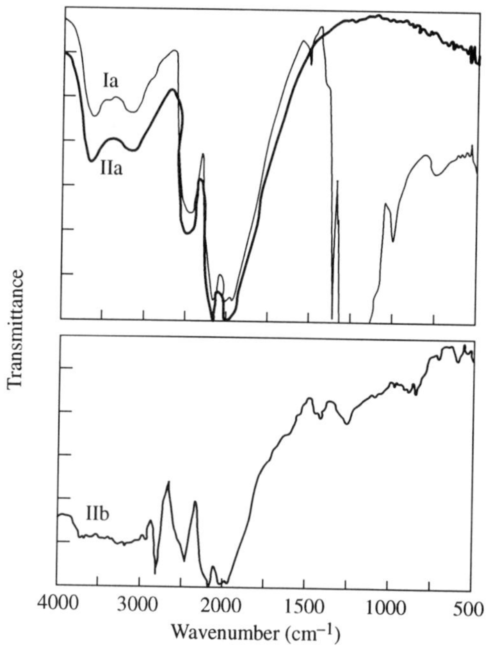

Fig. 1.24 The optical transmittance of diamond in the near infrared[25]. (Reprinted with permission from *Appl. Spertvos.*, **37**, 284, (1983).)

stones with a low optical birefringence, which is an indication of low internal strains and is said to ensure better mechanical resistance.

3.1.2 Seating of the anvils

The anvil seats transmit a force to the anvil tables of the order of 10 kN. Thus the seats are subjected to a normal stress of some 2 GPa, for a 3 mm diameter table. This value may well be exceeded for ultrahigh-pressure work. Thus seats are most often made of tungsten carbide with an optical finish. In some X-ray measurements, beryllium seats have to be used, for their transparency. The mechanical performance is then drastically decreased, and the beryllium must be machined in specialized workshops. Single-crystal sapphire has been used, to provide an increased optical aperture.[26]

The anvil seats are provided with a cylindrical or conical hole for optical access, on which the anvils must be centred. The anvils may be held in place with epoxy glue, which is simple, but is unsuitable for variable-temperature work when the anvils must be set with a soft copper ring which is held in place by screws onto the seat.[23] Since the parallelism and centring of the two anvils is a critical parameter, the shape of the seats usually provides tilt and *x*–*y* alignment (Fig. 1.25). Hemispherical shapes are currently the most widely used, although, for moderate pressures, quite

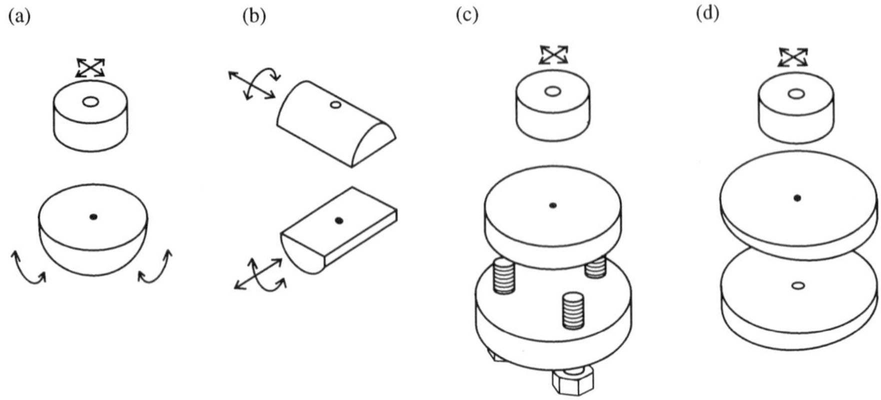

Fig. 1.25 Tilt and *x–y* alignment of seats: (a) hemispherical design, (b) semicylindrical rockers, (c) stand-off screws, and (d) rotating wedges. (From reference 23 and the references therein. (Reprinted with permission from *J. Phys. E.*, **22**, 913, (1989).)

simple devices such as Merrill–Basset cells[27] do not have any tilt adjustment. Alignment procedures are performed using a microscope or a monocular. The parallelism of the diamond culets is adjusted by observing the interference fringes between them, and by tilting them to reach the *flat-fringe pattern*.

3.1.3 Gaskets

The gaskets serve two purposes: they are pierced with a hole which provides the high-pressure chamber for the experimental volume (Fig. 1.26) and also give lateral support to the conical faces of the anvils (Fig. 1.27, B). In the latter function, they perform in a way like the belts and girdles discussed in the previous section. Gaskets are usually made of stainless steel or beryllium copper. Tungsten or molybdenum sheet has also been used on occasion. Rhenium is suitable for work in the ultrahigh-pressure region, or at high temperatures. Normally, the metal sheet, 100–250 μm in initial thickness, is preindented to a typical value of 30–50 μm between the anvils. It is then pierced with a microdrill or a spark erosion machine. It is necessary to observe the drilling process with a suitable optical microscope to centre the hole with respect to the indentation polygon. The quality of the centring (a 10 μm precision is preferable) is important to ensure the stability of the experimental volume under load. When the gasket is squeezed between the anvils, it deforms plastically and its thickness decreases (Fig. 1.27). This brings about a reduction of the volume of the cylindrical hole, which results in a pressure build-up in the sample. Whether the sample diameter decreases, remains constant, or increases (unstable situation) in this process depends, among other factors, on the relative radii of the sample hole and culet, on the initial thickness and shear

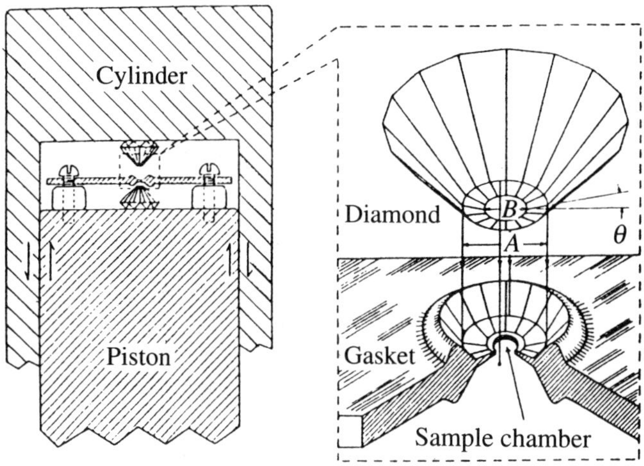

Fig. 1.26 (*Left*) Diamond anvils and (*right*) the gasket configuration[28]. (Reprinted with permission from reference 28. Copyright Plenum Publishing Corporation, 1989.)

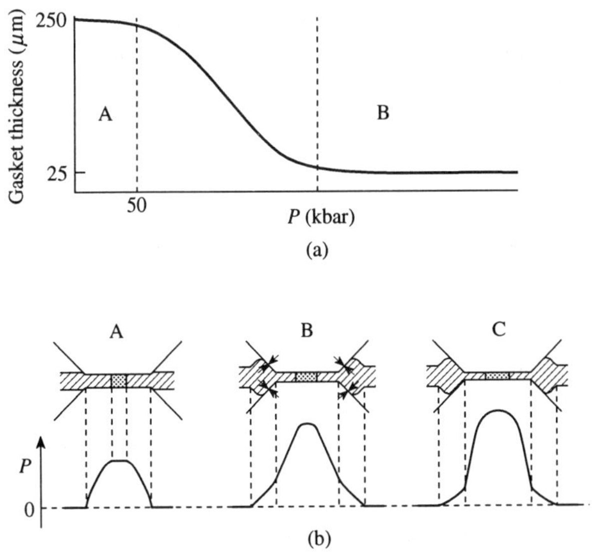

Fig. 1.27 (a) The relation of the gasket thickness to pressure, and (b) (hatched) a cross section of the gasket and (stippled) sample (stippled) for three pressure distributions: (A) $>$ 5 GPa, (B) 15–50 GPa and, (C) $>$ 50 GPa[28]. (Reprinted with permission from reference 28. Copyright Plenum Publishing Corporation, 1989.)

strength of the gasket, on the friction coefficient between the gasket and the anvil faces, and most importantly on the compressibility of the sample. No convincing analysis of this problem, which has seldom been addressed theoretically[29], has been made at the present time.

The following four factors of the gasket geometry should be kept in mind.

1. The gasket hole will decrease in diameter under load for a highly compressible sample. This is evident in the extreme case of an empty hole: the gasket material will flow in, and decrease the inner diameter. This is frequently used after indentation of the gasket and drilling of the hole to decrease the initial sample diameter further and to increase the work hardening of the gasket material.
2. Conversely, for an incompressible sample, the diameter will increase under load: when the thickness decreases under load the diameter must increase.
3. It is always advantageous to use the smallest possible sample size, relative to the culet's diameter. A very small sample will ensure negligible perturbation to the full gasket, and this will be stable at all loads.
4. In practice, up to ~ 50 GPa, gaskets are usually drilled to one-third of the culet diameter, for thicknesses of ~ 40 μm after indentation.

3.1.4 Pressure media

If the sample is a liquid or a gas at ambient conditions, it can be loaded directly, or condensed into the sample hole directly. If it is a single crystal, a fluid-pressure-transmitting medium will be used, and the procedure is the same. A real problem exists with powder samples, which are widely used for X-ray diffraction. These can be loaded alone and compressed as sintered powders, or a pressure-transmitting fluid (an ethanol–methanol mixture, a silicone oil or a rare gas) may be used to improve the hydrostatic conditions. The result may be disastrous, especially when using very compressible media, such as helium or neon. Fluid helium has a volume decrease of 1:0.2 between ambient pressure and 10 GPa, so that a powder sample with a compactness of 50 per cent at ambient pressure will have a compactness of 83 per cent at 10 GPa due to the compression of the interstitial helium. Highly nonhydrostatic conditions (shear components) will be generated at the grain contacts, which may be worse than if the powder had been compacted alone under pressure. Thus, if hydrostatic conditions are required in a powder sample, the only solution is to use a sample-to-pressure-medium volume ratio which remains smaller than ~ 0.3 under pressure, which means drastically reducing the initial volume of the sample. However, pressure homogeneity in a solid powder will be greatly enhanced by wetting it with some adequate liquid (preferably of a comparable compressibility) which will solidify as a glass. Such a liquid will *not* act as a hydrostatic-pressure transmitting medium, but it will fill in the voids between the powder grains and ensure that there is a constant stress between them (the Reuss condition), so avoiding local shear at the grain contacts.

The most widely used compressing medium is a 4:1 methanol–ethanol mixture which remains tolerably viscous up to ~ 10 GPa.[30] Above this, it becomes a *very* hard glass which will easily sustain pressure gradients of 3 GPa over 100 μm at an average pressure of 50 GPa (300 GPa cm^{-1}). The liquid-to-glass transition time in this mixture is sufficiently slow (minutes) that it may be overcome by fast (a few seconds) pressurization, and this has been used to create homogeneous—not hydrostatic—pressure conditions[31] in DACs up to 35 GPa. This technique does not allow variation of the pressure once the maximum pressure has been reached and is, therefore, of limited interest. Above ~15 GPa, a commercial silicone oil is preferable to an ethanol–methanol mixture since it will have a lower shear strength under pressure. Rare gases are the best possible quasihydrostatic pressure media, especially at low temperatures. Helium and neon can be loaded under moderate (150 MPa) pressures, at ambient temperature, by enclosing the DAC in an adequate loading cell,[32] but the most frequently used rare-gas medium is argon,[33] which can be loaded at liquid-nitrogen temperature. Gases which can be condensed under moderately low temperatures and at a pressure of a few megapascals are loaded by enclosing the anvils[34] with a compressible dam as shown in Fig. 1.28.

3.1.5 Pressure measurement

Since diamond is transparent to visible light, the easiest method of measuring the pressure is with the ruby-fluorescence (Al_2O_3:Cr^{3+}) scale.[35] The fluorescence associated with the R_1 and R_2 transitions of the Cr^{3+} ion around 14400 cm^{-1} at ambient pressure shifts by -7.57 cm^{-1} GPa^{-1} under pressure, and pressure variations of ~100 GPa can be measured with a modest spectrometer. Excitation-argon-laser powers of ~10 mW are sufficient to obtain measurable signals from ruby chips with a volume of ~1000 μm^3. A chromium content in the ruby of anywhere between 500 and 5000 p.p.m.

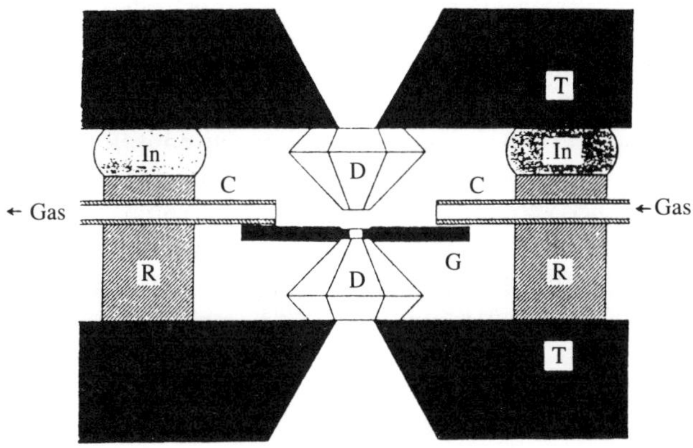

Fig. 1.28 A cross section of a DAC for loading condensable gases. The chamber is sealed with indium: (D) diamonds, (G) gasket, (R) metal ring, (In) indium ring, (C) capillary tubing, (T) tungsten carbide seats[34]. (Reprinted with permission from the publisher.)

(parts per million) will have no influence on the pressure coefficient. At low temperatures, the precision of the measurements is increased because of the narrowing of the emission lines with decreasing temperature, and the temperature itself can be estimated from the ratio of the intensities of the R_1 and R_2 lines.[36] At high temperatures, the ruby lines broaden excessively, and other fluorescent materials which have smaller temperature coefficients and less temperature broadening have been proposed for high-pressure metrology, such as samarium YAG crystals[37] or Sm^{2+}-doped SrB_4O_7.[38]

The ruby-fluorescence scale is only a secondary scale for pressure measurements. It has been calibrated against absolute measurements with dead-weight piston gauges up to 1.4 GPa, with the Decker equation of state[39] for NaCl up to 29 GPa, and with isothermal equations of state from shock-wave measurements at around 100 GPa. From this, interpolated calibration laws for the wavelength shift with pressure have been derived,[40] and are generally used. Nevertheless, above 50 GPa, there are large differences between calibrations under 'hydrostatic' or 'nonhydrostatic' conditions, so that when quoting pressure values for ultrahigh-pressure measurements it is advisable to indicate both the shift of the luminescence lines and their broadening under pressure. For energy-dispersive X-ray measurements at synchrotron sources in DACs, the ruby fluorescence may be excited by a white X-ray beam, so that laser is necessary. An alternative method is to use a volumetric calibrant (NaCl, Au, etc.) together with the sample, and to derive the pressure from the known equation of state of the calibrant. Further details are given in Chapter 2.

In any case, whatever method is used to determine the pressure in the DAC, under pressures of several tens of gigapascals, the main problem does not lie in obtaining a precise pressure reading from the calibrant, but in relating the pressure in the calibrant to the pressure in the sample, and this deserves a further brief comment. Above 11.5 GPa, at 300 K, the last material to solidify is helium. Above this pressure quasihydrostatic conditions are obtained in a solid sample and calibrant mixture. Indeed, solid helium has been shown to be a homogeneous pressure transmitter[33] up to some 50 GPa, but it is not in general use. Even the use of argon does not entirely solve the problem of pressure homogeneity. At 25 GPa, pressure differences of 2 GPa can be observed across the 200 μm diameter of an argon-filled DAC. A second problem arises when using a mixed powder of sample and calibrant, with or without an interstitial pressure medium: the problem of stress or strain continuity at the grain boundaries. With stress continuity (Reuss conditions), the pressure is the same in the calibrant and the sample. With strain continuity (Voigt conditions), the pressure is not the same when the sample and calibrant have different elastic constants, which is generally the case. Moreover, for a given system, the grain-boundary conditions have recently been shown[41] to evolve under pressure in a way which is not understood at the present time. Thus in a powder-sample experiment, there is no assurance that the pressure read by the calibrant, whatever the method, represents the pressure which is experienced by the powder sample itself.

3.2 Squeezers

In the previous subsection it was noted that a force of 10 kN or, at most 20–30 kN for very-high-pressure measurements has to be applied to the anvil seats to generate the pressure in the sample–gasket assembly. This can be done in a variety of ways and several hundreds of devices or variations on a given device have been described since 1965 in the literature. Only a few characteristic examples will be given here of the numerous drive mechanisms which have been devised.

3.2.1 Characteristics of drive mechanisms

When selecting a drive mechanism for a DAC, the following three points should be considered:

1. *Stability of the thrust axis.* The first cause of failure of diamond anvils is shear which is normal to the load. Moreover, nonparallelism of the diamond culets causes instability of the gasket through asymmetrical deformation of the sample hole. Nevertheless, if a piston carrying the anvils has a length of 30 mm and a tolerance for the bore of 0.02 mm in the diameter, its maximum tilt will be 0.7 milliradians, which amounts to a nonparallelism of less than half a micrometre on a 600 μm culet. Thus, very long guiding bores such as that shown in Fig. 1.29 are not really necessary.

2. *Continuous variation of pressure.* Drive mechanisms may, or may not, allow continuous variation of pressure. Screw systems such as that shown in Fig. 1.29 normally have to be removed from the optical system, from the X-ray system, or, even more so, from the cryostat to apply the thrust needed to vary (increase) the pressure. Thus realignment procedures must be performed. In contrast, some mechanical-drive or pneumatic-drive systems allow the load to be varied without moving the cell from its experimental stand. This is a most important factor to take into account, especially for work with large instruments, where the beam time is limited.

3. *Opening of the cell.* The larger is the angular access to the sample, the better. This is true for optical, infrared, and X-ray measurements, for different reasons. Obviously, large angular access can be achieved only at the expense of the mechanical strength of the anvils and seats, and of the quality of guiding of the piston, and thus of the maximum attainable pressure.

3.2.2 Mechanical-drive systems

One of the simplest and most successful DACs has been the squeezer[20] of G. J. Piermarini and S. Block. A much improved version[24] has been used to generate pressures well above 100 GPa (Fig. 1.29). The shape of the fulcrum

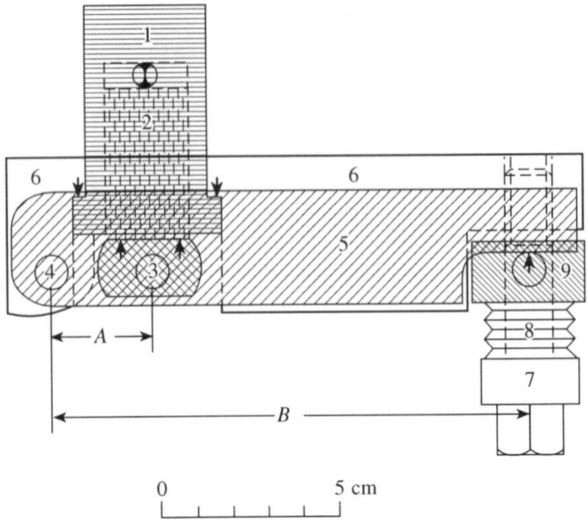

Fig. 1.29 A cross section of a megabar pressure cell: (1) cylinder and diamond anvils (cf. Fig. 26, *left*), (2) piston, (3) thrust block, (4) fulcrum, (5) lever arms, (6) body of the lever assembly, (7) torque nut, (8) Belleville springs, (9) spring holder. The ratio *B*/*A* results in a 5:1 force multiplication on the piston[24]. (Reprinted with permission from *Rev. Sci. Instr.*, **50**, 1002, (1979).)

and thrust block in this cell have been designed to minimize nonaxial components of the thrust on the piston. A completely different design[42] is shown in Fig. 1.30. This cell is designed to provide a 2 × 45° access for incident and diffracted X-rays through beryllium seats, and the load multiplication of the thrust on the thrust block is through a toggle arrangement. The force is transmitted through screws so that it may be varied without moving the sample much.

Much simpler cells have been used for moderate-pressure generation (~ 10 GPa) with clamped systems which allow the mass of the device to be

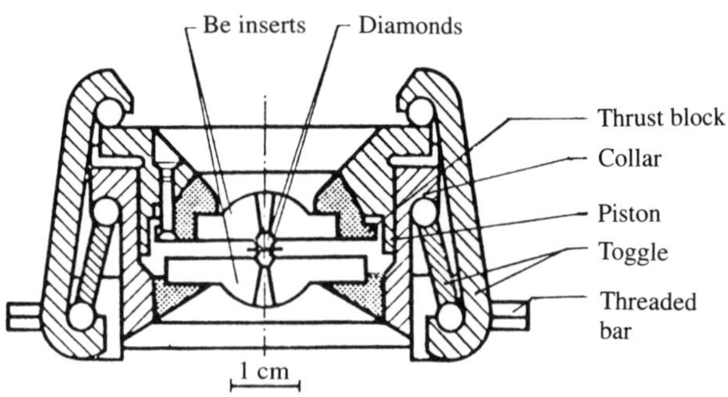

Fig. 1.30 An X-ray-diffraction DAC with beryllium seats[42]. (Reprinted with permission from *Rev. Sci. Instr.*, **48**, 517, (1977).)

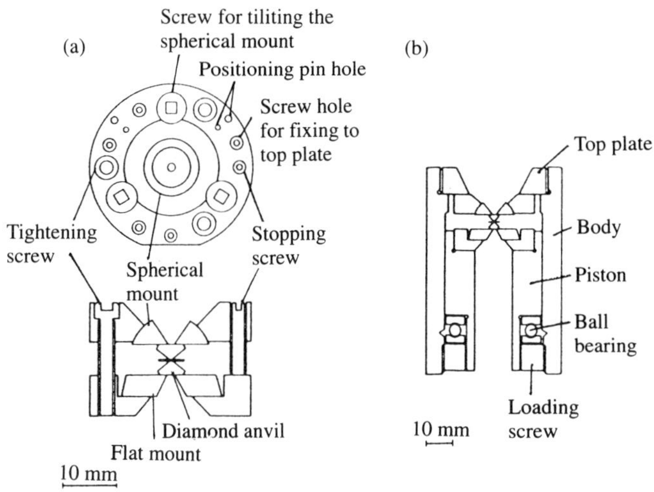

Fig. 1.31 (a) A compact DAC for low-temperature measurements, and (b) the loading device[43]. (Reprinted with permission from *Rev. Sci. Instr.*, **57**, 2993, (1986).)

reduced. This is quite important for low-temperature work, or when the device is to be placed on a goniometer head for X-ray diffraction. The DAC shown in Fig. 1.31 is a clamp cell where the necessary force is applied through a loading screw of an auxiliary device (Fig. 1.31(b)).[43] The cell itself is less than 100 g in mass, and thus it can easily be cooled down to low temperatures (15 K) once the pressurizing load has been fixed by the tightening screws.

An even simpler system used for X-ray measurements is the Merrill–Bassett cell shown in Fig. 1.32. No alignment is allowed for the anvils, and the thrust is provided by successively screwing down the three retaining bolts. This very light and cheap device has been used, on occasion, above 10 GPa. An extreme in miniaturization[44] has been reached with the cell shown in Fig. 1.33, which is a clamp cell with diamond alignment by wedge seats, made of nonmagnetic material and weighing less than 10 g.

3.2.3 Pneumatic-drive systems

The force to be applied to the anvils can be generated by some kind of hydraulic or pneumatic ram. This ram may be a sliding piston with O-ring sealing[45] or it may make use of bellows. The former can normally only be used at ambient temperature, whereas the latter can also be used at low temperatures by using fluid helium as a hydraulic medium. A number of such devices have been described and successfully used[46] at low temperatures. The problem with multiple bellows is that they drastically reduce the optical opening, whereas there is a need in a number of cases for the largest possible optical (or X-ray) access on both sides of the cell. For this reason membrane-

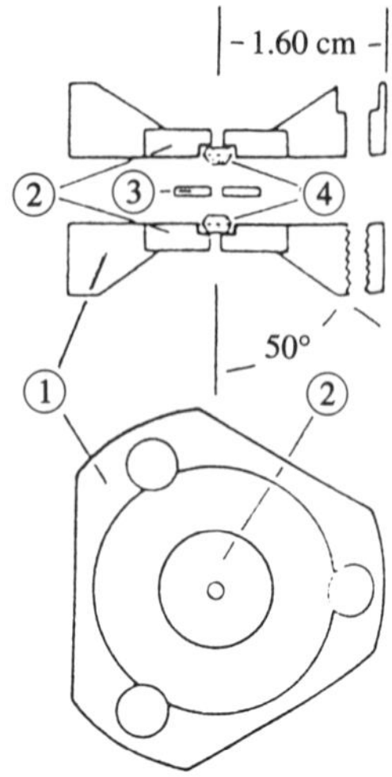

Fig. 1.32 A Merrill–Basset cell: (1) stainless-steel platens, (2) beryllium anvil seats, (3) gaskets, and (4) diamond anvils[27]. (Reprinted with permission from *Rev. Sci. Instr.*, **45**, 290, (1974).)

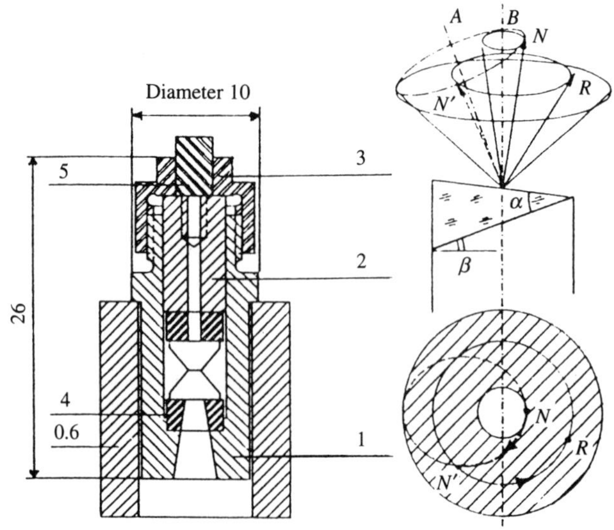

Fig. 1.33 A diamond-anvil microcell: (1) body, (2) piston, (3) screw, (4) anvil support, (5) pusher piston, and (6) support for loading. (Reprinted with permission from reference 44. Copyright Gordon and Breach Science Publishers, 1991.)

type systems are now preferred; they use a single (pressurized) metallic diaphragm to generate the thrust.

The first such system to be used practically[47] had a 0.2 mm thick Cu–Be, or stainless-steel, membrane with an area of 10 cm^2 providing a thrust of

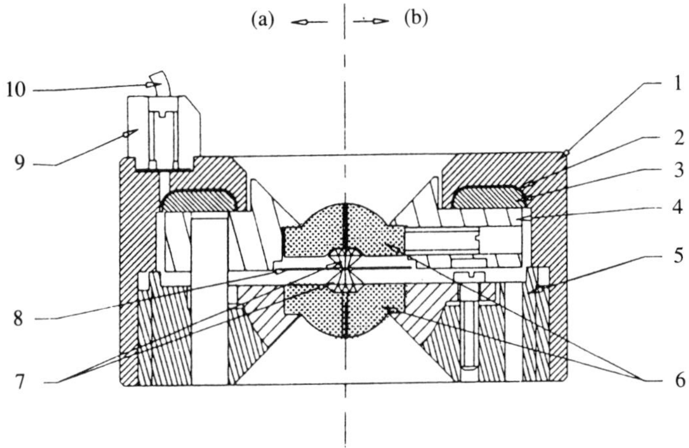

Fig. 1.34 A membrane DAC used for X-ray diffraction for half cuts through (a) the locating pins and through (b) the alignment screws: (1) retaining nut, (2) membrane, (3) push piece, (4) piston, (5) lower platen (cell body), (6) beryllium seats, (7) diamond anvils, (8) gasket, (9 and 10) gas inlets. The cell dimensions are 50 × 25 mm ^{2}. (Reprinted with permission from reference 48. Copyright Gordon and Breach Science Publishers, 1993.)

~ 15 kN for a gas pressure of 20 MPa. The optical aperture was, nevertheless, only 2 × 15°, and later designs such as those shown in Figs 1.34 and 1.35 allow the reconciliation of most of the requirements for X-ray diffraction and optical measurement.

The X-ray cell[48] of Fig. 1.34 has two 2 × 45° accesses for X-rays through beryllium seats. The reduction in area of the membrane compared to that used by Toullec *et al.*[47] necessitates a pressure of 10 MPa to produce a force of 4 kN, but this is not a limitation since the membranes will support

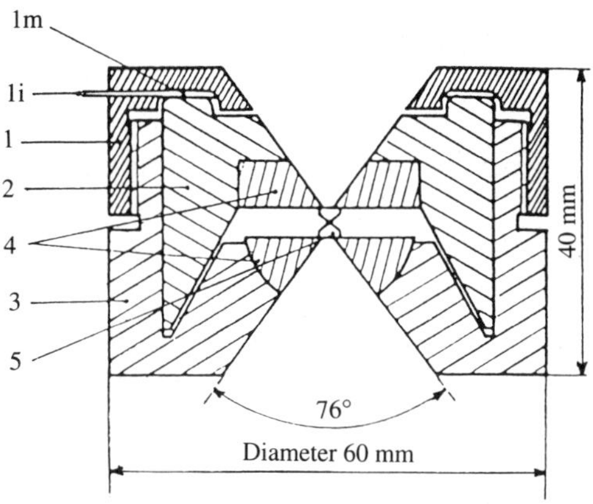

Fig. 1.35 An optical DAC: (1) cap, (1m) membrane ram, (1i) gas inlet, (2) piston, (3) cell body, (4) tungsten-carbide diamond seats, (5) diamond anvils[50]. (Reprinted with permission from *Rev. Sci. Instr.*, **66**, 2595, (1995).)

pressures well above 30 MPa. The beryllium seat is the limiting piece in the set-up: under thrusts of ~8 kN, it irreversibly flows under the diamond table, leading to anvil failures. A very recent advance has been to replace the beryllium seat by sintered boron carbide, which is transparent to X-rays but has a much higher hardness. With this material, single-crystal X-ray diffraction has been performed[49] up to 130 GPa. The optical cell shown in Fig. 1.35 has been designed[50] to produce the maximum optical aperture ($2 \times 38°$), taking into account the refractive index of diamond (~2.4). With standard anvils (with a diameter of 3 mm, a height of 2 mm, and a culet diameter of 0.4 mm) such an optical cell is not meant to be used above 50 GPa, due to the flexing of the anvil seats; it is, of course, usable at low temperatures. The performance of DACs is rather difficult to assess exactly, because so many variables are involved in a given experiment that quoting meaningful numbers for a given set-up would require careful specification—in addition to the maximum pressure claimed—of:

(1) the pressurizing fluid;
(2) the pressure homogeneity and the precision of the pressure;
(3) the pressurized volume and the sample volume;
(4) the nature of the sample;
(5) the temperature homogeneity;
(6) the nature and quality of the physical measurements performed.

Thus, reports of ultrahigh-pressure performance alone are seldom significant: any modern DAC with tolerably well-aligned parts will allow the compression of a 10 μm diameter sample of hard metal to 200 or 300 GPa, using selected bevelled anvils. However, at the time of writing, no one has succeeded in pressurizing hydrogen to 250 GPa in a DAC. Another striking example is laser heating in a DAC. The heating of samples to 5000 K under 100 GPa was reported over one decade ago, but, at the present time, whether reliable measurements can be performed under such conditions is still an open question. Thus, in Section 3.3 we shall necessarily only give incomplete information on the present pressure and temperature characteristics of DACs.

3.3 The pressure and temperature characteristics of DACs

3.3.1 Pressure generation

Diamond anvil cells are usually equipped with stones weighing between 0.2 and 0.45 carats (that is, with table diameters between 2.6 and 3.6 mm in the Drukker standard cut). Thus the linear dimensions are rather well constrained to within 30 per cent, and this puts rather precise limits on the size of the samples that can be pressurized with DAC technology. The maximum volume, V (in μm^3), that can be pressurized to a pressure P (in GPa) is given by[28]

$$P^3V = 10^{10}\ \text{GPa}^3\ \mu\text{m}^3. \tag{1.3.2}$$

This puts stringent limits on the type of measurement that can actually be performed in the high-pressure range, say above 100 GPa. At that pressure, the sample has a volume of 10^4 μm^3 at most, and it is surrounded by the highly strained material of the anvils whose volume is 10^6–10^7 times larger than the sample. With these caveats in mind, the maximum pressures that have been generated in DACs are in the range 300 GPa[51] to 500 GPa.[52] Routine experiments (for which anvils can safely be reused) can be performed in the range of 40–60 GPa.

A most serious problem in the ultrahigh-pressure range is that of pressure homogeneity. Figure 1.36 shows the pressure gradient[53] in a solid medium between the anvils; it is representative of the pressure gradients in an actual DAC experiment—the pressure variation in the central part of a 100 GPa cell may be as large as 1 GPa μm^{-1}.

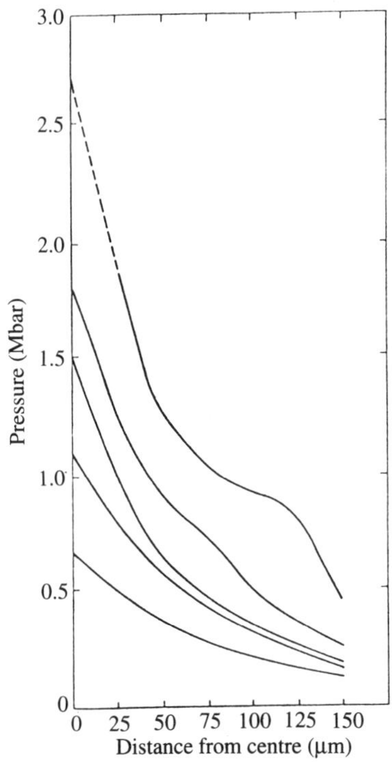

Fig. 1.36 The radial pressure distribution in the gasket of a bevelled anvil system, measured by ruby fluorescence[53]. (Reprinted with permission from *Rev. Sci. Instr.*, **56**, 1422, (1985).)

3.3.2 Low-temperature operation

Most DACs can be operated at low temperatures. Even lever-arm systems can be used if some kind of mechanical drive can be introduced into the cryostat. Nevertheless, bellows or membrane systems are much easier to operate since only a small (1.5 mm outer diameter) capillary tubing is needed to increase or decrease the load. Such a system is shown in Fig. 1.37; it can be used to perform optical measurements[54] at high pressures and variable temperatures down to 10 K. When using bellows, the minimum temperature may be limited by the freezing point of the ^4He gas, which is 3.3 MPa at 2 K. Thus, to generate the required ~ 10 kN thrust, the area of the ram must be at

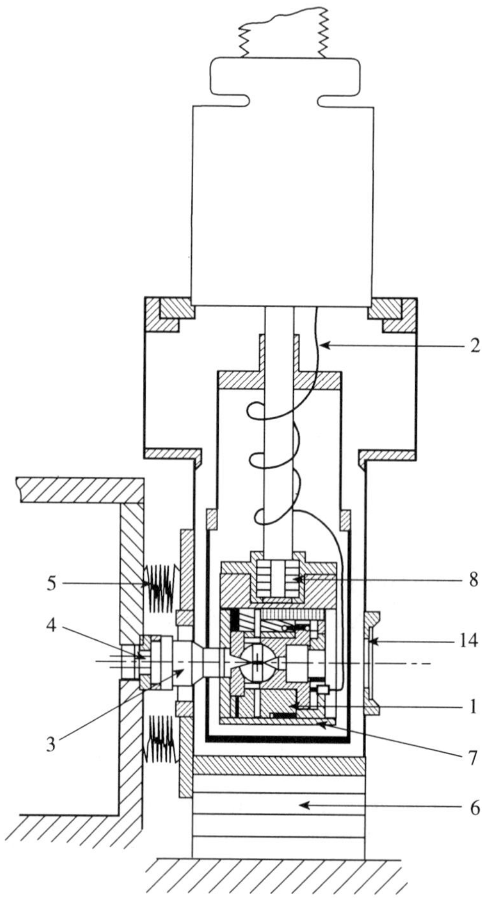

Fig. 1.37 A membrane DAC (after reference 47) in a helium-flow cryostat: (1) DAC, (2) gas feedthrough, (3) optical objective, (4) window, (5) bellows, (6) X–Y–Z–Φ micropositioners, (7) copper sleeve, (8) thermal exchanger, (14) optical window[54]. (Reprinted with permission from *Rev. Sci. Instr.*, **64**, 203, (1993).)

least 30 cm^2. An alternative to this, which allows a more compact construction, is to use Bowden cables to transmit the required force.[55] A common feature of all low-temperature systems is that they must be thoroughly cleaned to remove any grease or humidity, to prevent sticking of the piston at low temperatures. Failure to do this results at best in jamming of the piston, which prevents the build up of pressure. At worst, the piston may suddenly be liberated, and the ensuing shock may break the anvils.

3.3.3 High-temperature operation

A number of different techniques have been used to operate DACs at temperatures above 300 K, depending on the temperature to be reached. In the 'low-temperature' range, the cell can be placed in a high-temperature furnace so that the temperature is well controlled and homogeneous. Temperatures up to 1000 K have been reached[56] in this way. Even at these moderate temperatures, the components of the cell have to be modified: tungsten-carbide seats are sometimes replaced by boron carbide, and heat-resistant alloys have to be used. Rhenium gaskets have been found to be preferable to stainless steel. In any case, the cell has to be placed in an evacuated oven. To get to higher temperatures, resistive heating around the diamonds is used (Fig. 1.38), sometimes together with external heating. With local heating, temperatures up to 1300 K may be reached[57]; and with both local and external heating,[56] temperatures up to 1500 K have been reported at 10 GPa.

Problems occur with diamond anvils at high temperatures. Diamond oxidizes in air at temperatures above 900 K. At temperatures above 1200 K, diamond tends to weld to the gaskets. In practice, such cells are used to ~20–30 GPa and 900–1100 K under a vacuum or in a nonoxidizing atmosphere.[59]

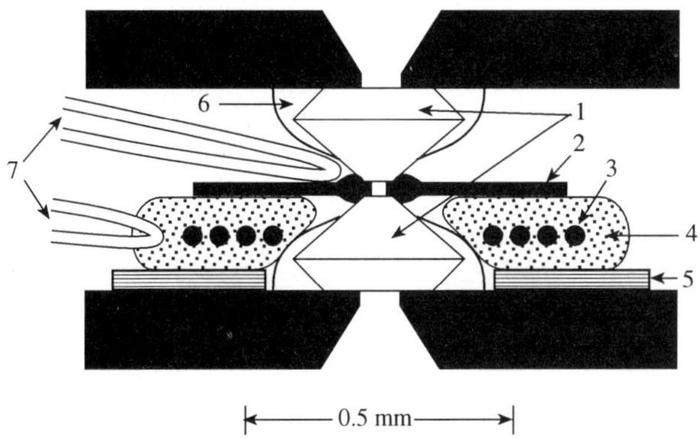

Fig. 1.38 A micro-heater assembly for a DAC: (1) anvils, (2) gasket, (3) platinum wires, (4) alumina cement, (5) mica, (6) high-temperature adhesive, (7) thermocouples[58]. (Reproduced with permission from Kluwer Academic Publishers.)

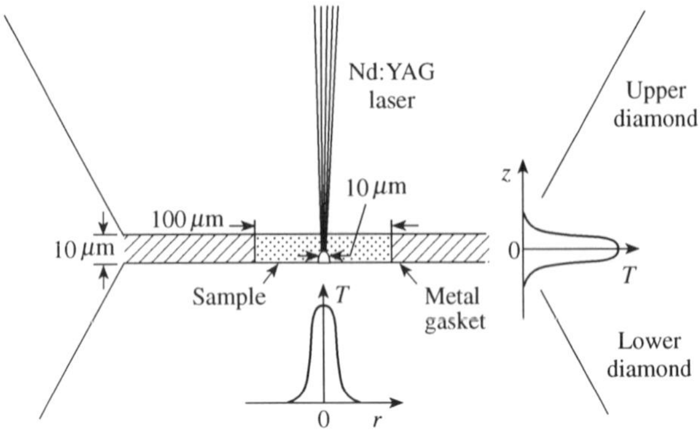

Fig. 1.39 A schematic geometry of a laser heating arrangement. Representations of the horizontal and vertical temperature distributions are given, with typical dimensions of the sample and the YAG-laser focal point (reference 60).

Laser heating is preferable to reach temperatures well above 1500 K. YAG lasers were first utilized[60] but CO_2 lasers are now also used. Type Ia diamond has a rather high absorption coefficient for the 10.6 μm wavelength of CO_2 lasers, so it is better to use type IIa diamonds for this purpose. An advantage is that many samples (silicates, glasses) which are transparent at the YAG-laser wavelength (1.06 μm) are absorbent at 10.6 μm and they therefore readily absorb the heat load of the laser. Whatever laser is used, the main problem with laser heating is the measurement of the temperature[60] and the temperature–pressure gradients in the cell (Fig. 1.39). The temperature is measured by optical pyrometry, so this technique is more adequate for very high temperatures (for example 3000 K), where the blackbody-radiation maximum is within the range of silicon-photodiode detectors, than for lower temperatures. Temperatures exceeding 4000 K at pressures above 100 GPa have been reached,[60] but the problems of stabilization and measurements of the thermodynamic conditions remain severe.[60,61]

3.4 Experimental considerations

Caution! For safety procedures see Appendix 2. Diamond-anvil cells are now the most practical devices for work above 2 GPa. They should be preferred to other set-ups in this pressure range, whenever large ($\geqslant 1$ mm^3) volumes of sample are not needed. The sample volume in DACs nevertheless decreases very quickly with increasing pressure. Between a maximum pressure of 5 and 100 GPa the available sample volume decreases by a factor of 10 000. Moreover, whereas experiments up to pressures of 40–60 GPa can be repeatedly and routinely performed, measurements above 120 GPa frequently lead to diamond failure. Except when the maximum pressure is a vitally important requirement of the experiment, DAC measurements should be planned for the lowest possible pressure, and preferably not above ~ 50 GPa.

3.5 Protocols

Protocol 1. Alignment of DACs

Caution! For safety procedures see Appendix 2.

Equipment

- A microscope with 10–100× magnification and a large working distance to allow for insertion of the DAC
- Cotton buds
- Clean alcohol or acetone
- A dry-air duster
- Sharp tweezers, preferably with bent ends
- A sharp steel needle mounted in a handle
- A few chips of transparent plastic smaller than the diameter of the DAC piston and typically 100 μm thick

1. Clean both diamond anvils with cotton buds slightly moistened with alcohol or acetone; however, do not use excess liquid if the diamonds are glued by a soluble cement onto their supports! Remove any remaining dust with a dry-air duster.
2. Check the alignment of the anvils after perfect cleaning.

   Attention! Inexperienced users often damage the anvils when they try to check the alignment. Therefore, it is highly recommended that beginners first insert a thin (100–200 μm thick) transparent plastic sheet on top of the lower anvil to avoid hard contact on first testing the free movement of the piston when it is lowered carefully (!) under the microscope to make a first contact. When both anvils start to make contact with the plastic, the contact should be allowed to occur without much deformation of the plastic and over the whole anvil face. The flat face of the lower anvil must match with all its edges precisely aligned with the flat face of the upper anvil.
3. Correct tilt or shift any misalignment with the adjustment screws (which should be accessible even when the DAC is mounted on the microscope: otherwise, each small change must be checked under the microscope before these screws are again altered).
4. Remove the chip after perfect alignment. Bring the anvils, with extreme care, into very slight contact. Remove any Newton fringes, which indicate misalignment, by further slight adjustments of the tilting screws (take care not to increase the misalignment by maladjustment).

   Recheck perfect alignment of the edges of both the diamond flats after each adjustment.

Protocol 2. Preparation of the gaskets

Caution! For safety procedures see Appendix 2.

Equipment

- All the equipment listed in Protocol 1
- An optoelectronic micrometer with sharp needles which have tip diameters smaller than 100 μm
- Wax or plasticine
- Chips of steel (or any other suitable gasket material), typically 300 μm thick
- Drills, typically with 150 μm diameter anvils and 500 μm diameter flats

1. Align the DAC and clean the anvils and gasket. Centre the gasket chip on the lower anvil under the microscope. Mark the corner using a coloured mark, a scratch mark or a cut to allow for reproducible mounting of the gasket in the same place.
2. Insert the piston with the upper anvil (carefully) until it touches the gasket (care should be taken not to shift the gasket away from its centre position). Close the DAC carefully until the gasket material starts to flow. (Beginners must check the number of turns of the tightening screws, or the gas pressure in the case of membrane DACs, to avoid excessive pressing.) Make a 50 μm indentation in the gasket. Check the thickness of the indented area using an optoelectric micrometer. (Smaller thicknesses improve gasket stability but reduce the available experimental volume; this is particularly problematic for highly compressible samples.) Repeat the indentation procedure until the optimum thickness is obtained. Make a smaller indentation carefully exactly in the centre of the indented flat to guide the drill. (A 150 μm hole in the centre of the flat indentation can be drilled by hand. However, electric microdrill machines or spark erosion work more reliably.) Prepare a new gasket if the hole deviates noticeably from the centre of the flat.
3. Ensure that both rims of the hole are free of borings, using a larger drill (300 μm diameter) to smooth ragged edges. Clean the gasket again with alcohol and the dry-air duster and carefully insert it in exactly the same position on the lower diamond anvil as during the prepressing. Use wax or plasticine (not too close to the diamond) to help in the mounting of the gasket.

Protocol 3. Loading the sample

Caution! For safety procedures see Appendix 2. The loading of the sample depends to some extent on the properties of the sample. Gases may need special provisions if they cannot be loaded at lower temperatures in solid form. Solids should be loaded in most cases with a pressure-transmitting liquid such as the magic 4:1 methanol–ethanol mixture which remains fluid at pressures up to 10 GPa.

Equipment

- All the equipment listed in Protocols 1 and 2
- A pressure-transmitting fluid
- A syringe for the pressure-transmitting fluid
- Small ruby chips or spheres with diameters typically smaller than 10 μm to avoid contact with both the anvils at higher compression

1. Insert the (solid) sample and its pressure sensors (for instance one to three ruby spheres). Drop a very small amount of the pressure-transmitting liquid into the hole. Avoid dropping too much excess fluid since this would lead to a loss of the liquid by contact with the lower support area. Rapidly insert the upper anvil and make a first weak closure of the device to avoid the liquid evaporating. Ensure that no air bubbles are visible and that the liquid is present in the experimental space.
2. Make a slight further closure of the device (this should not lead to closure of the gasket hole, which will only occur if the space is not filled with liquid). Measure the pressure using the ruby-luminescence method and apply the desired pressure. (Instabilities of the gasket will be noticed predominantly at lower pressures (5 GPa) and in the upper working range of the device.)

   Note Experience should teach experimentalists to optimize the procedure for the loading of their special samples.

References

1. Sherman, W. F.; Stadtmuller, A. A. *Experimental Techniques in High Pressure Research*; Wiley: Chichester, **1987**.
2. Spain, I. L.; Paauwe, J., eds., *High Pressure Technology*; Marcel Dekker, **1977**.
3. Bridgman, P. W. *Proc. Am. Acad. Sci.* **1938**, *72*, 157. Bridgman, P. W. *Proc. Roy. Soc. London* A, **1950**, *203*, 1.
4. Bradley, R. S., ed. *High Pressure Physics and Chemistry*; Academic Press: London, **1963**, Vol. 1.
5. Lees, J. In *Advances in High Pressure Research*; Bradley, R. S., ed. Academic Press: London, **1966**, Vol. 1.
6. Balchan, A. S.; Drickamer, H. G. *Rev. Sci. Instrum.* **1961**, *32*, 308.
7. Fitch, R. A.; Slykhouse, T. E.; Drickamer, H. G. *J. Opt. Soc. Am.* **1957**, *47*, 1015.
8. Sherman, W. F. *J. Sci. Instrum.* **1966**, *43*, 462.
9. Kendall, D. P.; Dembouski, P. V.; Davidson, T. E. *Rev. Sci. Instrum.* **1975**, *46*, 629.
10. Fukunaga, O. *J. Physique* **1984**, *45*, C8 suppl. 11, 315.
11. Khvostantsev, L. G.; Vereshchagin, L. F.; Novikov, A. P. *High Temp-High Press.* **1977**, *9*, 637.
12. Besson, J. M.; Nelmes, R. J.; Hamel, G., Loveday, J. S.; Weill, G.; Hull, S. *Physica* B, **1992**, *180*, *181*, 907.
13. Klotz, S.; Besson, J. M.; Hamel, G.; Nelmes, R. J.; Loveday, J. S.; Marshall, W. G.; Wilson, R. M. *Appl. Phys. Lett.* **1995**, *66*, 1735.
14. Gontcharenko, I. N.; Glazkov, V. P.; Irodova, A. V.; Lawova, O. A.; Somenkov, V. A. *J. Alloys Compounds* **1991**, *97*, 122.

15. Wakutsuki, M.; Ichinose, K. In *High Pressure Research in Geophysics—Advances in earth and planetary sciences*; Akimoto, S.; Manghnani, M. M., eds., D. Reidel: Dordrecht, **1982**, Vol. 12, p. 5.
16. Kawahara, N.; Imuna, T. *Jpn. J. Appl. Phys.* **1989**, *28*, L615.
17. Liebermann, R. C.; Wang, Y. *High Pressure Research Application to Earth and Planetary Sciences*; Syono, Y.; Manghnani, M. M., eds., Terra Scientific: Washington, DC, **1992**, p. 5.
18. Kawai, N.; Endo, S. *Rev. Sci. Instrum.* **1970**, *41*, 425.
19. Ohtani, E.; Irifune, T.; Hibberson, W. O.; Ringwood, A. E. *High Temp-High Press.* **1987**, *19*, 523 (1987).
    Walker, D.; Carpenter, M. A.; Hitch, C. M. *Amer. Miner.* **1990**, *75*, 1020.
20. Piermarini, G. J.; Block, S. *Rev. Sci. Instrum.* **1975**, *46*, 973.
21. Jayaraman, A. *Rev. Mod. Phys.* **1983**, *55*, 65.
22. Jayaraman, A. *Rev. Sci. Instrum.* **1986**, *57*, 1013.
23. Dunstan, D. J.; Spain, I. L. *J. Phys.* E, **1989**, *22*, 913, 923.
24. Mao, H. K.; Bell, P. M.; Dunn, K. J.; Chrenko, R. M.; de Vries, R. C. *Rev. Sci. Instrum.* **1979**, *50*, 1002.
25. Wong, P. T. T.; Klug, D. D. *App. Spectros.* **1983**, *37*, 284.
26. Hirsch, K. R.; Holzapfel, W. B. *Rev. Sci. Instrum.* **1981**, *52*, 52.
27. Merrill, L.; Bassett, W. A. *Rev. Sci. Instrum.* **1974**, *45*, 290.
28. Mao, H. K. In *Simple Molecular Systems at High Density*; NATO ASI Series B: Physics, Vol. 186; Polian, A.; Loubeyre, P., eds, Plenum Press: New York, 1989.
29. Dunstan, D. J. *Rev. Sci. Instrum.* **1989**, *60*, 3789.
30. Piermarini, G. J.; Forman, R. A.; Black, S. *Rev. Sci. Instrum.* **1978**, *49*, 1061.
31. Besson, J. M.; Pinceaux, J. P. *Rev. Sci. Instrum.* **1979**, *50*, 541.
32. Mills, R. L.; Liebenberg, D. M.; Bronson, J. C.; Schmidt, L. C. *Rev. Sci. Instrum.* **1980**, *51*, 891;
    Schouten, J. A.; Trappeniers, N. J.; van den Berg, L. C., *Rev. Sci. Instrum.* **1983**, *54*, 1209.
    Besson, J. M.; Pinceaux, J. P. *Science* **1979**, *206*, 1073.
33. Bell, P. H.; Mao, M. K. *Carnegie Inst. Yearbk.* **1991**, *80*, 404.
34. Shimizu, H.; Shimazaki, I.; Sasaki, S. *Jpn. J. Appl. Phys.* **1989**, *28*, 1632.
35. Barnett, J. D.; Block, S.; Piermarini, G. J. *Rev. Sci. Instrum.* **1973**, *44*, 1.
36. Weinstein, B. *Rev. Sci. Instrum.* **1986**, *57*, 910.
37. Hess, N. J.; Exarhos, G. J. *High Press. Res.* **1989**, *2*, 57.
38. Léger, J. M.; Château, C.; Lacam, A. *J. Appl. Phys.* **1990**, *68*, 2351.
39. Decker, J. L. *J. Appl. Phys.* **1971**, *12*, 3239.
40. Mao, H. K.; Xu, J.; Bell, P. M. *J. Geophys. Res.* B, **1986**, *91*, 4673.
41. Funamori, N.; Yagi, T.; Ushida, T. *J. Appl. Phys.* **1994**, *75*, 4327.
42. Keller, R.; Holzapfel, W. B. *Rev. Sci. Instrum.* **1977**, *48*, 517.
43. Tanino, H. *Rev. Sci. Instrum.* **1986**, *57*, 2993.
44. Eremets, M. I.; Timofeev, Yu. A. *High Press. Res.* **1991**, *8*, 510.
45. Webb, A. W.; Qadri, S. B. *Rev. Sci. Instrum.* **1982**, *53*, 1796.
46. Shields, A. J.; Wooley, R. A.; Klystein, P. C.; Simmons, J.; Friend, R. M. *Semicond. Sci. Technol.* **1989**, *4*, 301.
    Kawamura, H.; Tachibawa, K.; Shimomura, O. *Rev. Sci. Instrum.* **1989**, *56*, 1903, and the references therein.

47. Le Toullec, R.; Pinceaux, J. P.; Loubeyre, P. *High Press. Res.* **1988**, *1*, 77.
48. Le Toullec, R.; Loubeyre, P.; Pinceaux, J. P.; Mao, H. K.; Hu, J. *High Press. Res.* **1993**, *8*, 691.
49. Loubeyre, P.; Le Toullec, R.; Datchi, F.; Hu, J., private communication.
50. Chervin, J. C.; Canny, B.; Besson, J. M.; Pruzan, Ph. *Rev. Sci. Instrum.* **1995**, *66*, 2595.
51. Mao, H. K.; Wu, Y.; Hemley, R. J.; Chen, L. C.; Shu, J. F.; Finger, L. *Science* **1989**, *246*, 649.
52. Ruoff, A. L.; Xia, H.; Luo, H.; Vohra, Y. K. *Rev. Sci. Instrum.* **1990**, *61*, 3830.
53. Goettel, L.; Mao, H. K.; Bell, P. M. *Rev. Sci. Instrum.* **1985**, *56*, 1422.
54. Chervin, J. C.; Canny, B.; Gauthier, M.; Pruzan, Ph. *Rev. Sci. Instrum.* **1993**, *64*, 203.
55. Dunstan, D. J.; Schena, W. *Rev. Sci. Instrum.* **1988**, *59*, 627.
56. Schiferl, D.; Katz, A. I.; Mills, R. L.; Schmidt, L. C.; Skelton, E. F.; Webb, A. W.; Quadri, S. B.; Schaeffer, M. *Physica.* **1986**, *139*, *140*, 897.
57. Ming, L. C.; Manghnani, M. M.; Balogh, J. In *High Pressure Research in Mineral Physics*, Manghnani, M. M.; Syono, Y., eds, Terrapublishers: Tokyo, **1987**, p. 69.
58. Shimomura, O.; Yamaoka, S.; Nakazawa, H.; Fukunaga, O. *High Pressure Research in Geophysics*; Akimoto, S.; Manghnani, M. M.; Syono, Y. D., eds. Reidel Dordrecht, **1982**, p. 50.
59. Wu, T. C.; Bassett, W. A.; Burnley, P. C.; Weathers, M. S. *J. Geophys. Res.* **1993**, *98*, 19 767.
    Furnish, M. D.; Basset, W. A. *J. Geophys. Res.* **1983**, *88*, 10 333.
60. Heinz, D. L.; Jeanloz, R. reference 57, p. 113.
61. Yagi, T.; Susaki, J. In *High Pressure Research. Application to Earth and Planetary Sciences*; Syono, Y.; Manghnani, M. M., eds. Terrapublishers: Tokyo, **1992**, p. 69.
    Boehler, R.; Chopelas, A., *ibid.*, p. 55.
62. Eemets, M. *High Pressure Experimental Methods*, Oxford University Press, **1996**.

2

Pressure determination

W. B. HOLZAPFEL

1. Measurements in hydraulic systems

1.1 The U-tube

Modern methods of pressure measurement began as early as 1643 with the development of the mercury-column manometer by Torricelli, which are still used in some antique domestic barometers. The unit 1 Torr = 1 mmHg = 133.322 Pa (by definition exact) is a reminder of this almost outdated technique.

Nevertheless, for very accurate calibration work at moderate pressures (typically below 100 kPa), these U-tube manometers are still being used, both with mercury and with other suitable liquids.[1] The U-tube manometer shown schematically in Fig. 2.1 illustrates that the measurement of the height, h, between the two liquid menisci determines the pressure difference, $\Delta p = p_1 - p_0$, between the external pressure, p_0, and the internal pressure, p_1, of the high-pressure tube going to the (high) pressure system. If such an instrument is not calibrated by the manufacturer, one has to know the density, ρ_b, of the barometer liquid, and for small corrections the density, ρ_g, of the external gas must also be known, together with the value of the local acceleration due to gravity, g_{loc}, to calculate the pressure difference given by

$$\Delta p = g_{loc}\,(\rho_b - \rho_g)\,h.$$

When either the column is very high (columns more than 10 m high have been used in calibration work) or the external pressure, p_0, varies over a wide-pressure range, which occurs in stepwise calibration work, ρ_b must be known not just at the ambient pressure but also as a function of pressure. Details of all the special precautions for accurate calibration work with U-tube manometers are given in the specialist literature.[1,2] Here, the primary intention is not to give a detailed discussion of absolute-pressure calibration work but rather to give a survey of the different techniques used for pressure measurements.

1.2 Piston–cylinder pressure balances

The most accurate (absolute) pressure measurements—especially for the

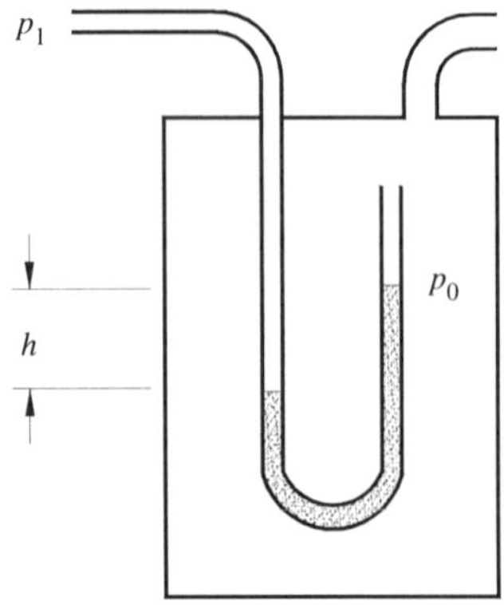

Fig. 2.1 A U-tube manometer with an external gas pressure of p_0 and an internal pressure of p_1.

realization of a pressure scale and for the calibration of other secondary pressure sensors (or pressure transducers)—are obtained with well-designed piston–cylinder devices, which establish a pressure scale from measurements of the force per unit area according to the equation $p = F/A$.

Usually, the force, F, is generated by the dead weight on a piston with well-defined area, A, as illustrated in Fig. 2.2. Accurate determinations of the effective area, and corrections for friction and for the buoyancy of the piston require great care if the best possible (absolute) accuracies are required (that is, accuracies of the order of 10 p.p.m (parts per million)).[1–3] For practical applications, other secondary pressure gauges (manometers) are used; they are usually commercially calibrated, or they can be recalibrated, if necessary, by comparison with transfer standards.[3]

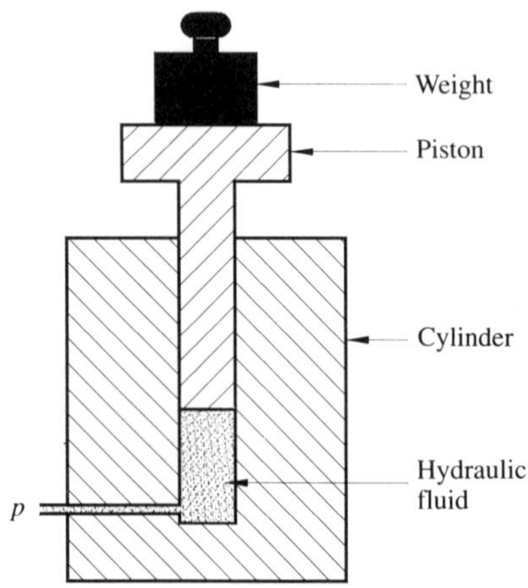

Fig. 2.2 Pressure determination by a piston–cylinder device.

1.3 Bourdon gauges

Good precision and simple operation are provided by commercial Bourdon gauges which mechanically amplify the deformation of a spiral tubing into the deflection of a large pointer, so giving, at their best, a resolution of 1/1000 of the maximum deflection. For modern applications, electrical signals can be provided by potentiometers which are coupled to the axis of the pointer.

1.4 Strain gauges

With the development of strain-gauge techniques, the deformation of specially designed, small, pressure vessels with suitably attached strain-gauge sensors are used in commercial strain-gauge manometers. To reduce the sensitivity of these sensors with respect to temperature variations, four strain gauges (resistors) are connected in the form of a Wheatstone-bridge circuit: two of the sensors measure the deformation and the other two sensors compensate for the zero-pressure temperature drift. At present, the electronic circuitry required for strain-gauge manometers seems to be more expensive than the mechanical systems used in Bourdon gauges. However, a detailed comparison of their actual costs may well be justified, especially if an electronic output is necessary in any case.

1.5 Manganin gauges

Since the temperature coefficient for the electrical resistance of manganin is almost zero at room temperature, the pressure coefficient of about 2.4×10^{-11} Pa^{-1} is sufficient for accurate pressure measurements, and calibrated manganin gauges are commercially available. However, the precise value of the pressure coefficient depends not only on the composition of the alloy but also on its aging, and, in addition, the stresses which arise in the preparation of a resistor coil also effect the actual resistance. Therefore, special care seems to be necessary when one wants to obtain very good reproducibility with manganin gauges over extended periods of time.[3]

1.6 Other gauges

Since many other properties of liquids and solids depend on pressure, one could think of many more different types of sensor for pressure measurement. However, the sensors discussed so far seem to meet most of the demands of typical hydraulic systems.

2. Measurements with piston–cylinder and anvil techniques

2.1 The similarity technique

If two piston–cylinder devices are coupled as shown in Fig. 2.3, where one larger piston acts on a smaller piston to convert the pressure on the low-pressure

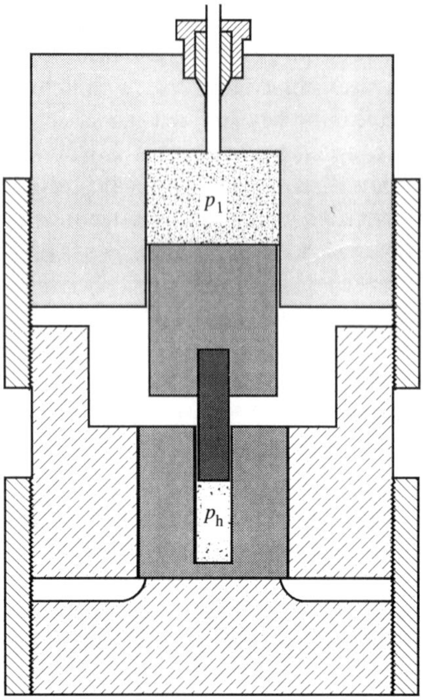

Fig. 2.3 A coupled piston–cylinder device for upward conversion of calibrated pressures by a constant factor.

side, p_l, into a higher pressure, p_h, in the smaller piston–cylinder device; one can provide similar deformations of the effective areas for both pistons by careful design of the shape of both of the piston–cylinder devices together with the selection of appropriate piston–cylinder materials. With carefully planned similarity, the effective areas of both pistons change in the same way, so that an accurate pressure measurement on the low-pressure side provides accurate calibration on the high-pressure side if the effects of friction can be controlled by rotation of the pistons and by cycling the pressure.[2,3] In this way, absolute calibrations can be carried out in piston–cylinder devices up to 5 GPa; however, due to problems from friction, the (absolute) accuracy of such calibrations decreases rapidly above 1 GPa.

At higher pressures, the failure of simple piston–cylinder devices necessitates that other techniques are used for pressure measurement.

2.2 Equation-of-state calibrants

Pressure measurements above 5 GPa are presently based on a pressure scale which is derived from a combination of theoretical (or semiempirical) equation-of-state calculations and multiple comparisons with results obtained from shock-wave methods.

Details on the evaluation of shock-wave data are given in many places in the literature; therefore, it is sufficient to note here that the determination of a pressure–volume (p–V)-isotherm from shock-wave data is affected not only by uncertainties in the shock-wave measurements but also by problems in the data reduction. Thus, (p–V)-isotherms from shock-wave measurements can be trusted only to within a 5 per cent accuracy at intermediate pressures; and only a rigorous intercomparison with first-principle calculations leads to confidence that the pressure calibration realized by various (p–V)-isotherms of specially simple solids represents an absolute scale with an accuracy of better than 5 per cent (in the pressure) up to 1 TPa.[4]

For practical applications in X-ray and neutron-diffraction studies under pressure (and in volumetric or, more, recently in extended X-ray-absorption fine-structure spectroscopy (EXAFS) studies), a single (p–V)-isotherm at ambient temperature is insufficient, so low- and high-temperature isotherms are also required. Thus, the complete (thermal) equation of state (EOS), $p = p(V,T)$ or $V = V(p,T)$, must be known for at least one solid to establish uniquely a pressure scale (for the range above 5 GPa).

For this purpose, one can recommend[4–6] the use of the EOS

$$p = \frac{3K_0\,(1-x)}{x^5}\cdot\exp\left[c_0\,(1-x)\right] \tag{2.2.1}$$

where $x = a/a_0 = (V/V_0)^{1/3}$ and $c_0 = \frac{3}{2}(K_0'-3)$.

Given the reference data shown in Table 2.1 for the lattice parameter at ambient pressure and temperature, a_{0R}, the linear thermal expansion, α_{0R}, the bulk modulus, K_{0R}, and its pressure derivative K'_{0R}, and given the temperature derivative dK_0/dT ($= -3\alpha_{0R}K_0\delta_{\alpha R}$), where $\delta_{\alpha R}$ is the dimensionless parameter $\delta_{\alpha R} = \partial(\ln\alpha)/\partial(\ln V)_{T_R}$, the EOS data can be easily computed for temperatures T around or above $T_R = 300$ K, when the linear relations

$$\begin{aligned}
a_0\,(T) &= a_{0R}\,[1+\alpha_{0R}\,(T-T_R)],\\
V_0\,(T) &= V_{0R}\,[1+3\alpha_{0R}\,(T-T_R)],\\
K_0\,(T) &= K_{0R}\,[1-3\alpha_{0R}\,\delta_{\alpha R}\,(T-T_R)],
\end{aligned}$$

are used with a constant value K'_{0R}. Only when the highest possible accuracies and complete consistencies at lower and higher temperatures are required should the precise temperature dependence of $\alpha_0(T)$ and $K'_0(T)$ and the very weak temperature dependence of $K_0(T)$ also be taken into account.[6] A similar analysis for the EOS data for NaCl has not yet been performed, and, therefore, the NaCl scale may ultimately have minor revisions with respect to the data for copper, gold, and silver. For further details it is recommended that recent reviews are consulted.[4–6]

2.3 Luminescence sensors

Red-ruby luminescence is not only valued in gemstones and used in laser action, it also plays a part in high-pressure measurements, where diamond

Table 2.1. Parameters recommended for pressure determinations by EOS measurements using various face-centred cubic (f.c.c.) and body-centred cubic (b.c.c.) metals as calibrants with the EOS given by (2.2.1) and the reference temperature T_R = 300 K, as discussed in the text; the numerals in parentheses represent the uncertainties in the last significant digit obtained by inspection of the relevant literature[6–13]

Metal	a_{0R} (pm)	K_{0R} (GPa)	K'_{0R}	$\alpha_{0R}(\times 10^6$ K)	$\delta_{\alpha R}$
Al	404.98(1)	72.5(4)	4.8(2)	23.0(4)	5.5(11)
Cu	361.55(1)	133.2(2)	5.4(2)	16.6(3)	6.1(6)
Ag	408.62(1)	101.0(2)	6.2(2)	19.2(4)	7.1(6)
Au	407.84(1)	166.7(2)	6.3(2)	14.2(2)	7.2(6)
Pd	388.99(1)	189(3)	5.3(2)	11.6(4)	6.0(11)
Pt	392.32(1)	277(5)	5.2(2)	8.9(4)	5.9(11)
Mo	314.73(1)	261(5)	4.5(5)	5.0(4)	5.2(14)
W	316.47(1)	308(2)	4.0(2)	4.5(4)	4.7(11)

anvils allow optical access. In fact, the application of red-ruby luminescence and improvements in gasket techniques have led to the widespread application of the diamond-anvil technique.

Typically, a small splinter or sphere of ruby with a diameter of about 10 μm and containing about 0.3 per cent chromium in the Al_2O_3 host crystal is excited by light with a power of 50 mW or less from the blue–green lines of an argon ion laser focused through diamond anvils on to the ruby splinter. The emitted luminescence is then focused on to the entrance slit of a grating monochromator, which is used in scanning mode either with a simple photomultiplier detector or with fixed gratings and a modern charge-coupled-device-camera detector in multichannel mode. The red $R_{1,2}$ luminescence doublet is observed at 300 K and ambient pressure at $\bar{\nu}_{01}$ = 14 402(2) cm^{-1} and at $\bar{\nu}_{02}$ = 14 432(2) cm^{-1}, with slight variations in the line positions from sample to sample, probably due to internal strains. For precise pressure measurements ($\Delta p < 0.3$ GPa) one has thus to determine the initial line positions for each grain in each experiment. Within the accuracy of the existing calibrations both the lines shift linearly with pressure at the same rate, −7.56(2) cm^{-1} GPa^{-1}, although the uncertainty quoted does not include the likely uncertainty in the NaCl pressure scale (which is the original calibrant) of ±3 per cent. Over the range 200–500 K, both lines also shift almost linearly with temperature, at a rate of −0.126(9) cm^{-1} K^{-1}, and this temperature correction seems to be independent of the pressure within the (limited) accuracy of the available calibrations, as discussed previously.[4] The ruby-line shift has also been calibrated over an extended pressure range up to 180 GPa, by X-ray diffraction with respect to EOS data for copper, silver, platinum, molybdenum, and tungsten. The nonlinear shift in this extended region has been represented by

$$p = 1904 \text{ GPa } [(\lambda/\lambda_0)^B - 1]/B$$

Table 2.2. A comparison of various luminescence pressure sensors[15] with singlet and doublet lines (the numerals in parentheses are the uncertainties in the last significant digit)

Material	λ_0 (nm)	$d\lambda/dp$ (nm GPa^{-1})	$d\lambda/dT$ ($\times 10^{-3}$ nm K^{-1})	$d\lambda/dp)/\Gamma$ (nm GPa^{-1})	$\frac{d\lambda/dT}{d\lambda/dp}$ (10^{-3} GPa K^{-1})	Transition
$Cr^{3+}:Al_2O_3$	694.2	0.365(9)	6.2(3)	0.49	17.0	$^2E \rightarrow {}^4A_2$ Doublet
$Sm^{2+}:SrB_4O_7$	685.4	0.255	−0.1	1.7	−0.4	$^5D_0 \rightarrow {}^7F_0$ Singlet
Sm^{2+}:BaFCl	687.6	1.10	−1.6	4.8	−1.5	$^5D_0 \rightarrow {}^7F_0$ Singlet
Sm^{2+}:SrFCl	690.3	1.12(3)	−2.36(3)	5.8	−2.1	$^5D_0 \rightarrow F_0$ Singlet
Eu^{3+}:LaOCl	578.7	0.25	−0.5	1.0	−2.0	$^5D_0 \rightarrow {}^7F_0$ Singlet
Eu^{3+}:YAG	590.6	0.197	−0.5	0.7	−2.5	$^5D_0 \rightarrow F_1$ Doublet

where $B = 5$ for nonhydrostatic measurements, and $B = 7.665$ for quasi-hydrostatic measurements.[14] For many modern spectrometers with wave-number scales, $\bar{\nu}$, instead of wavelength scales, λ, the relation $\lambda = 1/\bar{\nu}$ gives $\lambda/\lambda_0 = \bar{\nu}_0/\bar{\nu}$. Thereby λ and $\bar{\nu}$ represent the line positions under pressure and the two line positions for the ruby doublet at ambient pressure, $\lambda_{01} = 1/\bar{\nu}_{01} =$ 694.3(1) nm and $\lambda_{02} = 1/\bar{\nu}_{02} = 692.9(1)$ nm, must be determined again with higher precision for each ruby splinter, when pressure measurements with accuracies of better than ±0.3 GPa should be made.

In addition to the shift of the ruby-luminescence lines other optical sensors have also been discussed in the literature.[4,15,16] A comparison of the relevant parameters is given in Table 2.2, which shows that the sharp line width Γ for the luminescence of the $^5D_0$–$^7F_0$ transition for Sm^{2+} in either SrB_4O_7 or BaFCl or SrFCl always gives a much larger sensitivity, $(1/\Gamma)(d\lambda/dp)$, than ruby does. At the same time these sensors also show much weaker temperature dependences, $(d\lambda/dT)/(d\lambda/dp)$. While $Sm^{2+}:SrB_4O_7$ seems to work up to 200 GPa, both the other hosts exhibit quenching at much lower pressures, and, therefore, similar luminescence from the electronically equivalent Eu^{3+} in different hosts may be more favourable. At elevated temperatures or pressures, the precision of the ruby sensor is further reduced due to the merging of the doublet lines, which does not occur with the singlets of the other sensors. Nonlinearities and quenching, which have also been observed with Sm^{2+} sensors, could possibly be avoided with other sensors in the future.

2.4 Fixed-point calibrations

For a long time fixed-point calibrations were much used in high-pressure measurements, especially with piston–cylinder, belt, and multianvil apparatus, where optical or X-ray access was difficult but electrical wires could be introduced easily. The useful fixed points were related to sharp changes in the electrical resistance at first-order phase transitions. Sharp transitions with minor hysteresis (for instance, in resistance measurements) were considered to be the most useful. For this almost historical reason, some recommended fixed points[4] are given in Table 2.3.

Table 2.3. Pressure fixed points at the ambient temperature

p (GPa)	Transition	Remarks
2.550(6)	Bi I–II	Accurance of thermodynamic average values (centre of hysteresis)
0.68(3)	Tl II–III	
5.5(1)	Ba I–II	
7.7(2)	Bi III–IV	
9.4(3)	Sn I–II	Precision for onset of forward transition only
12.3(5)	Ba II–III	
13.4(6)	Pb I–II	

References

1. Jäger, J. In *Landoldt–Börnstein New Series: Units and fundamental constants in physics and chemistry*; Bortfeldt, J.; Kramer, B., eds. Springer Verlag: Berlin, **1991**, Section 2.3.1.7, pp. 2.165–2.176.
2. Dadson, R. S. In *High Pressure Measurement Techniques*; Peggs, G. N., ed. Applied Science: London, **1983**, pp. 3–67.
3. Heydemann, P. L. M. In *High Pressure Chemistry*; Kelm, H., ed. D. Reidel: Dordrecht, **1978**, pp. 1–49.
4. Holzapfel, W. B. In *Landoldt–Börnstein New Series: Units and fundamental constants in physics and chemistry*; Bortfeldt, J.; Kramer, B.; eds. Springer Verlag: Berlin, Section 2.3.1.8, **1991**, pp. 2.177–2.184.
5. Holzapfel, W. B. *High Press. Res.* **1991**, *7*, 290–293.
6. Holzapfel, W. B. *J. Phys. Chem. Solids*, **1994**, *55*, 711–719.
7. Gray, D. E., ed. *American Institute of Physics, Handbook*; McGraw Hill: New York, **1972**.
8. Wyckoff, R. W. G. *Crystal Structures*. Interscience: New York, **1963**, Vol. 1, p. 10.
9. Gschneidner, K. A. *Solid State Phys.* **1964**, *16*, 275–446.
10. Simmon, G.; Wang, H. *Single Crystal Elastic Constants and Calculated Aggregate Properties: A handbook*; MIT Press: Cambridge, MA, **1971**.
11. Guinan, M. W.; Steinberg, D. J. *J. Phys. Chem. Solids*, **1974**, *35*, 1501–1512.
12. Weast, R. C.; Astle, M. J. *CRC Handbook of Chemistry and Physics*, 59th ed; CRC Press: West Palm Beach, **1979**, p. D-225.
13. Young, D. A. *Phase Diagrams of the Elements*; University California Press: Berkeley, **1991**.
14. Mao, H. K. In *Simple Molecular Systems at Very High Density*; Polian, A.; Loubeyre, P.; Boccara, N.; eds. Plenum Press: New York, **1989**, pp. 221–236.
15. Shen, Y. R.; Gregorian, T.; Holzapfel, W. B. *High Press. Res.* **1991**, *7*, 73–75.
16. Lorenz, B.; Shen, Y. R.; Holzapfel, W. B. *High Press. Res.*, **1994**, *12*, 91–99.

3

Techniques for the study of physical properties

1. Electrical measurements

J. M. Besson

Electrical measurements are among the most ancient methods used to characterize samples at high pressures. Indeed, before the advent of diamond-anvil cells (DACs), they were the most often used characterization methods in belt-type or multianvil devices at pressures above 2 GPa for the detection of solid–solid phase transitions—upon which the so-called fixed-point pressure scale was based (see Section 2.2 in Chapter 2).

Electrical methods actually include a variety of techniques in addition to measurements of the resistivity of the sample under study, or of resistive pressure gauges. These include:

(1) magnetoelectrical properties (Hall effect, magnetoresistance);

(2) optoelectronic properties (photoconductivity–photovoltaic effect, emission from diode lasers);

(3) temperature measurements by thermocouples, or differential thermal analysis.

In all cases, the commonest problem is the introduction of electrical leads, or thermocouples, into the high-pressure cell. Different techniques are used in large-volume fluid cells at pressures up to ~ 2 GPa and in multianvil systems (including DACs) above 2 GPa.

1.1 Electrical measurements in large-volume fluid cells

The most reliable and simplest type of electrical leadthrough, shown in Fig. 3.1, was developed by P. W. Bridgman. Its main components are a metallic cone made of beryllium–copper and treated to its full hardness (42 HRc) or of maraging steel (52 HRc), and an insulating sleeve which is a machined cone made of (pipestone) pyrophyllite or plastic, preferably Makrolon. In the former case, the porosity of pyrophyllite is dealt with by wetting the insulating sleeve with silicone oil.[1] The initial sealing can be made during the first

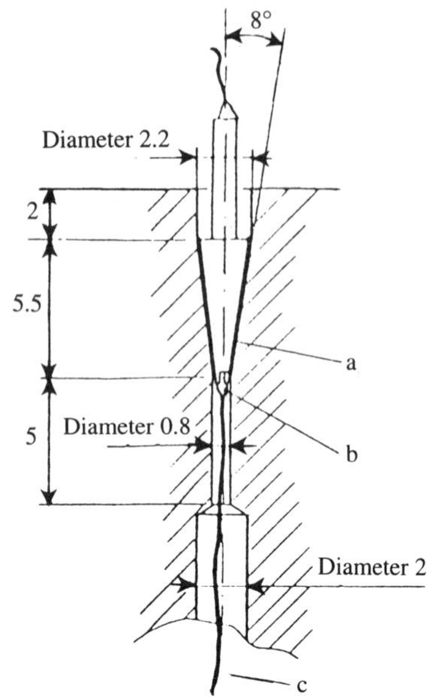

Fig. 3.1 A conical electrical leadthrough. (After P. W. Bridgman: (a) a pyrophyllite (pipestone) or plastic sleeve, (b) solder, and (c) an insulated electrical wire. The dimensions are in millimetres. The cone is made of either steel or hardened beryllium–copper. It is set into the plug under a force of 2–3 kN. *Proc. Am. Acad. Arts Sci.* **74,** 11 (1940).)

pressurizing of the leadthrough, or, preferably, by applying a force of ~ 4 kN to the electrode. Simple as these leadthroughs are, they will withstand helium-gas pressures up to 1.8 GPa at room temperature[1] and helium-gas pressure above 1 GPa at 77 K with no measurable leak over a three week period.[2] In addition[3] they are suitable for use at high frequencies (above 10 MHz) as a short element in a high-frequency circuit with a matched impedance, without causing too much reflection or attenuation. Some authors favour the use of a retaining nut on top of the cone, which is screwed into the electrical plug to keep the cone in place. This may be useful if the angle of the cone is much larger than 8°.

The same type of conical leadthrough may be used for swaged thermocouples (Fig. 3.2) the outside sheath of which is either silver soldered to the cone or simply glued with epoxy resin. If the outer sheath is cut and only the centre wire is used, porous MgO or Al_2O_3 powder insulator will be in contact with the pressurized fluid, and some kind of impervious filler, such as uncured epoxy resin, has to be forced under pressure into the insulator.

Only a limited number of such leads can be mounted on an electrical high-pressure plug. Alternative methods can be used to introduce many leads on a small-diameter plug, such as that shown in Fig. 3.3; more than ten individual enamelled wires can be embedded in an insulating conical shell made of an epoxy and elastomer resin which has been used[4] above 3 GPa. Special lead-

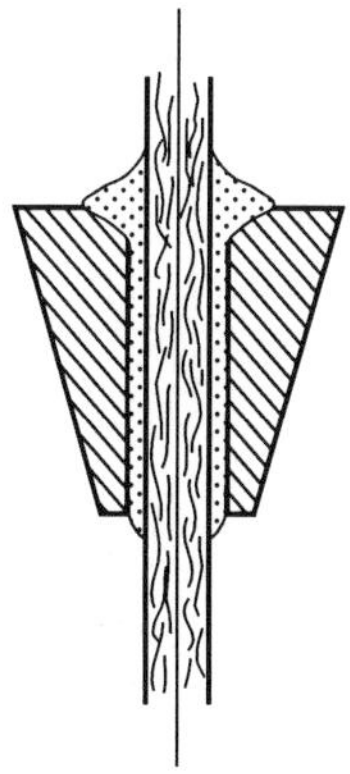

Fig. 3.2 A swaged thermocouple leadthrough. The thermocouple is silver soldered on to the cone. (From reference 5.)

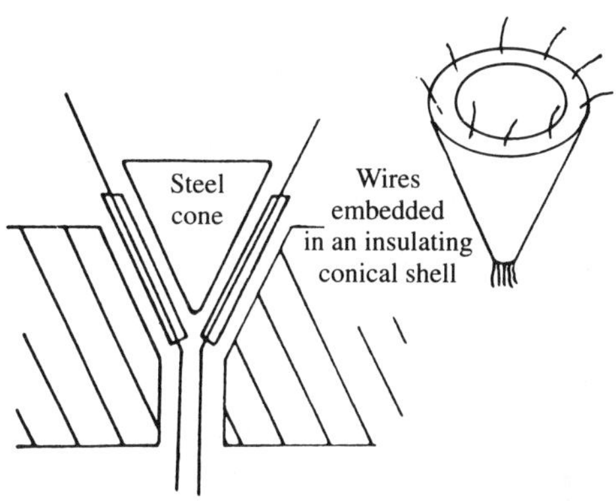

Fig. 3.3 A multiwire high-pressure leadthrough. The conical shell is made of epoxy resin. (Reprinted with permission from L. G. Blosser and H. S. Young, *Rev. Sci. Instr.* **33**, 1007 (1962), quoted in reference 5.)

throughs with very high leakage resistances, or small capacitance, have been described and reviewed.[5] They usually involve considerably more work, and they are less reliable than the simple conical types described here, which should be satisfactory in 90 per cent of cases.

When very high insulation is required ($R \gg 10^8\ \Omega$), the electrical leadthrough can be mounted on a pierced sapphire plug using a double Poulter-seal arrangement (cf. Chapter 3, section 5), as shown in Fig. 3.4.

1.2 Multianvil systems

The common features of the electrical leadthrough of multianvil systems are:

(1) the use of a solid pressure-transmitting medium which avoids the difficulties encountered with leaks in fluid media;

(2) extrusion of the gasket material, which may cause a mechanical failure of the electrical leadthrough.

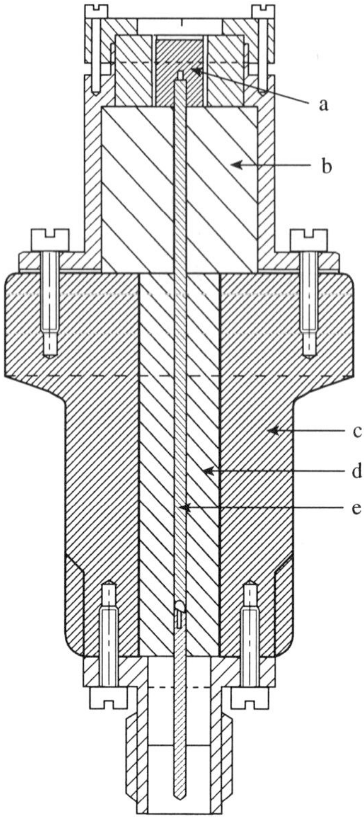

Fig. 3.4 A high-frequency, high-insulation leadthrough: (a) beryllium–copper, (b) sapphire plug, (c) maraging-steel plug, (d) Teflon cylinder, and (e) brass electrical lead. (Designed by the Département Hautes Pressions, Université Paris 6–4, place Jussieu, F-75252, Paris, France.)

For these reasons electrical leads or thermocouples are currently sheathed with alumina[6] and stainless-steel tubing (Fig. 3.5) in belt-type or large volume-multianvil systems.

The case of DACs is quite different since their gasket materials are usually metallic, and they thus have to be electrically insulated from the leads. In addition, at pressures up to ~10 GPa, fluids such as ethanol–methanol or helium are used in DACs, which results in rather large deformations of the gasket. Several methods have been proposed[7,8] to allow multiprobe electrical measurements to be performed in DACs, on single crystal samples in a (quasi) hydrostatic environment. The example[8] shown in Fig. 3.6 has been tested to ~20 GPa. Simpler arrangements are quoted in reference 8. Arrangements[9] which use metal leads deposited on the anvils permit measurements to be made on highly resistive media at very high pressures.

Although electrical measurements in DACs are indeed possible and have

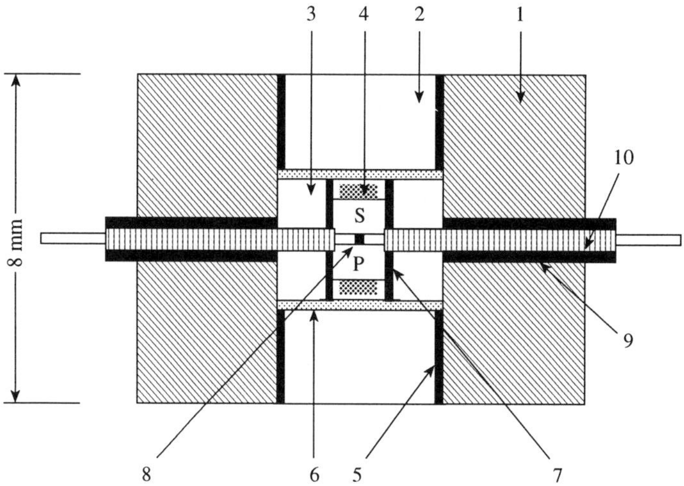

Fig. 3.5 A thermocouple leadthrough for a high-pressure, high-temperature assembly in a cubic-anvil device. (From [6].) (1) boron-epoxy, (2) end plug, (3) sleeve, (4) cap, (5) heater-current leadthrough (gold), (6) molybdenum, (7) graphite heater, (8) platinum–rhodium thermocouple, (9) stainless-steel protective tube, (10) alumina. S and P are the sample and pressure marker (NaCl), respectively.

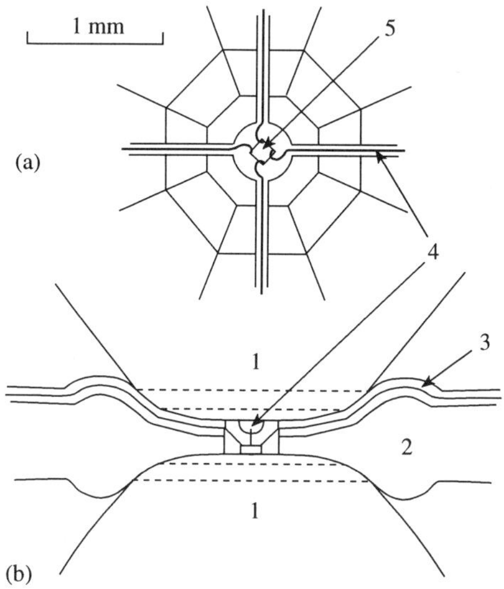

Fig. 3.6 A set-up for electrical measurements in a gasketed DAC: (a) a top view of the sample and leads, and (b) a cutaway view; (1) anvil culets, (2) metallic gasket, (3) alumina layer, (4) grooves with leads, and (5) sample[8]. (Reprinted with permission from *Rev. Sci. Instr.*, **57**, 160, (1986).)

been performed, they usually require fastidious work due to the small size of the samples. They are not commonly used, for this reason, as is shown by the very small number of published results using these methods.

1.3 Experimental considerations

Caution! For safety procedures see Appendix 2. Conical Bridgman lead-throughs (Fig. 3.1) should be preferred for electrical measurements in fluid cells, for their simplicity, performance, and reliability.

References

1. Langer, D.; Warschauer, S. M. *Rev. Sci. Instrum.* **1961**, *32*, 32.
2. Fournier, J. Radiative emission of a semiconductor laser under pressure; *Mémoire du Conservatoire National des Arts et Métiers*, Paris, **1969** (unpublished). Fournier, J.; Besson, J. M. *Revue de Physique Appliquée* **1970**, *5*, 333.
3. Besson, J. M.; Calawa, A. R.; Paul, W. *Phys. Rev.* **1968**, *173*, 699.
4. Boehler, R.; Getting, I. C.; Kennedy, G. C. *J. Phys. Chem. Solids* **1977**, *38*, 233.
5. Sherman, W. F.; Stadtmuller, A. A. *Experimental Techniques in High Pressure research*, Wiley: Chichester, **1987**.
6. Shimomura, O.; Utsumi, W.; Taniyuchi, T.; Kikegawa, T.; Nagashima, T. *High Pressure Research*; Syono, Y.; Manghnani, M. H., eds, Terra Scientific: Washington, DC, **1992**, p. 6.
7. Tozer, S. W.; King Jr, H. E. *Rev. Sci Instrum.* **1985**, *56*, 260.
8. Gonzalez, J.; Besson, J. M.; Weill, G. *Rev. Sci Instrum.* **1986**, *57*, 106.
9. Grzybowski, J. A.; Ruoff, A. L. *Phys. Rev. Lett.* **1984**, *53*, 489.

2. Volumetric techniques

W. B. Holzapfel

A complete thermodynamic description of a material under pressure requires knowledge of the (thermal) equation of state (EOS), for example, in the form $V = V(p,T)$ or $\rho = \rho(p,T)$, for the specific material, where V is usually the specific volume (the volume per unit mass) or the molar volume and ρ is the specific density $\rho=1/V$. For a rigid cylinder with a fixed amount of a sample enclosed by the piston, the relative change in volume $V(p,T)/V_0$ is directly related to the movement of the piston if deformations of the piston and cylinder, and leakage and friction, can be neglected. For pressures below 1 GPa and moderate temperatures these techniques are well developed for liquids and gases. However, for higher pressures and variable temperatures the encapsulation of the sample and the deformation of the pressure vessel need special attention. The state of the art for encapsulated solids[1] corresponds to a precision of only 500 p.p.m. (parts per million) in V/V_0 or 3 per cent in pressure (whichever is the larger value) in the temperature range 4 K $< T <$ 300 K for pressures, p, below 2 GPa.

For higher pressures optical-length measurements in DACs have been tested either on single crystals[2] or by measuring the thickness and area of the sample space.[3] Useful results are obtained for highly compressible samples such as[3] H_2. However, problems with shadowing and with the interferometric measurement of the sample thickness result in a precision in V/V_0 of only 3 per cent, which would be considered rather poor in most other cases. Nevertheless, one can envisage some interesting high-pressure applications of this technique for pressures and temperatures below 50 GPa and 1000 K. For example, thermal expansion could be determined by measuring the thermal pressure variation at constant volume, using a digital video system to check the area of the sample hole and using an optical interferometer to control the thickness, making use of the fact that the refractive index also remains constant at constant density. Similarly, the pressure and temperature changes upon melting at constant volume, $\Delta p_m|_V$ and $\Delta T_m|_V$, could be determined much more precisely than in earlier attempts,[4,5] and more accurate values could be obtained for the data at constant pressure, $\Delta V_m|_p$ and $\Delta S_m|_p$, which are the crucial parameters for melting theories. For this purpose, temperature-insensitive pressure sensors (such as the new lanthanide-luminescence sensors) need to be developed further, as discussed in Section 2.2 in Chapter 2.

2.1 Experimental considerations

Caution! For safety procedures see Appendix 2. For volumetric measurements with commercial equipment (at pressures less than < 1 GPa), information on sample preparation, precision, and accuracy should be supplied by the manufacturer.

For volumetric measurements with piston–cylinder devices at pressures up to 2 GPa much experience and special calibrations are required.[1]

For volumetric measurements with DACs high precision can only be obtained using interferometric and video control, and the primary applications may be for thermal pressure measurements at constant volume. The techniques need still further development.

References

1. Anderson, M. S.; Swenson, C. A. *J. Appl. Phys.* **1984**, *56*, 2697–2702.
2. Polian, A.; Besson, J. M.; Grimsditch, M.; Vogt, M. *Phys. Rev.* B, **1982**, *25*, 2767.
3. Van Straaten, J.; Wijngaarden, R. J.; Silvera, I. F. *Phys. Rev. Lett.* **1982**, *48*, 97.
4. Hirsch, K. R.; Holzapfel, W. B. *J. Phys.* E, **1983**, *16*, 412–417.
5. Yagi, T.; Hirsch, K. R.; Holzapfel, W. B. *J. Phys. Chem.* **1983**, *87*, 2272–2273.

3. X-ray diffraction

W. B. Holzapfel

Developments in DAC techniques have strongly promoted the widespread application of X-ray diffraction to solids under pressure.[1–6]

With this technique lattice parameters and structural changes are studied almost routinely in polycrystalline solids at ambient temperatures and at pressures up to 80 GPa, and even up to 500 GPa and above in some laboratories.[5,6] With cryogenic systems, temperatures down to 4 K (and even lower if required) can also be obtained,[7,8] and elevated temperatures have been applied in more limited pressure ranges, by placing conventional external heaters around either the whole DAC[9] or around the internal diamond mounting;[3,10] more recently, well-controlled laser heating has been used to focus on a small part of the sample which is embedded in a transparent, inert, pressure-transmitting medium just within the gasket hole.[11] However, under these extreme conditions, useful X-ray diffraction signals are only obtained with the very intense X-ray beams which have recently become available from modern synchrotron-radiation sources.

Since various reviews of the different techniques used for X-ray diffraction on single crystals and polycrystalline samples under pressure give very detailed introductions into this field,[1–4] only the most common techniques and recent developments are illustrated here.

The basic condition for the observation of the maxima in the scattered intensity is given by the Bragg equation

$$\lambda = 2d_{h\,k\,l} \sin\theta,$$

whereby λ is the wavelength of the (elastically) scattered radiation, $d_{h\,k\,l}$ is the spacing of the lattice planes with the Miller indices h, k, l, and θ is the angle of specular reflection with respect to these planes. The diffraction angle, defined as the angle between the incident primary beam and the diffracted beam, is therefore equal to 2θ. For monochromatic radiation (radiation with a fixed λ), maxima in the scattered intensity are observed at some discrete values of $2\theta_{h\,k\,l}$ corresponding to the lattice spacing $d_{h\,k\,l}$. For ideal polycrystalline samples, where all possible orientations of the individual microcrystalline grains are equally probable, the radiation is scattered symmetrically around the primary beam into discrete diffraction cones, which can be registered photographically as concentric circles on a flat film, or, more commonly, as Debye–Scherrer rings on a cylindrical film wrapped around the sample with the primary beam passing through the film on one side and through the collimator hole and on the other side through the beam stop. In high-pressure applications, these conventional film techniques are usually replaced either by a scanning diffractometer with a single slit in front of the detector or with position-sensitive linear or curved detectors. Image-plate techniques and area detectors with a scintillation screen coupled optically to a charge-coupled-device (CCD) camera are now coming into use.

In contrast to these angular dispersive X-ray diffraction (ADXD) techniques, which require monochromatic X-rays and rather large exit slits on the DAC, most high-pressure X-ray measurements made with DACs are currently made using a complementary technique, which works with white or polychromatic X-rays, typically in the range 10 pm $< \lambda <$ 100 pm, and with a small collimator slit at a predetermined fixed diffraction angle 2θ in front of a solid-state detector which performs the energy analysis. The best compromise between quantum efficiency and resolution is commonly provided by high-purity germanium detectors for this energy dispersive X-ray diffraction (EDXD) technique. Each individual X-ray photon absorbed in the detector produces a charge pulse, which is then amplified and routed into one specific channel of a multichannel analyser (MCA) according to its pulse height. When the channel numbers of the MCA are calibrated with respect to the energy of the incident X-ray photons the MCA registers a diffraction pattern in the form of an energy spectrum, and the Bragg equation for the fixed angle θ now determines the exact positions, $E_{h\,k\,l}$, of the diffraction maxima on the energy scale:

$$E_{h\,k\,l} = hc/(2d_{h\,k\,l} \sin\theta)$$

A conversion from wavelength to energy using the equation $E = hc/\lambda$, requires the constant $hc/2 = 619.691$ keV pm, which is needed in the evaluation of the lattice spacings $d_{h\,k\,l}$ from the measured energies $E_{h\,k\,l}$.

For a comparison of the ADXD and EDXD techniques Figs 3.7 and 3.8 present two diffraction patterns for the same substance, NaCl, from two typical high-pressure measurements.[2]

With the use of MoKα radiation (17.5 keV) from a conventional X-ray generator, rather large diffraction angles ($5° < 2\theta < 40°$) are necessary to observe the first few diffraction lines from simple solids with the ADXD technique. With the EDXD technique on the other hand, using a fixed angle of around 14°, even a conventional X-ray generator with a tungsten tube and a maximum energy below 60 keV will result in a few more lines with higher h, k, l values, while modern synchrotron-radiation sources sometimes even extend the useful energy range to 100 keV and above.

A more quantitative comparison of these two techniques is illustrated in Fig. 3.9 where the solid curves represent the Bragg relation for various fixed values of the lattice spacing, $d_{h\,k\,l}$. If one uses a fixed energy in this diagram as indicated by the dotted vertical line, the intersections with the curves determine the diffraction angles $2\theta_{h\,k\,l}$ in the ADXD technique for the selected d-values. On the other hand, the intersections with the dashed horizontal line locate the energy values $E_{h\,k\,l}$ for the EDXD technique (the fixed scattering angle $2\theta = 10°$ here). When the dominant diffraction lines correspond to d-spacings below 400 pm, it is convenient to work with diffraction angles of 12°

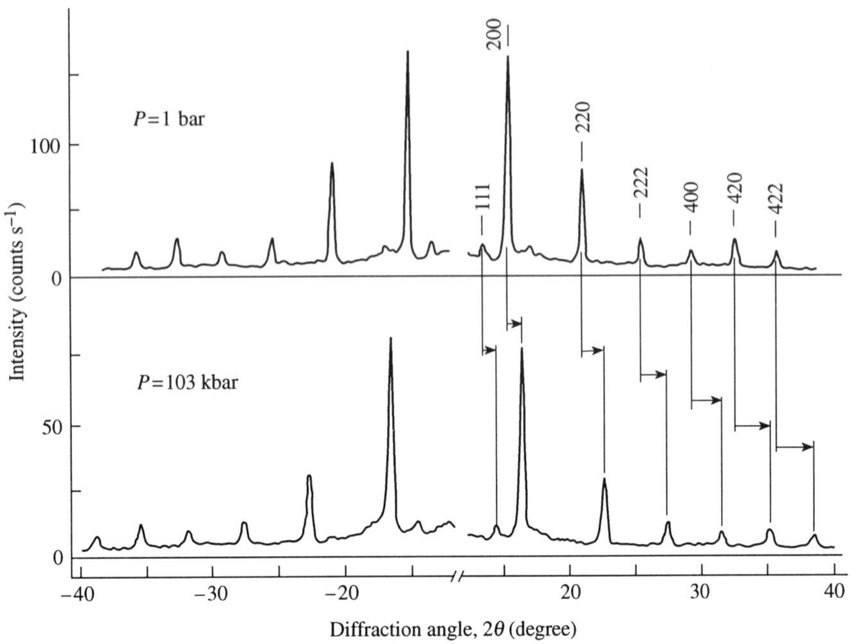

Fig. 3.7 An angular dispersive X-ray diffraction (ADXD) pattern for NaCl at the ambient and a high pressure.[2] (Reprinted with permission from Kluwer Academic Publishers.)

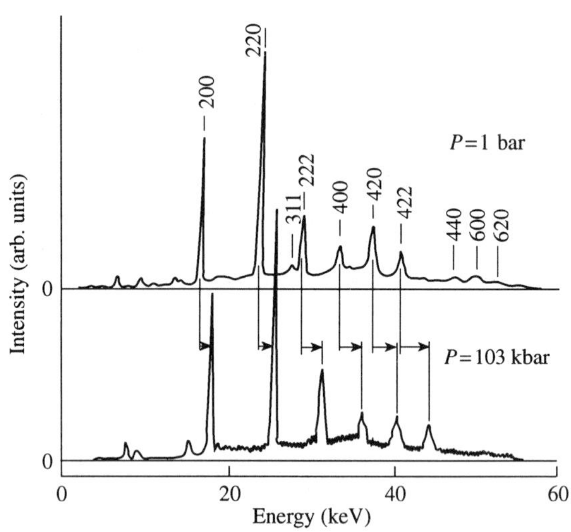

Fig. 3.8 An energy dispersive X-ray diffraction (EDXD) pattern for NaCl at the ambient and a high pressure.[2] (Reprinted with permission from Kluwer Academic Publishers.)

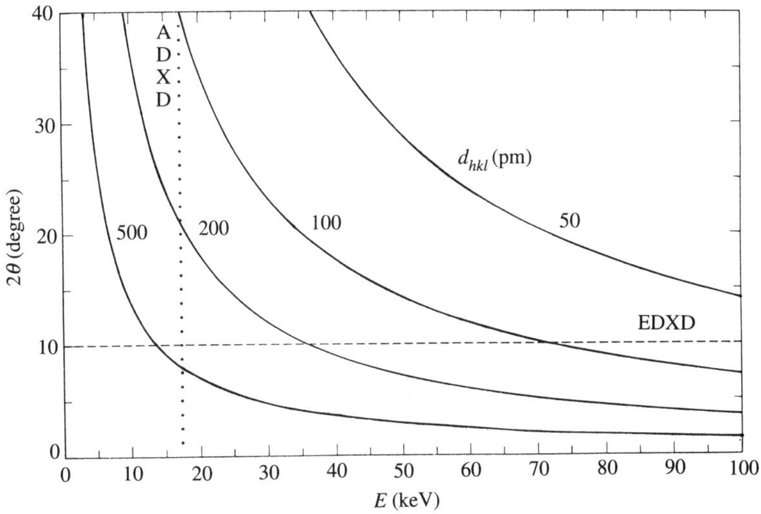

Fig. 3.9 A comparison of the accessible lattice spacings d_{hkl} which can be obtained using the ADXD and EDXD techniques.[8] (Reprinted with permission from Plenum Press.)

or less. In this case, a complete conical-slit system with a large-area detector can be used to optimize the counting rate, as is realized by the very compact EDXD system illustrated in Fig. 3.10, whose cone also includes an optical system for simultaneous optical observation and pressure determination by the ruby-luminescence technique during the X-ray measurements.[8]

A more detailed comparison of the ADXD and EDXD techniques must take into account many different parameters. The EDXD technique using conventional 2 kW X-ray tubes and a conical slit system permits, for instance, kinetic studies with time resolutions of a few minutes on simple samples under pressure,[10] whereas the ADXD technique is several orders of magnitude slower even with a position-sensitive detector. On the other hand, the resolution of possible overlapping lines, especially with the large-area detector used with the conical slit system, is much poorer in the EDXD than in the ADXD technique.

For a quantitative comparison of the resolution $d/\Delta d$, where Δd is the full width at half maximum (FWHM) of the diffraction peaks, one can estimate the ultimate optimum for EDXD by the equation

$$\frac{d}{\Delta d} = \left(\frac{350\ \mathrm{nm}}{d \sin\theta}\right)^{1/2} = 24\left(\frac{E}{\mathrm{keV}}\right)^{1/2},$$

and for ADXD by the equation

$$\frac{d}{\Delta d} = \frac{1\ \mathrm{rad}}{\Delta\theta}\tan\left[\arcsin\left(\frac{\lambda}{2d}\right)\right] \approx \frac{\lambda}{2d}\frac{1\ \mathrm{rad}}{\Delta\theta},$$

when the angular resolution, $\Delta\theta$, is measured in radians. With typical values

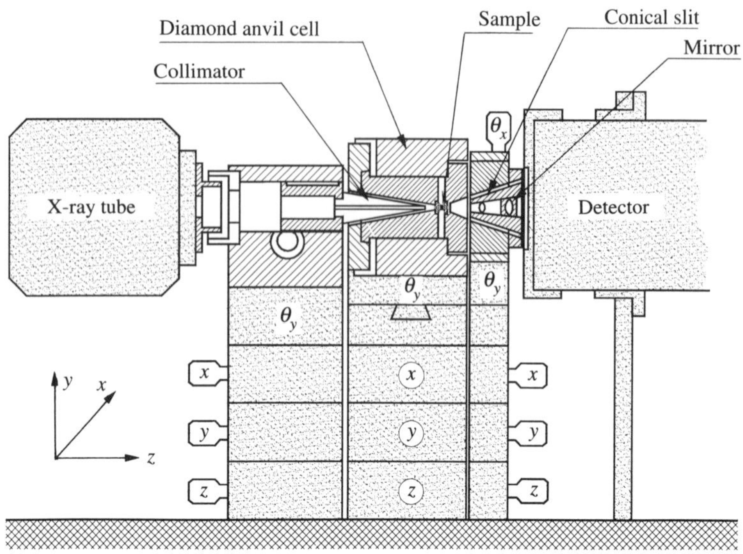

Fig. 3.10 A compact EDXD system with a conical slit and a large-area germanium detector.[8] (Reprinted with permission from Plenum Press.)

for the wavelength λ in the ADXD technique of 50–100 pm and d values of 100–400 pm one obtains a resolution of better than 1/500 with well-collimated synchrotron radiation if the film or area detector is positioned at a distance of 20 cm or more from the sample. However, with conventional sources, the finite divergence and shorter sample-to-film distances lead typically to resolutions below 1/100.

Energy dispersive X-ray techniques result in roughly the same resolution when conventional sources (10 keV $< E <$ 50 keV) are used with a typical value for $\theta \approx 7°$. With synchrotron radiation (10 keV $< E <$ 100 keV), small receiving slits ($<$100 μm) and large detector distances ($>$30 cm) allow resolutions above 1/200 can be obtained if broadening of the lines due to nonhydrostatic conditions can be avoided.

In addition, nonhydrostatic stresses can also lead to anisotropic variations in the d-spacings (this is known as the Singh–Kennedy effect)[12] with a further reduction in the accuracy. Thus full use of the high resolutions of these advanced techniques can only be made with nearly hydrostatic pressure conditions, which are mostly obtained at pressures below about 20 GPa. At higher pressures the EDXD techniques therefore commonly give just the required resolution but much shorter recording times than those typical of ADXD techniques. Also, digital evaluation of the recorded spectra is much better developed at the present time for high-pressure EDXD data. However, current developments in ADXD techniques at large synchrotron-radiation laboratories may rapidly change this situation.

3.1 Single-crystal X-ray diffraction

Single-crystal X-ray diffraction is usually necessary if one wants to determine, under pressure, not just the crystallographic unit cell and its dimensions but also the atomic positions within the unit cell, which corresponds to a complete *structural determination*. In simple or favourable cases structural refinements of powder-diffraction data also give some information about atomic positions under pressure,[13,14] but only with a limited accuracy and only for a small number of free parameters, because of the (usually) small number of reflections which are registered and because of the uncontrolled texture effects which are produced during pressure loading. Thus many applications *require* high-pressure single-crystal-diffraction techniques.

Due to the complexity of the problems involved in the use of this technique, the excellent (and still very comprehensive) monograph by Hazen and Finger on this technique should be consulted.[3] Significant progress in the subject matter of this monograph has only been made in recent years by the application of synchrotron radiation to high-pressure single-crystal X-ray diffraction. Typical examples include on the one hand[15–19] very light gases such as H_2 and He, and on the other hand[20] the first tests on very tiny single crystals in hydrostatic surroundings. With future developments in X-ray area detectors, together with improvements in synchrotron radiation, one can imagine many new opportunities for these techniques.[21–23]

3.2 Experimental considerations

Caution! For safety procedures see Appendix 2. Special hardware, software, and experience is usually needed for high-pressure X-ray diffraction. Therefore, anyone interested in this field should first read some of the extensive literature[1–12,15–23] and preferably contact one of the groups working in this field for possible cooperation at the outset.

References

1. Banus, M. D. In *High Temp.–High Press.* **1969**, 1, 483–515.
2. Holzapfel, W. B. In *High Pressure Chemistry*; ed. Kelm, H. Reidel: Dordrecht, Holland, **1978**, pp. 177–197.
3. Hazen, R. M.; Finger, L. W. *Comparative Crystal Chemistry*; Wiley: New York, **1982**.
4. Holzapfel, W. B. *Revue Phys. Appl.* **1984**, *19*, 705–713.
5. Ruoff, A. L.; Xia, H.; Luo, H.; Vohra, Y. K. *Rev. Sci. Instrum.* **1990**, *61*, 3830.
6. Ruoff, A. L.; Xia, H.; Xia, Q. *Rev. Sci Instrum.* **1992**, *63*, 4342.
7. Skelton, E. F.; Spain, I. L.; Yu, S. C.; Liu, C. Y.; Carpenter Jr, E. R. *Rev. Sci Instrum.* **1977**, *48*, 879–883.
8. Holzapfel, W. B. In *Simple Molecular Systems at Very High Density*, Polian, A.; Loubeyre, P.; Boccara, N.; eds. Plenum Press: New York, **1989**, pp. 257–276.
9. Schiferl, D. A.; Katz, I.; Muls, R. L.; Schmidt, L. C.; Vanderborgh, C.; Skelton, E.

F.; Elam, W. T.; Webb, A. W.; Quadri, S. B.; Schaffer, M. In *Proc. Xth AIRAPT International High Pressure Conference*; Trappeniers, N. J.; Biswas, S. N.; van den Berg, H. R.; Prins C.; Prins, K. O.; Schouten, J. A., eds. North Holland: Amsterdam, **1986**, pp. 897–900.
10. Krüger, T.; Merkau, B.; Grosshans, W. A.; Holzapfel, W. B. *High Press. Res.* **1989**, *2*, 193–235.
11. Boehler, R. *Nature*, **1993**, *363*, 534–536.
12. Singh, A. K.; Kennedy, G. C. *J. Appl. Phys.* **1974**, *45*, 4686–4691.
13. Akahama, Y.; Kobayashi, M.; Kawamura, H. *Phys. Rev.* B, **1993**, *47*, 20–47.
14. Holzapfel, W. B.; Krüger, T.; Sievers, W.; Vijaya-kumar, V. *Jpn. J. Appl. Phys.* **1993**, *32*, 16–21.
15. Hazen, R. M.; Mao, H. K.; Finger, L. W. and Hemley, R. *Phys. Rev.* B, **1987**, *36*, 3944–3947.
16. Mao, H. K.; Helmey, R. J.; Wu, Y.; Jephcoat, A. P.; Finger, L. W.; Zha, C. S.; Bassett, W. A. *Phys. Rev. Lett.* **1988**, *25*, 2649–2652.
17. Mao, H. K.; Jephcoat, A. P.; Hemley, R. J.; Finger, L. W.; Zha, C. S.; Hazen, R. M.; Cox, D. E. *Science* **1988**, *239*, 1131–1134.
18. Hemley, R.; Mao, H. K.; Finger, L. W.; Jephcoat, A. P.; Hazen, R. M. and Zha, C. S. *Phys. Rev.* B, **1990**, *42*, 6458.
19. LeToullec, R.; Loubeyre, P.; Pinceaux, J. P.; Mao, H. K.; Hu, J. *High Press. Res.* **1992**, *6*, 379–388.
20. Skelton, E. F.; Ayers, J. D.; Quadri, S. B.; Moulton, N. E.; Cooper, K. P.; Finger, L. W.; Mao, H. K.; Hu, Z. *Science* **1991**, *253*, 1123–1125.
21. Piltz, R. O.; McMahon, M. I.; Crain, J.; Hatton, P. D.; Nelmes, R. J.; Cernik, R. J.; Bushnell-Wye, G. *Rev. Sci Instrum.* **1992**, *63*, 700–703.
22. Nelmes, R. J.; Hatton, P. D.; McMahon, M. I.; Piltz, R. O.; Crain, J.; Cernik, R. J. Bushnell-Wye, G. *Rev. Sci. Instrum.* **1992**, *63*, 1039–1042.
23. Nelmes, R. J.; McMahon, M. I.; Hatton, P. D.; Piltz, R. O.; Crain, J. *High Press. Res.* **1992**, *8*, 677–684.

4. Neutron scattering

J. M. Besson

4.1 Introduction

The elastic scattering of slow (thermal) neutrons from crystals is an important tool for the study of condensed matter. It complements the information obtained from other physical probes such as X-ray diffraction or the Mössbauer effect. In addition, because of the very low energy of neutrons, compared with X-rays of the same wavelength, inelastic processes due to interactions with excitations of the system (phonons or magnons) can readily be detected and studied. Nevertheless, neutron scattering under high pressure has not been used for the study of highly condensed matter on the same scale as X-ray scattering or optical measurements. This is due to the specific properties of neutrons. Thus, although neutron scattering has, in many respects, advantages over X-rays as a physical probe of condensed matter, it also has drawbacks, the principal one being the requirement for large sample volumes. This has somewhat restricted the use of neutron-scattering methods under high pressure. In contrast to X-rays, neutrons are massive particles and their associated de Broglie wavelength is

$$\lambda = h(2mE)^{-\frac{1}{2}},$$

with $h(2m)^{-\frac{1}{2}} = 0.028\ (nm)(eV)^{-\frac{1}{2}}$ in the common laboratory units.

Thus neutrons with wavelengths of 0.1 nm which are suitable for diffraction measurements on typical crystals have a kinetic energy of 80 meV, whereas X-rays of the same wavelength have an energy of 12.4 keV. Neutrons can thus exchange a large fraction of their energy with excitations (phonons or magnons) of the lattice by inelastic processes, and thus they can give direct information on the dispersion curves of these quasiparticles.

Neutrons have a magnetic moment of 1.94 Bohr magnetons and therefore they are sensitive to magnetic interactions in solids, for which they traditionally have been a favourite probe.

Since neutrons are uncharged their main interaction with matter (in contrast to electromagnetic waves) is not with the electrons in atoms, but with the nuclei through the nuclear interaction. This has a number of consequences.

1. In an elastic-scattering experiment (such as diffraction) the actual position of the atomic nuclei is given by the neutron scattering. This may be quite different from the barycentre of the electronic environment observed using X-rays, for example in C_6H_6, the electronic cloud around the hydrogen atoms is shifted some 10 pm with respect to the position of the proton.

2. The magnitude of the interaction depends on the nucleus and it can be quite large even for light elements (such as hydrogen or lithium) and so it depends on the *isotope*: neutron scattering is much more efficient than X-rays for probing the location of the all-important hydrogen sites in condensed matter. Actually, deuterium is preferred to hydrogen in coherent scattering experiments because of its small incoherent scattering cross section.

3. The nucleus is much smaller than the electronic cloud, so the effective absorption cross section of neutrons is orders of magnitude smaller than that of X-rays. Neutrons can easily penetrate matter, which is an obvious advantage for high-pressure studies. Thermal neutrons will easily go through the walls of high-pressure cells which are several centimetres thick. Moreover, the absorption corrections in a given sample are minimal, and structural studies using Rietveld profile-refinement procedures can be performed with a high accuracy.

4. The price to pay for these favourable features is high. The coherent scattering cross section of neutrons is *quite small* in comparison with X-rays, and much larger samples are required. In addition, the flux of neutron sources (reactors or spallation sources) at the sample position is of the order of

10^5–10^8 neutrons cm^{-1} s^{-1}, which is many orders of magnitude smaller than that of an X-ray synchrotron source, or of a rotating anode for that matter. In practice, an X-ray diffraction pattern can be obtained from a single crystal some 10^{-3} mm^3 in volume. For neutrons, 10 mm^3 crystals are required. This ratio of 10^4 also applies to powder diffraction patterns. With the most advanced neutron sources which exist at the present time (the Grenoble ILL reactor, or the ISIS spallation source in the UK), neutron scattering experiments require sample volumes of 10–100 mm^3 for powders, and 1–10 mm^3 for single crystals, for both elastic (diffraction) and inelastic scattering.

Neutron diffraction experiments with much smaller sample volumes (~ 0.1 mm^3) are indeed possible, but they only yield values for the lattice parameters, and not for the interatomic distances in a unit cell. Neutron scattering methods have no advantage over X-ray methods for measuring these interatomic distances, for which X-rays should be preferred.

4.2 Neutron sources

Unlike X-rays, which can be produced in a laboratory, neutrons are only available in sufficient quantities at large facilities. Two main types of such facilities must be distinguished, reactor sources and spallation sources, because they have completely different characteristics in high-pressure experiments. The principle of reactor sources need not be described since they are by far the most common sources and have been used for several decades for solid-state studies. The flux from a reactor is usually continuous—the pulsed DUBNA source is an exception—and reactor fluxes are used mainly in the monochromatic angle dispersive mode (as ADXD) for X-ray diffraction.

Spallation sources, which tend to be preferred for neutron diffraction at high pressures are quite different (Fig. 3.11). They give a pulsed flux of neutrons, for a few microseconds, with a repetition rate of 10–50 Hz. Since thermal neutrons are slow particles with velocities of a few kilometres per second, they have flight times of several milliseconds to the sample position, which is typically 10–100 m from the target, and thus their kinetic energy $E = \frac{1}{2}m^{v2}$ can be measured by time-of-flight methods. This method has the same advantages as EDXD for X-rays (see Section 3), by allowing work in a fixed 2θ diffraction mode. In high-pressure studies, the diffraction parallelogram can be maintained within the sample by appropriate collimation of the ingoing and outgoing beams (Fig. 3.12), and the background signal from the wall of the high-pressure cell can be almost eliminated, although the cell itself will be orders of magnitude more bulky than the sample. Figure 3.13 shows how important this effect may be[1,2] for measurements on powder samples.

Elastic or inelastic scattering measurements on single-crystal samples are much less affected by background problems because of the high intensity of

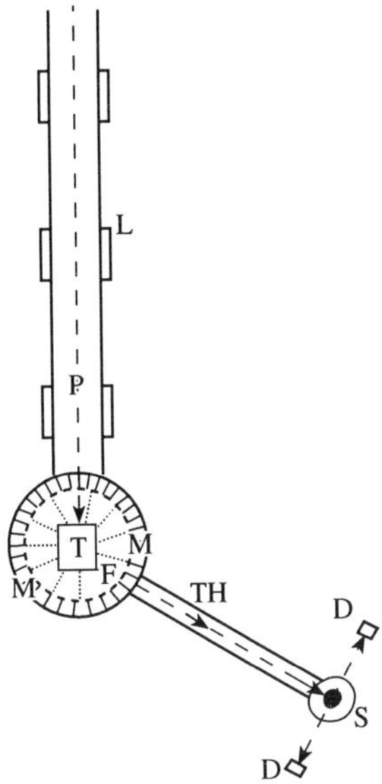

Fig. 3.11 The schematics of a spallation source: (L) a proton linear accelerator (linac), (P) pulsed proton beam (800 MeV), (T) heavy-metal target (tantalum, uranium, etc.), (F) fast neutrons, (M) moderator (H_2O,CH_4,), (TH) thermal neutrons, (S) sample location, (D) 90° detectors.

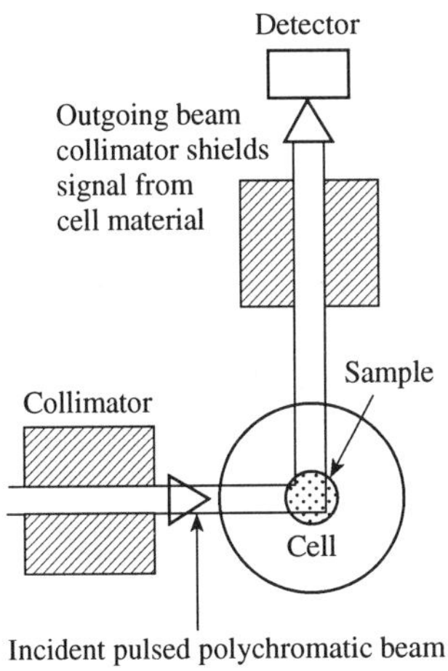

Fig. 3.12 Shielding of the incident and scattered neutron beams with a fixed-angle ($2\theta = 90°$) geometry.

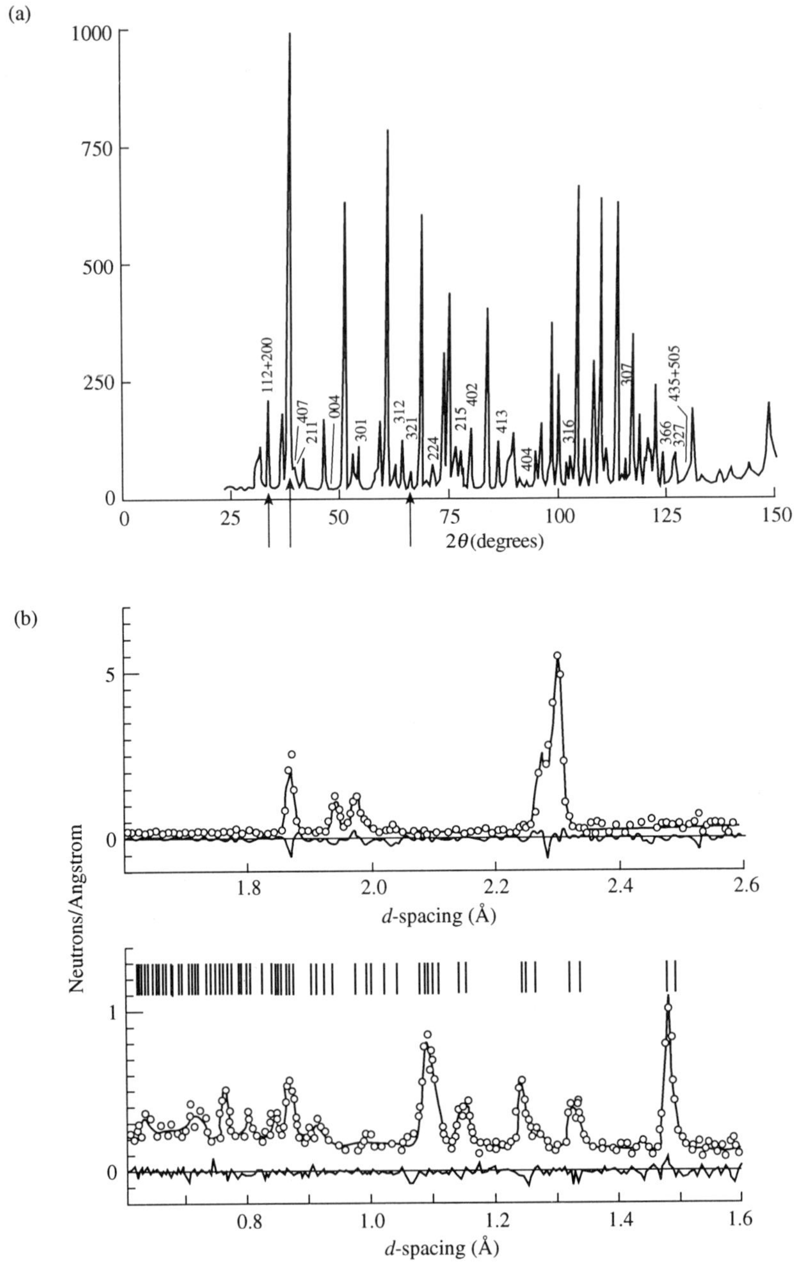

Fig. 3.13 A comparison of (a) angle-dispersive (2.5 GPa) and (b) energy-dispersive (7.8 GPa) set-ups for shielding the background signal. (a) A D_2O ice VIII pattern from reference 1. The cell was the alumina bicone arrangement shown in Fig. 3.15. The large diffraction peaks all are background signals from the Lucalox (Al_2O_3) cylinder. The sample's signal is the very small black peaks indicated by the arrows. (b) A D_2O ice VIII pattern from reference 2. The set-up is shown in Fig. 3.16 and 3.17, and it uses the shielding procedure shown in Fig. 3.12. No background signal from the tungsten carbide anvils is detectable.

the individual reflections from single crystals, and both types of source can be used in high-pressure work.

The high-pressure cells currently used for neutron-scattering experiments fall into three well-defined categories

(1) quasihydrostatic cells for measurements below 1 GPa, preferably using helium as the pressure medium;
(2) clamp-type supported piston–cylinder devices for measurements up to 2.5 GPa;
(3) opposed-anvil systems, which have been used, at the time of writing, up to 25 GPa.

4.3 Measurements below 1 GPa

Below a pressure of 1 GPa most measurements can be made using single-crystal samples with helium as the pressure-transmitting medium; this ensures excellent pressure homogeneity, even at low temperatures, since helium is a soft solid. Simple cylindrical cells with a capillary-tubing inlet for pressurization (see Section 1) are in frequent use (Fig. 3.14). They have been

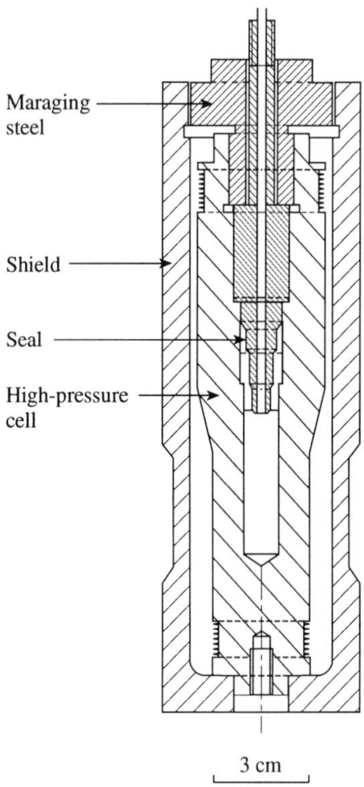

Fig. 3.14 A high-pressure helium-gas cell. This cylindrical-window aluminium-alloy cell will support 800 MPa at 4.2 K. (Reprinted with permission from ref. 3. Copyright 1975 Pion Limited, London.)

manufactured from low-absorption materials, such as aluminium alloys,[3] but they now tend to be made of a Ti–Zr null-scattering alloy or zero-matrix materials, which have zero (or small) coherent scattering cross sections and thus do not introduce the background scattering shown in Fig. 3.13(a). The properties of such alloys stem from the fact that the terms in the Fermi scattering length in isotopic species can be positive or negative. Thus the scattering cross section, which is proportional to the square of the scattering length, may be made to have a zero coherent term by choosing an appropriate isotopic composition. This is the case for an approximately 50–50% alloy of natural titanium and zirconium. In addition, such alloys have a compressive strength of 700–900 MPa, and thus they are tolerably strong construction materials. Unfortunately, if their coherent scattering cross section can be made equal to zero, their incoherent scattering and absorption will remain quite normal and thus, in order to reduce the attenuation of the neutron beams, the thickness of the walls must be reduced.

Cells with a capillary fluid inlet, such as that shown in Fig. 3.14, when used at low (liquid ^4He) temperatures have to be heated while the pressure is varied until the helium in the capillary becomes fluid. This can be avoided by directly compressing the cell with a piston: a number of such devices exist. Although the devices described in this subsection have usually been tested above 1 GPa, they are, in practice, only used between 500 and 800 MPa, for safety reasons.

4.4 Measurements up to 2.5 GPa

A number of devices[7] have been described for the quasihydrostatic compression of monocrystals or powders up to pressures of 2–3 GPa, but the biconical clamp-type cell is now the most commonly used.[5,6] In this type of device, the inner cell is an alumina (Lucalox) bicone which is externally supported by matching steel conical rings under an axial compression of 200–300 kN (Fig. 3.15). A gap between the rings gives a full 360° angle access to the sample in the equatorial plane. The sample is compressed using tungsten-carbide pistons. In the original design, the supporting external thrust was matched to the internal pressure, and pressures up to ~ 4 GPa were generated. Currently, clamped devices with a fixed supporting thrust are used, and they are generally capable of pressures of ~2.5 GPa. The cell can readily be cooled to 4.2 K, especially in the compact version.[6] When used with a liquid (Fluorinert or CS_2) pressure-transmitting medium, encapsulation of the sample volume is necessary because of the porosity of alumina. This can be avoided by replacing the alumina of the bicones by a Ti–Zr alloy, which also allows work to 2 GPa, with no background scattering, but with the drawback of somewhat larger attenuation.

4.5 Pressures above 3 GPa

Carlile and Slater have given an account of early attempts at designing devices for pressures above 3 GPa.[4] Since 1990, two new developments have

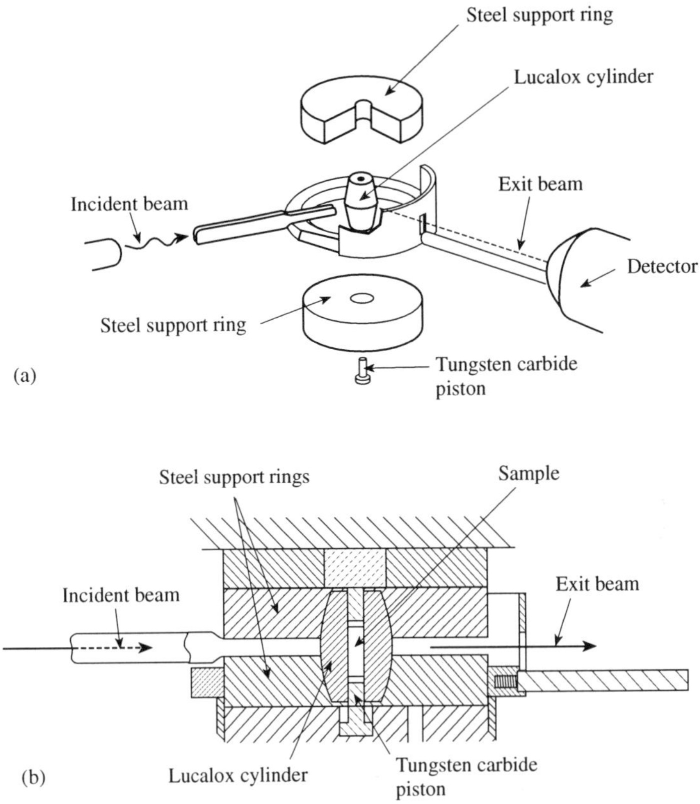

Fig. 3.15 A supported alumina bicone cell:[5] (a) an expanded view, and (b) a cross section[5]. (Reprinted with permission from *Rev. Sci. Instr.*, **45**, 643, (1974).)

allowed studies well above 2.5 GPa. One is the use of sapphire-anvil cells, which allow refinable patterns to be obtained up to 7 GPa. The other is the development of Paris–Edinburgh cells with toroidal anvils, which allow refinable patterns of powder samples to be obtained at pressures up to 25 GPa.

Sapphire-anvil cells are based on the use of sample volumes smaller than 1 mm^3; they have a high detector sensitivity and excellent background-rejection levels (for which dedicated wide-angle detectors have been designed[7]). The capabilities of such methods have been demonstrated by a remarkable measurement of the equation of state of deuterium[8] at pressures above 30 GPa, in a single-crystal experiment with large diamond anvils. In this type of experiment, the volume ($< 10^{-2}$ mm^3) is much too small to yield powder patterns of refinable quality. This means that only lattice parameters can be extracted, and not interatomic distances, and this can actually be done much more easily by X-ray diffraction. Since large diamond anvils of a quality

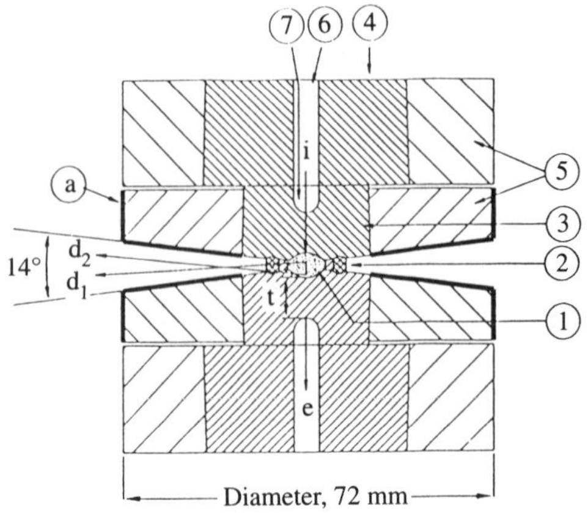

Fig. 3.16 A cross section of the toroidal anvil arrangement used in the Paris–Edinburgh cell:[2] (1) sample (~ 100 mm^3), (2) gaskets (null-scattering Ti–Zr alloy), (3) tungsten-carbide anvils, (4) tungsten-carbide seats, (5) steel binding rings, (6) hole for the incident neutron beam, (7) recess at the back of the anvils to reduce absorption, (a) cadmium plating, (e) transmitted neutron beam, (d_1, d_2) diffracted neutron beams, and (i) incident neutron beam.

adequate for high-pressure experiments are, and will remain (see Chapter 3, section 5) unavailable, sapphire anvils have to be used. The order of magnitude of the sample volume for this type of apparatus is $V(\text{mm}^3) = 1000 \times P^{-4}$ (GPa).

The advantage of such devices is that they are light (< 10 kg) and can easily be cooled to cryogenic temperatures. They also allow a large-angle access in the equatorial plane, and in directions normal to this plane. Their main drawback is the limited volume of the samples, which, at the present time does not allow profile-refinement procedures to be carried out above 10 GPa. Nevertheless, they are extremely useful for low-temperature, high-pressure work, such as the study of magnetic phase transitions.[9]

A different approach has been used in the Paris–Edinburgh cells and presses shown in Figs 3.16 and 3.17. The compressed sample volume is enclosed in a spheroidal chamber at the centre of a tungsten-carbide or a sintered diamond anvil, and it is contained by an assembly of null-scattering gaskets. The sample and gasket assemblies are compressed by a hydraulic ram[2] with a capacity of 2.5 MN. The press and ram assembly of the original design, shown in Fig. 3.17, has a mass of only 50 kg and it is thus portable. With this type of press and anvil profile, the available sample volume is $V(\text{mm}^3) \approx \frac{1}{2}d(\text{mm})^3 \approx 3000\, P^{-3/2}$ (GPa), that is, about 100 mm^3 at 10 GPa. Tungsten carbide is used for these anvils for pressures up to 12 GPa, and sintered diamond is used for higher pressures, up to at least 25 GPa.

For powder samples, the compressed volume will contain the sample, a calibrant (NaCl) and a pressure medium, such as Fluorinert, to provide homogeneous—if not hydrostatic-strain conditions. Paris–Edinburgh cells can be loaded with samples which are liquids (H_2O), or condensable gases (NH_3 or

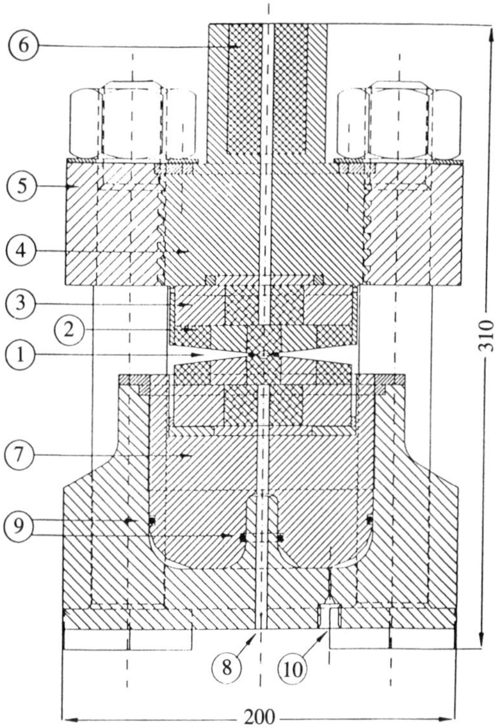

Fig. 3.17 The Paris–Edinburgh V4 press; it has a nominal working capacity of 250 tonnes (2.5 MN):[2] (1) sample and gasket arrangement (see Fig. 3.16), (2,3) anvil and seat binding rings, (4) breech, (5) upper platen, (6) B_4C neutron collimator, (7) piston, (8) hole for transmission measurements, (9) O-rings, and (10) hydraulic-fluid inlet. The dimensions are in millimetres.

H_2S). An *ad hoc* loading system[11] allows room-temperature high-pressure loading of noncondensable gases, such as N_2, H_2, or He. The cell can be operated at low temperatures by replacing the hydraulic fluid by a pentane–isopentane mixture, which is suitable down to 115 K, and by helium for temperatures down to 20 K. This cell was designed specifically for time-of-flight measurements at a fixed 2θ scattering angle of 90° for the POLARIS station of the ISIS spallation source at the Rutherford Appleton Laboratory (Didcot, UK). The incident beam goes through a few millimetres of the anvil material, and the diffracted beam, which exits at $2\theta = 90° \pm 7°$ is detected at the POLARIS station by four ZnS detector modules covering ~ 140° in the plane perpendicular to the incident beam. The diffracted beam (Fig. 3.16) traverses the gasket material and, in part, the anvil material. Thus, in order to obtain precise intensity values for the diffraction patterns, wavelength-dependent attenuation corrections have to be performed; those corrections should include absorption and incoherent scattering from the various materials and

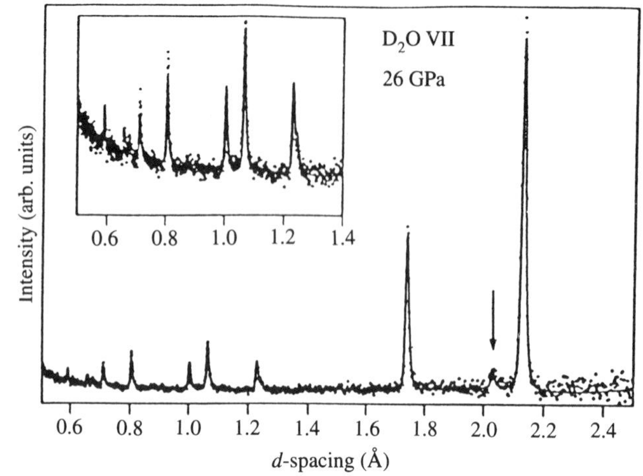

Fig. 3.18 A high-pressure neutron-diffraction pattern of D_2O ice VII collected with a Paris–Edinburgh cell. The arrow indicates the background signal from the sintered diamond anvils which could not be entirely removed by the collimator (from reference 10).

from their Bragg edges. In addition, careful collimation and shielding of the ingoing and outgoing neutron beams is needed, in order to obtain clean and refinable patterns at the highest pressures, such as that shown in Fig. 3.18.

4.6 Experimental considerations

Caution! For safety procedures see Appendix 2. Neutron scattering under high pressure is a very powerful but comparatively new method in condensed-matter studies. It can only be performed at a limited number of neutron facilities, such as the reactors in Grenoble (ILL), Orsay (LLB), or Dubna (Russia), at the spallation sources in the UK (ISIS, Rutherford Appleton Laboratory, Didcot) or the USA (LANSCE, Los Alamos). Potential users of these facilities are encouraged to submit requests for beamtime. Requests should only be made after gathering information on the experimental facilities available and after careful examination of the published results from individual sources.

References

1. Kuhs, W. F.; Finney, J. L.; Vettier, C.; Bliss, D. V. *J. Chem. Phys.* **1984**, *81*, 1220.
2. Besson, J. M.; Nelmes, R. J.; Hamel, G.; Loveday, J. S.; Weill, G.; Hull, S. *Physica* B, **1992**, *180* & *181*, 907.
3. Paureau, J.; Vettier, C. *High Temp.–High Pressures* **1975**, *7*, 529.
4. Carlile, C. J.; Slater, D. C. *High Temp.–High Pressures* **1978**, *10*, 1.
5. McWhan, D. B.; Bloch, D.; Parisot, G. *Rev. Sci. Instrum.* **1974**, *45*, 643.

6. Bloch, D.; Paureau, J.;, Voiron, J.; Parisot, G. *Rev. Sci. Instrum.* **1976**, *47*, 296.
7. Glazkov, V. P.; Naumov, I. V.; Somenkov, V. A.; Shil'stein, S. Sh. *Nucl. Inst. Meth.* A, **1988**, *264*, 367.
8. Glazkov, V. P.; Gontcharenko, I. N.; Iridova, A. V.; Somenkov, V. A.; Shil'stein, S. Sh.; Besedin, S. P.; Makarenko, I. N.; Stishov, S. M. *Zeit Phys. Chem.* **1989**, *163*, 509.
9. Gontcharenko, I. N.; Mignot, J. M.; Somenkov, V. A.; Rosset-Mignot, J.; Vogt, O. *Physica* B, **1994**, *199*, *200*, 625.
10. Klotz, S.; Besson, J. M.; Hamel. G.; Nelmes, R. J.; Loveday, J. S.; Marshall, W. G.; Wilson, R. M. *Appl. Phys. Lett.* **1995**, *66*, 1735.
11. Marenko, I. N. private communication.

5. Optical measurements

J. M. Besson

Optical measurements at high pressures represent the majority of the papers which are published in the field of high pressure studies, whatever the range of pressure. Before going into a description of the techniques, apparatus, and methods which are used for this type of experiment, a brief review is in order of the different physicochemical parameters which can be probed by optical methods, that is, by electromagnetic radiation between a few electronvolts (the near ultraviolet (u.v.)) and a few millielectron volts (the far infrared (i.r.)). For convenience, the various regions of the spectrum in this broad range of photon energies will be examined with regard to the physical information that is most frequently yielded rather than the associated technical problems. When using this approach, optical measurements naturally fall into two broad categories: determination of optical constants by transmission and reflection measurements (Section 5.1), and inelastic scattering and luminescence experiments (Section 5.2) which probe excitations of the sample, in a broad sense.

5.1 Optical constants

Measurement of the optical constants, that is, ideally, of the real and imaginary parts of the dielectric response of a medium, gives full information about the electronic and vibrational transitions in condensed media, at least those which are optically active through dipole interactions. Three regions of the spectrum can be distinguished, although they often overlap; they more or less correspond to distinct physical phenomena.

(1) the near u.v., visible, and near i.r. which allow electronic transitions in solids to be probed by, for example, interband absorption edges in semiconductors;

(2) the near and medium i.r. regions, corresponding to intramolecular vibrational energies;

(3) the medium and far i.r. domain, which mostly arise from intermolecular or lattice vibrations (phonons) (this domain is much less studied at high pressures than the first two domains because of the experimental difficulties with which it is fraught).

The transitions between electronic levels that can be studied at high pressures mostly lie in the photon energy range from 0.5 to 5–6 eV; measurements are most often made using absorption, fluorescence, and excitation spectroscopy, or reflection in the case of transitions above the energy gap with large oscillator strength. One example of this type of study is the variation with pressure of the energy gaps in semiconductors; this has been the prime tool for the past three decades in the study of crystals at high pressures. Classical semiconductors belong to a homogeneous series of families from group IV (Si, Ge), group III–V (GaAs, InSb), or group II–VI (CdS, ZnSe) of the periodic table, which all have the quasi-isostructural networks of diamond, zincblende, or wurtzite. In the 1960s, K. Paul and his group experimentally established the phenomenological rule that, in a given structural series, electronic transitions between levels with the same symmetry have similar pressure coefficients. This means that the shift which is observed under pressure in an absorption edge has a pressure coefficient that is characteristic of the symmetry of the initial and final levels. For example, in the

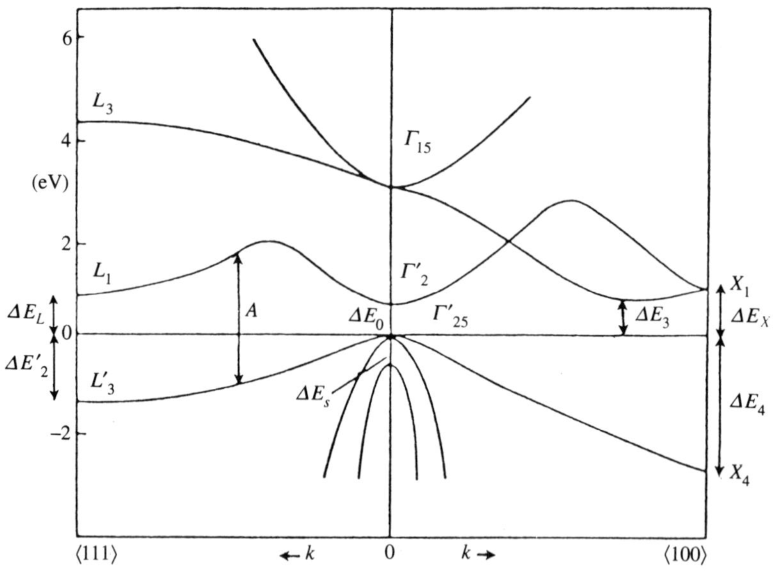

Fig. 3.19 The band structure of germanium, which is representative of that of the semiconductor families from groups IV, III, V and VI. (Reprinted with permission from Smith, R. A. Semiconductors. Cambridge University Press, **1978**.)

group III–V family, transitions (Fig. 3.19) between the top of the valence band ⟨000⟩ and extrema of the conduction band have the following values:

$$\Delta E_0 = \Gamma\text{–}\Gamma\,\langle 0\,0\,0\rangle \approx 100\text{–}140\ \mathrm{meV\ GPa^{-1}},$$
$$\Delta E_X = \Gamma\text{–}X\,\langle 1\,0\,0\rangle \approx -10\ \mathrm{meV\ GPa^{-1}},$$
$$\Delta E_L = \Gamma\text{–}L\,\langle 1\,1\,1\rangle \approx +50\ \mathrm{meV\ GPa^{-1}}.$$

Thus, the measurement of the pressure shift of an optically active electronic transition is a direct indication of its nature. Although this rule was originally validated for the group IV, and group III–V families, it was later extended to a number of other series—such as group IV–VI (PbSe, PbS), the chalcopyrites of the group I–II–VI_2 (Cu, Ga, S_2) families, or group III–VI layer compounds (GaSe, InSe)—with equal success. Measurements of such optical transitions under pressure are a powerful tool to test theoretical calculations on solids. Whichever method is used in such calculations, the first result is to predict the equation of state and the variation of the electronic levels with the density. Thus the variation of differences in energy (optical transitions) between electronic levels is a crucial test of the accuracy of calculations on condensed systems.

Since the relevant parameter in such measurements is the pressure coefficient of a given transition, this has often been simplified just to measurements of an absorption edge under pressure, that is, the shift under pressure of a transmission curve at experimentally accessible values of absorbancy. This, unfortunately, does not usually yield acceptable results. In general, *all* the parameters of an optical transition will vary with the pressure, and not just the characteristic energy. In particular, the intensity of the transition (oscillator strength) will vary; in the case of, for example, a direct transition between a valence and a conduction band, this is related to the density of states, to the effective mass, and to the dielectric constant, among other factors. Thus, an absorption edge may become more or less 'steep' under pressure, and measurements in the low-absorption region will yield pressure coefficients which are quite far from the values of the actual gap. There is no choice but to interpret a given absorption edge in terms of some physical model and to extract from it the variation of the energy gap(s) from which it originates. In the case of semiconductors, this implies, for instance, making use of the model of Elliott[1] and of Toyozawa[2] for direct transitions and that of Dumke *et al.* for indirect transitions.[3]

It has been assumed in the discussion that the absorption coefficient has indeed been extracted from transmission measurements under pressure. This is a far from trivial problem which is discussed in Section 5.4 in relation to the use of DACs, where this difficulty is most acute.

Reflectivity measurements are, in principle, necessary to obtain the absorption coefficient from the transmission data. Such measurements are usually too difficult to perform directly for precise absolute values under high pressures in regions where absorption measurements are carried out. They can be extrapolated from the known dispersion of the refractive index at ambient pressures, and from its variation with pressure in the region where interference fringes can be observed in plane-parallel Fabry–Pérot-type samples. High-energy transitions between deep levels are observed by following the peaks in the reflectivity spectra which are associated with singularities in the joint density of states corresponding to electronic transitions.

Infrared spectroscopy of condensed media under pressure is usually used to investigate the variation of the vibrational properties, which are a direct measure of the interatomic (intermolecular) restoring forces acting on the atoms, and thus of the effective interatomic interactions which are brought about by the electronic and ionic contributions. Ideally, the calculation of the vibrational properties of a solid should follow *ab initio* calculations of its pressure–temperature stability, of its equation of state, and of its electronic properties. This ideal is very seldom achieved, even for the $k = 0$ modes in the centre of the Brillouin zone which can be probed by optical measurements in the i.r.

Infrared measurements at high pressures are mainly used to study intramolecular vibrational modes in the near i.r. region, that is, of wavenumbers between 5000 and 500 cm^{-1}. This characteristic region for the intramolecular frequencies of oscillators is important in the chemistry of O–H, C–H, or C–C vibrations in organic molecules. Their variation at high pressures gives a direct measure of changes in chemical bonding for variable density and, in the case of complex molecules, of their deformation under pressure. Analogous results can be obtained from Raman scattering (see Section 5.2); but the variation in the intensity of active i.r. modes is information which can directly be interpreted as the variation of the associated dipole moment (effective charge), whereas variations in the intensities of Raman lines are of no direct significance. In crystals with inversion symmetry, both these methods are complementary, since they probe sets of frequencies which are mutually exclusive because of selection rules.

Such measurements at high pressures do not present any special difficulty since commercial spectrometers will accept most types of high-pressure cells in the experimental space with suitable high-pressure windows (see Section 5.3). Nevertheless, adequate pressure-transmitting fluids must be used for the study of solids, and this excludes organic fluids—such as the pentane–isopentane or the ethanol–methanol mixtures which are commonly used in the visible region—since their i.r. absorption bands preclude their use. For i.r. measurements, rare gases such as helium or argon must be preferred in both large-volume cells and in DACs.

The low-energy end of the optical spectrum, the mid and far i.r. region (wavenumbers of 1000 to 10 cm^{-1}), allows observation of phonon (intermolecular) modes in solids. This is a domain where high-pressure methods, although potentially quite useful, have been comparatively little used. The small number of experiments which have been performed in this region were mostly on molecular compounds[4] using large-volume, low-pressure cells with suitable windows, which present no particular difficulties for high-pressure studies (see Section 5.3). When using DACs attention must be paid to the adaptation of the cell to the optical instrument, which is usually a Fourier-transform interferometer. This is examined in Section 5.4.

In any case, at the present time, low-frequency vibrational modes are much more conveniently studied under pressure by Raman-scattering methods (see Section 5.2) than by i.r. methods, although the former only gives access to the g-modes when inversion symmetry allows to distinguish between u and g modes. In the far i.r., DACs are limited by the aperture of the gasket hole, which limits accurate measurements at long wavelengths because of diffraction effects. For this reason, reflectivity measurements under very high pressures in the *reststrahlen* region of ionic crystals (50–500 cm^{-1}) have not been performed, although they would yield a wealth of information on the evolution of ionic bonds at high densities.

Infrared reflectivity measurements can be used to measure the charge-carrier density in conducting media and to probe the occurrence of insulator–conductor transitions in dense matter. The real and imaginary parts of the dielectric function are given, respectively, by

$$\varepsilon_1 = n^2 - K^2 = \varepsilon_\infty - \frac{Ne^2}{m\varepsilon_0} \bullet \frac{\tau^2}{1+\omega^2\tau^2},$$

$$\varepsilon_2 = 2nK = \varepsilon_\infty - \frac{Ne^2}{m\omega\varepsilon_0} \bullet \frac{\tau}{1+\omega^2\tau^2}, \qquad (3.5.1)$$

where n is the refractive index, K is the extinction coefficient, m is the effective mass of the carriers, ε_∞ is the high-frequency dielectric constant, τ is the relaxation time, N is the number of carriers, and ω is the angular frequency of light.

Both ε_1 and ε_2 can be derived by the Kramers–Kronig inversion formulae from the reflectivity spectrum over a large enough region of frequencies and such measurements have been used to study, with DACs, the occurrence of conduction due to increasing band overlap in a number of systems, such as xenon and hydrogen. The accuracy of this approach is only moderate since it requires accurate absolute values for the reflectivity, which necessitates a knowledge of the optical properties of diamond anvils, which are strongly affected by the strains at the diamond–sample interface.

Measurements of the refractive index under pressure are necessary, in addition to the intrinsic refractive index as physical probes of condensed media, to interpret transmission measurements. This is most conveniently done under pressure by measuring Fabry–Pérot interference patterns in plane-parallel samples. The position of the interference maxima in transmission is given by

$$2\,nd\bar{\nu} = k, \qquad (3.5.2)$$

where n is the refractive index, d is the thickness of the sample, $\bar{\nu}$ is the wavenumber of light, and k is an integer.

So that the interfringe separation $\Delta\bar{\nu}$ is given by

$$2\,nd\Delta\bar{\nu}\left(1+\frac{\bar{\nu}}{n}\frac{dn}{d\bar{\nu}}\right)=1. \tag{3.5.3}$$

When the index dispersion $dn/d\nu$ is negligible, the interfringe separation gives the value of the refractive index directly, provided that the variation of d with pressure is known from separate measurements of the equation of state. This is, as a rule, not the case, and if the interference order, k, is not known and the variation of the interfringe separation is to be used on its own to derive the variation of n with pressure, some kind of measurements of, or assumptions about, the variation of $dn/d\nu$ with pressure must be made. The most direct and reliable method is to determine directly the interference order k for very small k in the i.r., and to use eqn (3.5.2) which unambiguously provides $n(P)$.

5.2 Scattering and luminescence

Raman scattering is the most widely used of the scattering and luminescence methods because it is quite well adapted to high- and very-high-pressure measurements. Visible light alone is used, and the sample does not need to be larger than 10–20 μm, that is, the size of the focused spot of the laser. Thus, at the present time, Raman scattering is the preferred method for the study of dense media, together with its complement, X-ray scattering. Raman scattering probes the restoring forces (vibrational frequencies) in Raman active optical modes, through the electron–phonon interaction. In addition, because of selection rules, it also gives information in molecular crystals about the molecular structure (molecular symmetry), the local static environment (site symmetry), and dynamical couplings (factor group). Thus, a complete interpretation of Raman spectra in polarized light should yield a wealth of information on condensed media. Such is, unfortunately, not the case in practice under high-pressure conditions. Polarized-light measurements are always very approximate—when performed—because of the depolarization induced by pressure-strained windows (see Section 5.3), and so variations in the intensity of Raman signals under pressure are not of any direct significance. Only the variation of the frequency of Raman active modes can easily be studied and interpreted under high pressures. Raman lineshapes and homogeneous or inhomogeneous broadening certainly contain precious information about lifetimes and decay processes, but they have not, yet, been studied under high pressures. The same is true for stimulated Raman scattering, which has been used in dynamic (shock-wave) measurements[5] but not used widely in static measurements.

Coherent anti-Stokes Raman spectroscopy[6] (CARS) is a potentially powerful method which is now emerging for high-pressure studies, both in large-volume cells and in DACs. Due to frequency coupling in the sample, CARS allows transitions with energies higher than the absorption edge of the window material to be probed; it thus overcomes this intrinsic limitation. Also, CARS signals are strong and free from luminescence or thermal backgrounds.[7]

Brillouin scattering is the inelastic scattering of light by thermal acoustical phonons in a material. Energy and wavevector conservation in the process are given, respectively, by

$$\begin{aligned} \hbar\Omega &= \pm\hbar(\omega_i - \omega_s), \\ \boldsymbol{Q} &= \pm(\boldsymbol{K}_i - \boldsymbol{K}_s), \end{aligned} \tag{3.5.4}$$

where Ω and $\boldsymbol{Q}$ are the frequency and the wavevector of the phonon, respectively, and ω and $\boldsymbol{K}$ are the frequency and wavevector of light; the subscripts i and s denote incident and scattered light, respectively.

The interaction occurs at small wavevectors, and thus at small energy differences. The frequency changes are of the order of a few per centimetre and Fabry–Pérot-type interferometers have to be used which are necessarily more complex than the simple spectrometers used for Raman scattering. The information obtained in Brillouin scattering is the elasticity tensor of the material, and this can be obtained with much higher precision by ultrasonic techniques which are usable in large-volume cells. Thus, for pressures below 2 GPa, ultrasonic techniques are preferred to Brillouin scattering, which, in turn, is the most straightforward technique for the measurement of elasticity coefficients at high pressures in DACs.

For high-pressure work, tandem-type Fabry-Pérot instruments with high contrast[8] should be used. The elasticity coefficients, C, are given by:

$$C = \frac{v^2}{\rho} = \frac{\Omega^2}{Q^2\rho}, \tag{3.5.5}$$

where

$$Q = f_n(K).$$

C is an effective coefficient which depends on the polarization and propagation direction of the phonon, and v is the speed of sound, ρ is the density, and $K \approx K_i \approx K_s$, and $f_n(K)$ is a function of the scattering geometry.

In the case of a cubic crystal with three independent coefficients, C has the following values for phonons propagating along the [1 1 0] direction:

for a longitudinal phonon

$$C = \tfrac{1}{2}(C_{11} + C_{12} + C_{44}), \tag{3.5.6}$$

for a transverse phonon with polarization along the [0 0 1] direction

$$C = C_{44}, \tag{3.5.7}$$

for a transverse phonon polarized along the [1 1 0] direction

$$C = \tfrac{1}{2}(C_{11} + C_{12}). \tag{3.5.8}$$

Now $f_n(K)$ is dependent on the scattering geometry through the refractive index, which is needed to interpret Brillouin shifts. This dependence can be solved either by performing measurements at different incident and scattered beam angles, or by having incident and scattered beams which are at right angles, with the sample in the DAC at 45° to both beams. Then the scattering equation is independent of the refractive index, n, since

$$Q = \sqrt{2}K. \tag{3.5.9}$$

In general, in order to obtain the full elasticity tensor from Brillouin measurements in a DAC, several incident and scattered beams must be used, and proper access ports must be machined into the diamond seats. In addition, the equation of state and the variation of the refractive index with pressure are necessary.

In practice, limiting values for combinations of elasticity coefficients versus pressure can be obtained using only the backscattering geometry and suitable hypotheses, and indirect evaluations of the equation of state of fluid media can be derived from Brillouin scattering measurements using the density dependence of the Brillouin shift.

A more elaborate method[10] allows the full elasticity tensor of cubic crystals to be determined, with arbitrary orientation, in a DAC. This requires measurements at several angles from the cell's axis over a large angular span (180°) and it therefore necessitates large (>90°) optical apertures. A fit of the observed velocities allows both the single-crystal orientations and the full set of elastic coefficients to be determined.

Observation of luminescent processes under high pressures does not present any particular experimental difficulties, other than those mentioned previously for the observation of the optical properties or of Raman scattering. Indeed, photoluminescence (fluorescence) of Al_2O_3:Cr^+ (ruby 'luminescence') is used as a standard high-pressure secondary gauge (see Section 2.2 in Chapter 2) and is therefore the most frequently used physical measurement under high pressures in DACs. Photoluminescence is quite a sensitive probe in semiconductors for band-crossing processes such as those that occur in GaAs, which is a direct band-gap semiconductor at ambient conditions and which is an indirect band-gap semiconductor above 4.1 GPa. The decrease in intensity, and the different pressure dependence of the luminescent levels below and above band crossing, have been the subject of a number of papers in the last decades.

Electroluminescence under high pressures is often used for the study of light-emitting diodes (LEDs) and semiconductor p–n diode lasers in large-volume cells. Moderate pressures can produce large variations in the band gap of semiconductors, and they may therefore yield large variations in the emitted wavelength of solid-state lasers.[11] These are the subject of numerous investigations on group III–V semiconductor heterostructures, where the detailed evaluation is difficult due to the large number of parameters which vary under pressure in such devices including: the band gap of the various components, the refractive indices, the band offsets, and the lifetimes in the different conduction-band minima.

The photoluminescent response is also used to study weak absorption coefficients due, for instance, to impurity states in the band gap of semiconductors, and it thus complements absorption measurements.

5.3 High-pressure optical instrumentation for the intermediate pressure range ($P < 2$ GPa)

A large number of different optical cells have been used in the past half-century for optical measurements at high pressures, and a description of them all would take up excessive space. A good summary of the subject has been given by Sherman and Stadtmuller.[12] Solid-media optical cells, such as Drickamer-type devices, although of great historical significance, are not in current use. This is also true of multiple-anvil optical cells.[12] Current high-pressure optical studies use:

(1) large-volume fluid cells with optical windows up to 1.5–2 GPa (at most);

(2) DACs above this value (sapphire-anvil cells may be an alternative for work in the u.v. or the near i.r., and for moderate ($P < 10$ GPa) pressures).

In this subsection we examine the design of intermediate-pressure cells, that is, the window materials, the window geometry, and the design of cells for the various measurements discussed in Sections 5.1 and 5.2.

More than two dozen different window materials have been studied for use at high pressure.[12–13] Sapphire (Al_2O_3) is the most reliable material for wavelengths in the range 0.17–5.5 μm (0.25–7 eV), and it should be used except when polarized-light experiments (see below) are necessary. For wavelengths longer than 5 μm a number of materials have been tested and used[13] such as polycrystalline germanium or silicon, various alkali halides or IRTRAN[13] materials. All these materials should be rejected for the following two reasons.

1. Polycrystalline silicon and germanium are difficult to obtain (monocrystalline materials, especially germanium, are much more readily available but they rupture by cleavage under stress). Moreover, since they are opaque in the visible, they do not allow safe seating of the windows (see below).

2. Soft materials such as ionic crystals flow quite easily and will rupture after extended periods of time in an unpredictable manner.[13]

The alternative to sapphire and its complement is diamond. Type II diamond is transparent for wavelengths of 0.4 μm and above, except between 4 and 6 μm where two-phonon absorption bands occur. Type I diamonds have an additional nitrogen absorption band at around 7.5 μm. The initial cost of diamond windows is more than offset by their performance, which allows[14] very-low-mass optical plugs and cells to be used, and thus there is a considerable saving on machining and cell material. In addition, diamond flats, by contrast with alkali-halide or semiconductor materials, cannot be scratched or damaged and so they are quite safe up to 2 GPa, for reasonable geometries.[14]

Float glass, as an alternative material to diamond or sapphire, is of practical importance when polarized light is used. Sapphire is birefringent and polarization scrambling by sapphire windows[12] is severe. Float-glass windows, by contrast, are probably the best choice for the wavelength range 0.35–2.5 μm, especially when only used in the Poulter geometry (see Section 5.4) in the centre of the window, where scrambling is minimal. Diamond windows should also be quite adequate for studies using polarized light in large-volume fluid cells, but this has not yet been documented.

After the choice of the window material, the window geometry is the next most important factor in the high-pressure performance. A review of the various geometries which have been tested has been given by Sherman and Stadtmuller.[12] In practice, only the three window arrangements shown in Fig. 3.20 need to be considered. For low pressures, that is up to 100 MPa, the opposed-force mount (Fig. 3.20(a)) or a transparent cylinder (Fig. 3.20(b)) can be used. The former simply uses the force exerted by a breech screwed onto some gasket to oppose the force exerted on the window surface by the inner pressure. Failure of such windows is usually not due to the pressure in the cell, but to the excessive stress gradients which are exerted on the edges of the window to ensure leak tightness. Indeed, failures frequently occur at zero pressure from excessive initial tightening of the window on its seat.

Cylindrical glass or sapphire cells are in common use when long samples have to be observed, as is the case in falling-sphere viscosimeters used in the petroleum industry (Fig. 3.20(b)). This geometry is not suitable for use above 100 MPa for safety reasons. The bursting pressure of such cylinders is very highly dependent on the surface quality, and on local stresses which are brought about by the end plugs or pistons used to generate the pressure.

In the pressure range 100 MPa–2 GPa, the Poulter window geometry

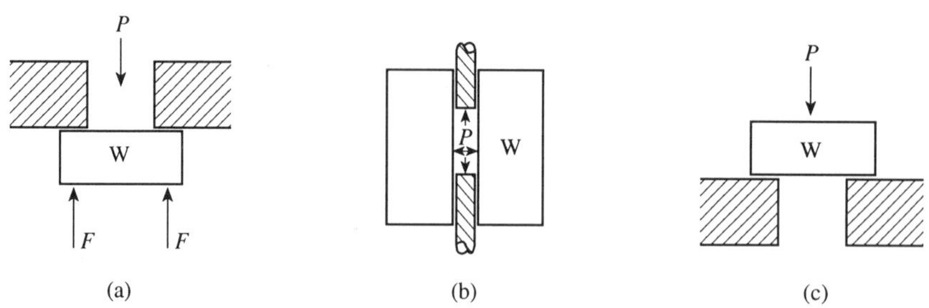

Fig. 3.20 Three common high-pressure window (W) arrangements: (a) an opposed-force mount, (b) a transparent cylinder, and (c) a Poulter unsupported-area window. (From reference 12).

should be used[15] (Fig. 3.20(c)). In such a geometry (Fig. 3.21), the force generated by the inner pressure, P

$$F = \tfrac{1}{4}\pi D^2 P,$$

is supported by the area of the window that rests on the window seat, that is, by

$$S = \tfrac{1}{4}\pi (D^2 - d^2).$$

Thus the pressure at the window–seat interface is

$$P_i = P\frac{D^2}{D^2 - d^2}. \qquad (3.5.10)$$

Thus, for a ratio for d/D of 1:2 and for an inner pressure of 1 GPa, the pressure at the window–seat interface is 1.33 GPa and the overpressure which opposes the leak is 333 MPa. In the same way as the Bridgman seals described in Section 1, increasing the pressure will tend to reduce leaks. Another advantage of the Poulter geometry is that the window material is subjected to the hydrostatic pressure it is meant to contain. This enhances its mechanical properties since the elastic coefficients of solids usually increase under high pressures.

In using unsupported-area Poulter-type ports (Fig. 3.21), several factors should be considered:

5.3.1 Seat materials

The window seat material is subjected to an overpressure of P_i (see 3.5.10). Obviously this pressure should not exceed the compressive strength of the window-seat material. Under high-overpressure conditions ($P_i > 1800$ MPa) a hard-steel or tungsten-carbide insert should be shrunk into the high-pressure optical plug for proper seating of the window.

5.3.2 Window seating

The window seating is shown schematically in Fig. 3.21. The initial seal on the window is made under moderate compression by a metallic cap, which simply keeps the window in place at low pressures.

The inner edge of the window seat must be rounded off to avoid any knife-edge effects on the window when it starts to bend in under pressure. The practical radius for sapphire windows with an unsupported-area diameter of 6 mm is 0.3 ± 0.1 mm.

The outer diameter of the window seat should be matched to the diameter of the window itself, and it must *not* be larger. At the highest pressures, the force acting on the window may be sufficient to imprint it slightly into the seat due to the *bending in* (concavity) under load of a Bridgman-type plug such as that shown in Fig. 3.22. This will cause failure of the window on the downstroke.

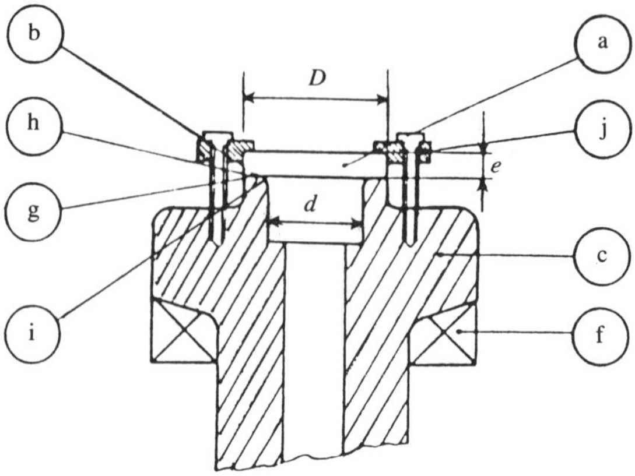

Fig. 3.21 A Poulter-type window on an unsupported-area Bridgman optical plug, used for window testing:[16] (a) window, (b) flat window seat, (c) hard-steel plug, (f) washer packing, (g and h) bevels on the outside of the plug and the window, (i) a bevel on the inside diameter of plug and, (j) a retaining brass cap; *d* and *D* are the inner and outer diameters of the window seat; *e* is the thickness of the window (from reference 16).

5.3.3 Selection and dimensioning of high-pressure windows

The behaviour under pressure has only been studied systematically for sapphire windows. Since these windows are by far the most frequently used, readers should refer to the full discussion of the geometries used in the gigapascal range given by Chervin *et al.*[16] Even when a commercial high-pressure cell is used for optical studies, it is often desirable to replace or modify one or several optical plugs or windows so that users could do it themselves.

Sapphire (Al_2O_3) ports are commercially available in (almost) any size. The surface polish should be specified as *optical quality, half fringe*. It is a good precaution to check for refractive-index inhomogeneities between crossed polarizers under sodium light. Although[16] there is no direct experimental evidence that sapphire fractures more easily because of internal strains, it certainly is safer not to use highly strained material. The orientation of the *c*-axis with respect to the window plane does not influence the mechanical properties much, and if the *c*-axis is selected as being perpendicular to the

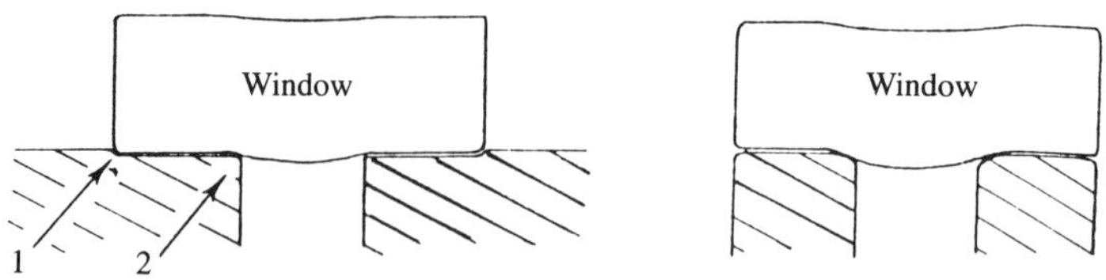

Fig. 3.22 The seating of a window: (a) improper seating (the window indents the seat at (1) and gets indented on the plug's sharp edge at (2)), and (b) proper seating. (From reference 12).

window plane, it should be in this direction for the optical requirements of the experiment (birefringence).

5.3.4 Mounting of the windows

After the flatness of the faces has been checked with an optical flat, the window should be placed on its seat, which must also be flat and polished to within ½. The quality of the seating can readily be checked under sodium light (Fig. 3.23).[12] When proper seating, such as that shown in Fig. 3.23(b), is achieved (that is, after cleaning of the window and seat), the window is fixed by screwing down its cap (Fig. 3.21). No intermediate glue or sheet, or 'lubricant' should be used between the window and the seat, although this is a common, and redundant, practice. A flat-to-flat seal has been shown to be quite leak-tight within the sensitivity of the measurement for a period of several weeks for helium in the gigapascal range, even at low temperatures.[17] The only exception to this is if the window has to be used at zero pressure (or at very low pressures) where the overpressure might not be sufficient to maintain the window firmly on its seat or to initiate the automatic sealing procedure.

5.3.5 Safety

Safety procedures are given in Appendix 2; however, it could be summed up by the sentence 'There is no such thing as a safe high-pressure window'. This means that operators should treat all high-pressure ports as though they might fail at any pressure, even at pressures well below the 'safe' working pressure. A sapphire window may well hold at, say, 1.2 GPa, and then fail at 0.5 GPa on a second run, because, for instance, the window seat was plastically deformed during the first experiment. Thus, never look at a high-pressure window directly with your eye. Using a mirror can be even more dangerous: sapphire fragments from an exploding window can be specularly reflected from a glass mirror into the eye of an overconfident observer. Using a total reflection prism is better, but, the use of a television monitor is the most sensible method of direct observation. Among other reasons it allows a safe distance to be maintained between the window and an experimenters' ears:

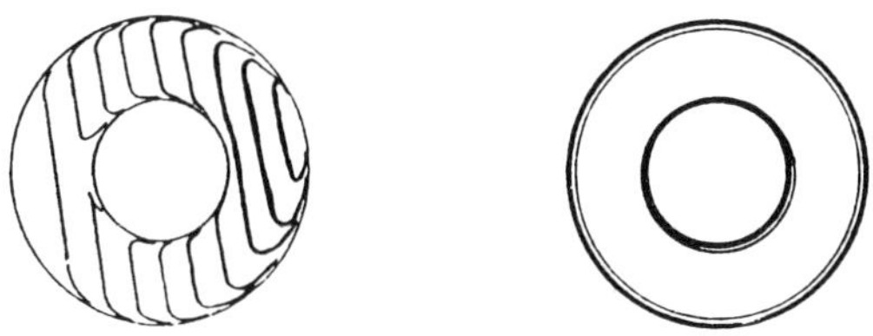

(a) (b)

Fig. 3.23 Interference fringes, observed under sodium light, of a window resting on its seat for: (a) improper seating with an angle between the window and the seat, and (b) proper seating with a flat fringe pattern (from reference 12).

failure of a window under a gas pressure of 1 GPa can be quite loud. Bullet-proof glass can and should be used to protect the optical equipment downstream, unless this is prevented by the alterations that it brings about to the optical path.

5.3.6 The design of optical cells with high-pressure ports

Ideally an optical cell should be designed specifically for one experiment. This is because the experimental conditions in optics involve so many variable parameters that no single cell design will provide the optimal conditions for different experiments. Among these parameters, the most important are:

(1) the wavelength range;

(2) the size of the sample/the nature of the sample (solid or fluid);

(3) the number of port holes, depending on the experiment;

(4) the minimum aperture of the windows;

(5) the minimum angular opening, which depends on the f-number of the instrument (spectrometer) used; and

(6) the pressure domain to be investigated.

The last factor is the most important. When designing, or purchasing, an optical cell, it is of the utmost importance that the required maximum pressure is kept as low as possible. In practice, designing a cell for a working pressure of 1.2 GPa is quite straightforward. In contrast, a working pressure of 1.7 GPa will take some parts of a multiwindow cell above the yield strength of the construction, which poses difficult construction problems.

With this in mind, starting from the sample, the characteristics of the light beams (focal spot, f-number) define the size of the windows and of the optical plug, which in turn define the size of the pressurized volume, and thus the size of the high-pressured cell. As a rule, for solid-state studies, the sample is small and most of the pressurized volume is taken up by the window material. Since the total volume (and mass) of a cell is more or less proportional to its inner volume, doubling the linear dimensions of the high-pressure windows will lead to an eight-fold increase in the mass of the optical device, with a corresponding increase in the cost. In addition, optical materials (sapphire, glass, or diamond) are always of higher quality (more homogeneous and lacking flaws) in small sizes. Finally, small devices are obviously easier to handle and install in a spectrometer or in a cryostat than bulkier devices. Thus, in selecting an optical cell design, the first criterion should be to use the smallest possible size which is compatible with the experimental requirements.

Cells may have any number of optical windows. Two examples are given in Figs 3.24 and 3.25 of a one-window cell and a four-window system. The windows themselves can be mounted, for example on wedge-type seal plugs, or on unsupported-area ones[17] (Figs 3.26 and 3.27)).

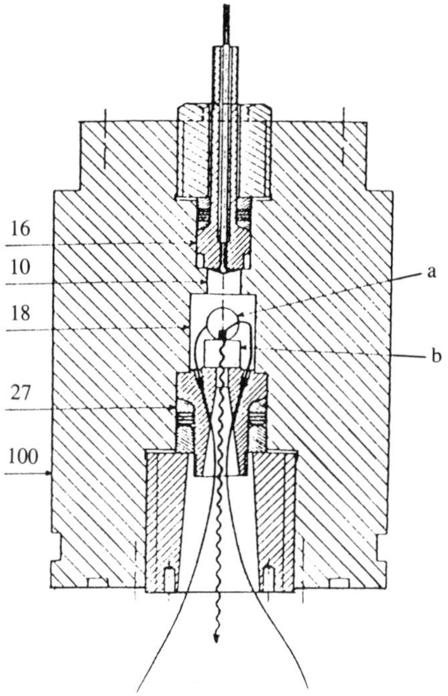

Fig. 3.24 A one-window cell, used here for observation of the emission of a diode laser under pressure: (a) diode laser, and (b) high-pressure window. The diameters are in millimetres. The body of the cell is made of maraging steel. This cell has been used at pressures up to 1.7 GPa.

Clearly, designs such as that shown in Fig. 3.25 are much weaker than simple cylinders with only one or two windows. Many ingenious devices have been devised to allow, for example right-angle scattering (Raman) experiments or reflectivity measurements within the cell, without any need for perpendicular bores, which weaken the cell's body. Examples are given in Figs 3.28, 3.29 and 3.30.

5.3.7 Cylindrical cells

Optical cylindrical cells can be manufactured from transparent materials (sapphire or glass) in a thick-cylinder geometry and used as piston–cylinder devices or connected to a high-pressure generating system by adequate plugs and plumbing. These devices, which are often used, for example, in petroleum research, may be quite advantageous for the observation of phase demixing in liquids (fluids) or for falling-sphere viscosimetry. They have not been used[14] above a fraction of a gigapascal. Simple glass or sapphire cylinders are only used up to 100 MPa for safety reasons: in such a geometry their behaviour is unpredictable, and the causes of failure are difficult to pinpoint (surface quality or crystalline defects may be involved).

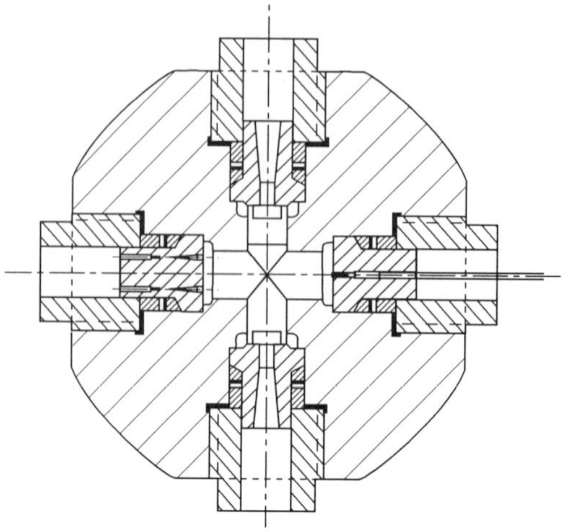

Fig. 3.25 A multiwindow cell with Bridgman plugs. (Département des Hautes Pressions, University of Paris 6, Pierre et Marie Curie).

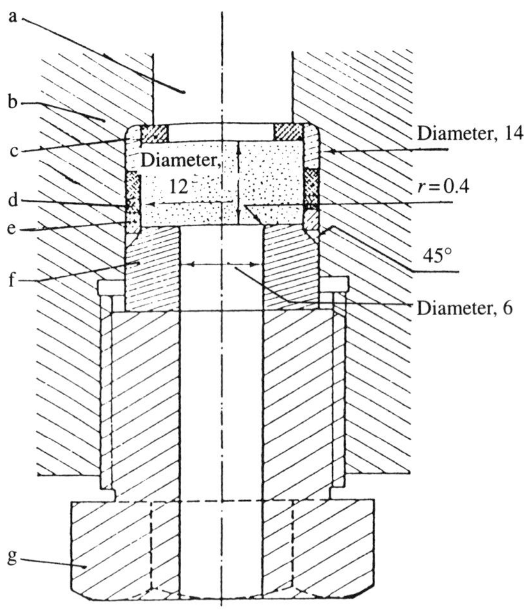

Fig. 3.26 A window seating with a square ring seal on the plug[17]. (This and Figs. 3.27, 3.28 reprinted with permission from *Rev. Phys. Appl.*, **5**, 333, (1970).)

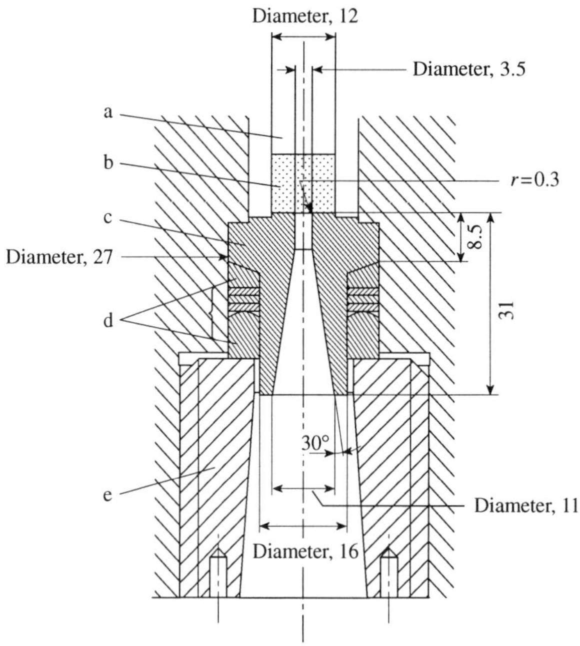

Fig. 3.27 A window seating with a Bridgman plug: (a) high-pressure medium, (b) a sapphire window, (c) a hard-steel plug (52 HRC), (d) washer packing, (e) a retaining breech. The dimensions are in millimetres[17].

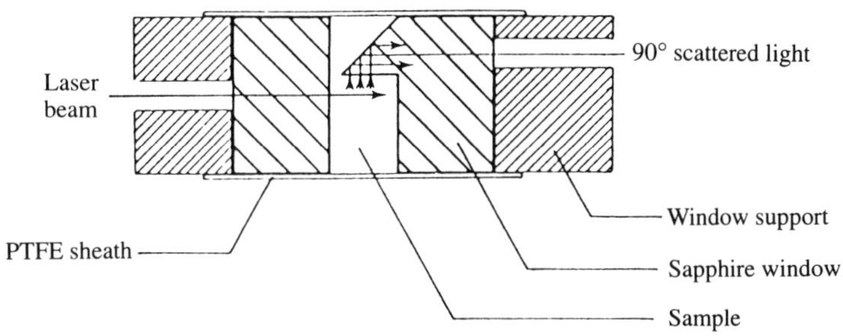

Fig. 3.28 A high-pressure arrangement which allows 90° Raman scattering from a fluid sample to be observed while maintaining in-line light paths. (Reprinted with permission from ref. 18. Copyright 1974 Pion Limited, London.)

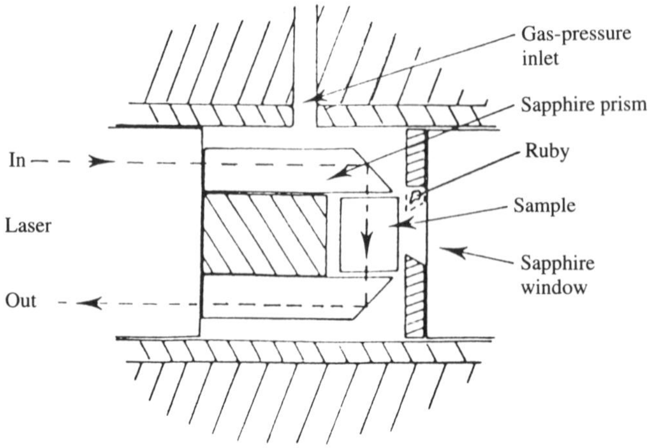

Fig. 3.29 An in-line window geometry permitting observation of 90° scattered Raman radiation from a solid sample[19]. (Reprinted with permission from *Rev. Sci. Instr.*, **50**, 340, (1979).)

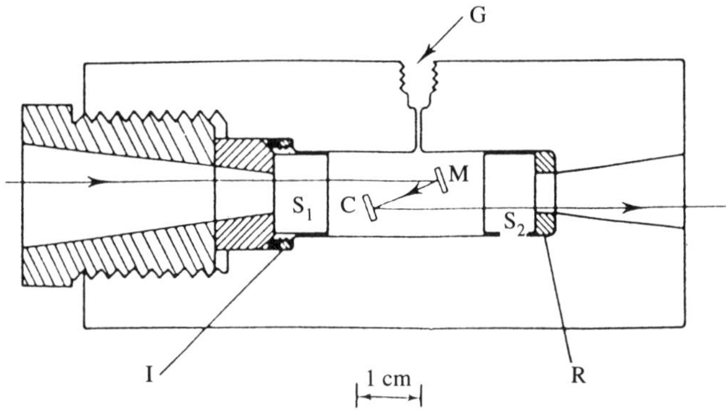

Fig. 3.30 An optical cell with in-line windows for reflectivity measurements: (C) sample, (I) seal, (G) gas inlet, (M) mirror, (R) ring seal, and (S_1, S_2) sapphire windows[20]. (Reprinted with permission from *The Institute of Physics*.)

5.3.8 Experimental considerations

Caution! For safety procedures see Appendix 2. High-pressure windows for measurements up to ~1.5 GPa should be used in the unsupported-area (Poulter) geometry. Sapphire, diamond, and, in a few cases float glass are the preferred materials. The high-pressure ports should always be shielded.

5.4 Optical studies in the higher-pressure range ($P \geqslant 2$ GPa)

This subsection deals with the specific problems associated with DACs and sapphire-anvil cells (which are analogous but are usable over a more modest pressure range). The other systems which have been used for studies above 2 GPa have been all but abandoned with the advent of DACs. These other systems include optical cells for solid media such as Drickamer-type staged NaCl windows, which are usable in girdle-type presses or Bridgman-anvil systems where part of the gasket is replaced by a transparent material. Even multiple-anvil systems (Chapter 2) have occasionally been provided with transparent windows in the anvils. These systems have been reviewed by Sherman and Stadtmuller.[12]

The general layout and principle of DACs is dealt with in Chapter 1.3. In this subsection only a few problems associated with the use of DACs in the study of optical properties will be examined.

The first difficulty which arises, particularly for luminescence and Raman measurements, is the selection of low-luminescence diamonds. The luminescence centres in diamonds are, even now, quite poorly understood, and it is adequate to select rough stones with a luminescence which is sufficiently low under the laser-light illumination which is used in the experiment: a given stone may have an excessive luminescence under 488 nm light and yet have a quite acceptable luminescence under 514.5 nm light. Stones with a uniformly low luminescence are, unfortunately, usually—but not always—of the extra-white or exceptional-white grades, and their cost is prohibitive. Since rough stones are irregularly shaped, the observed luminescent background should be scaled against the second-order Raman spectrum of diamond around 2660 cm^{-1}. A rule of thumb is that the peak luminescent intensity should be less than twice the second-order Raman feature under 514.5 nm illumination.

A second problem, which is specific to DAC use, is the interfacing of the cell with the spectrometer or interferometer. The actual problems arise from the size of the samples, and from the fact that the environment of the sample (diamonds, cell) precludes the use of the very short frontal distances which are in current use for the objective lenses used for magnification in optical spectroscopy.

The size of samples is determined by size of the anvils, which is seldom greater than 0.5 carats (100 mg). This is because the availability—and the cost—of diamond material scales as the seventh power of the linear dimensions if the anvil is large enough, that is, when the cost of cutting is not dominant (it increases 'only' as the cube of the linear dimensions). Thus, for a given geometry, for a given colour class, and for a given degree of quality (absence of flaws and defects), an anvil with a table diameter of 4 mm will weigh 0.35 carat and may cost 1000 ECU (European currency units). Enlarging all the linear dimensions by only 27 per cent will yield a 5.1 mm diameter anvil, weighing 0.70 carat and costing over 5000 ECU. Therefore, if you can work with crystal samples, for example 100 μm across, within a definite pressure range, then you should, because it would be *very* costly to design the experiments for samples 200 μm across.

The adaptation of a DAC to an optical system must interface with some

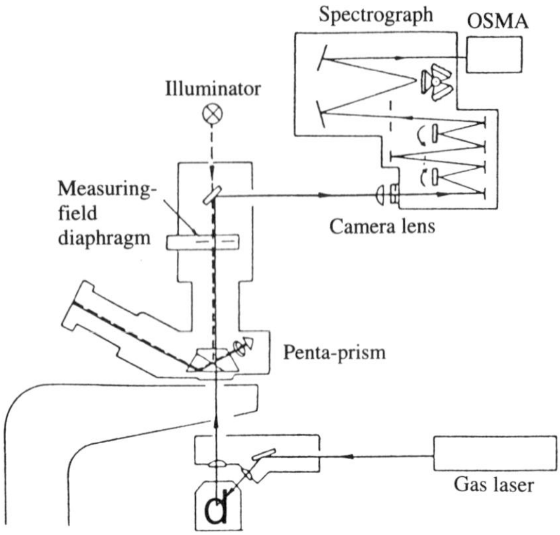

Fig. 3.31 A microscope set-up used to adapt a DAC (d) for a spectrograph with optical spectra multichannel analyser (OSMA)[10]. (Reprinted with permission from H. K. Mao in *Simple Molecular Systems at High Density,* Plenum Press.)

kind of a microscope system, which may simply be a commercial metallographic microscope such as that shown in Fig. 3.31. The two main factors to consider are: (i) the frontal distance of the objective, which must be as short as possible, depending on the geometry of the DAC, since the magnification is proportional to the f-number of the objective; and (ii) the focal-rejection iris (measuring-field diaphragm). The iris should be exactly the same size as the image of the active part of the sample in the focal plane of the objective. The active part of the sample is the part which is illuminated by the laser in luminescence or Raman measurements, or the portion of the area of the cell (sample or reference) which is explored in transmittance measurements. In the first type of measurements, the iris rejects any luminescence from the diamonds which does not originate from the plane of the sample. In the second type of experiment, the iris rejects stray light which has not gone through the sample itself (see below). In most cases the best results are obtained when the f-number of the objective is small, especially for luminescence measurements.

There are exceptions to this. Transmittance measurements cannot be made at an incidence which is too far from normal, since light rays incident on a sample of thickness d with an angle α to the normal go through a sample thickness of

$$d' \approx d\left(1 + \frac{\sin^2 \alpha}{2n}\right). \qquad (3.5.11)$$

This will cause errors of around 4 per cent for $\alpha = 30°$ and a refractive index of $n = 3$. Actually, this is an overestimate. Integration of eqn (3.5.11) for a homogeneous converging beam with a maximum incidence angle α

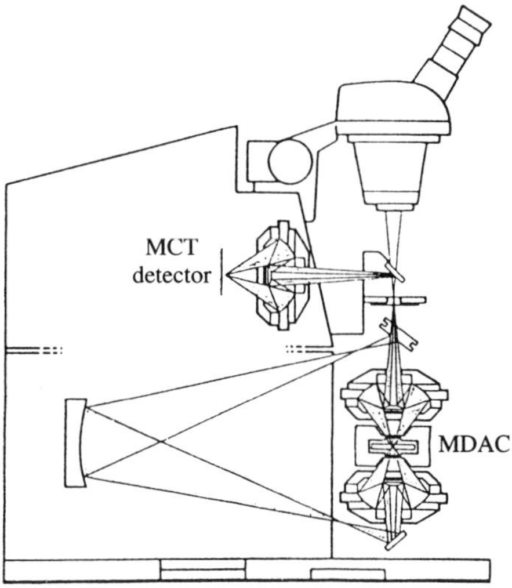

Fig. 3.32 An i.r. microscope arrangement with a membrane DAC (MDAC) on the sample stage, for interfacing with a Fourier-transform instrument using an MCT (i.r.) detector.[21] (Reprinted with permission from *Rev. Sci. Instr.*, **66**, 3, (1995).)

shows that the resulting error in d is only 2 per cent for $\alpha = 30°$, and it is less than 1 per cent for $\alpha = 20°$, for an assumed refractive index of 3 in the sample. The vast increase (proportional to α^2) in the efficiency of the set-up for such experiments makes it worthwhile to increase α and to accept the comparatively small error due to non-normal incidence measurements. This is not true for Fabry–Pérot-type experiments (Sections 5.1 and 5.2) for which sharp interference patterns require $\alpha/n \leqslant 5°$.

The same type of problems occur when making i.r. measurements with DACs. It is now preferable to use Fourier-transform (FT) interferometers, which are commercially available. In this case, the incident i.r. beam, which is a few millimetres in diameter, must be focused[21,22] to less than 100 μm by suitable mirror optics (see Fig. 3.32). This procedure can be avoided for studies of homogeneous samples, filling the DAC experimental volume, especially when the intense beam from synchrotron radiation is used, although the useful throughput is a very small fraction ($10^2 - 10^3$) of the available light. However, it cannot be avoided when solid samples embedded in a transparent pressure medium are studied.

The last point which will be examined in this subsection is that of extracting the true transmission values from a DAC experiment, in order to calculate the absorption-coefficient spectrum. In the case of a plane-parallel slab of sample with a thickness d, a refractive index n, and an absorption coefficient α which is immersed in a transparent medium with an index n' and negligible absorption, the theoretical transmission in the absence of interference fringes is

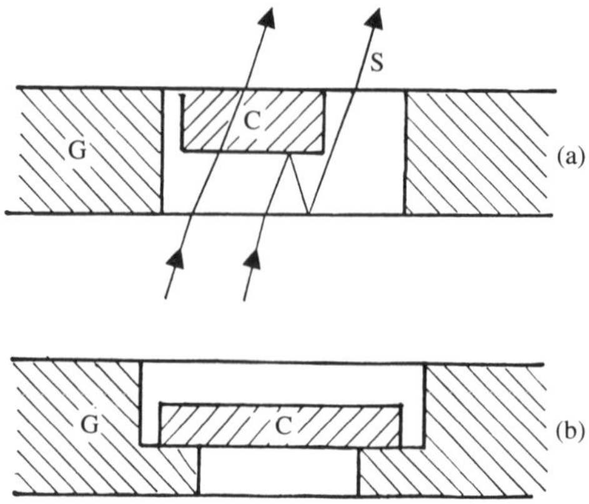

Fig. 3.33 The stray light in a DAC: (a) the ordinary gasket geometry, and (b) a two-aperture gasket which eliminates stray light. (G) gasket, (C) crystal, and (S) scattered light giving the background signal.

$$T = \frac{(1 - R)^2 \exp(-\alpha d)}{1 - R^2 \exp(-2\alpha d)}, \qquad (3.5.12)$$

where $R = (n - n')^2/(n + n')$.[2] We shall examine here the case of the observation of an absorption edge of a solid with direct, interband, electronic transitions giving absorption coefficients of some 10^4 cm^{-1} where Fabry–Pérot interference fringes disappear for $\alpha d > 1$.

The sample is located on one of the diamonds with the incident white light focused on it (Fig. 3.33). Small as the incident spot may be, some stray light (S) will go through the optical system and be detected. The solid line in Fig. 3.34(a) shows the transmission curve which is obtained from a 10 μm thick sample whose true absorption coefficient is shown by the dashed line in Fig. 3.34(b). The stray-light level at high pressures is taken to be 2×10^{-4} in this wavenumber range, which is quite good in practice. Now the actual transmittance spectrum which is expected with n–$n' = 3$, for argument's sake, is shown by the dashed line in Fig. 3.34(a). The low-frequency region is plagued by two effects: (i) small interference fringes appear, and (ii) the average transmittance is different from that expected by making α equal to zero in eqn 3.5.12. If the dependence of α on $\overline{\nu}$ is calculated from the solid line in Fig. 3.34(a) the solid curve in Fig. 3.34(b) is obtained which is wrong in both the high- and the low-frequency regions. The low-frequency transmission has been taken here to be half of the actual value. This often occurs because of scattering within or at the surface of, imperfect samples, or because of differences in the focusing between the I and I_0 measurements. As a matter of fact, transmittances which are larger than the theoretical transmittance for $\alpha = 0$ are 'observed' on occasion and must be corrected for. Thus, in order to

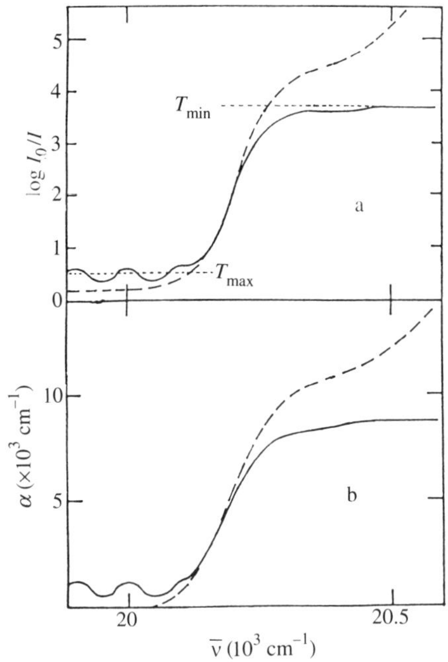

Fig. 3.34 High- and low-frequency corrections which are applied to transmittance measurements in DAC. (a) transmittance plotted against the wavenumber: (——) the uncorrected spectrum, which levels out at a transmittance of $T_{min} = I/I_0 = 2 \times 10^{-4}$ on the high-energy side, while the apparent transmittance on the low-energy side levels out at 30 per cent with interference fringes; (- - - -) the corrected transmittance, which takes into account the stray-light level of 2×10^{-4}, and which corrects the low-frequency values to the real value of 60 per cent with $\alpha = 0$ in eqn (3.5.12) and which eliminates interference fringes. (b) The absorption coefficient, (——) the uncorrected values of α from the uncorrected curve in (a), and (- - - -) the corrected absorption coefficient from the corrected curve in (a).

obtain sensible values for $\alpha(\bar{\nu})$ two corrections must be made before the transmission spectrum is analysed.

1. In the low-frequency region the asymptotic transmission must be made to fit with the theoretical value predicted by eqn (3.5.12) with $\alpha = 0$. A small and constant absorption coefficient is physically meaningless. If this type of attenuation is present, it cannot be ascribed to absorption due to the electronic transitions which are studied here. It can only have come from scattering and/or diffusion by point or extended defects, or, more usually, from the optical geometry (for example, focusing, the lensing effect due to the curvature of the anvils under pressure). Thus, the experimental transmission curve first had to be *multiplied* by some correction factor (larger or smaller than unity) to take these effects into account.

2. In the high frequency region the stray-light level (2×10^{-4} in this case)

has to be *subtracted* from the apparent transmittance. When this correction is made, a better approximation to the true absorption coefficient is obtained.

Actually, the two corrections are not independent, and an iterative procedure should ideally be used.

The stray light which spoils the high-energy region of the spectrum is not a constant, and its spectrum should be measured by replacing the crystal by an opaque (metallic) sample of the same size. An alternative is to use a gasket such as that shown in Fig. 3.33(b); this greatly decreases the level of the stray light. This type of arrangement is not in general use, possibly due to manufacturing difficulties and deformation under pressure.

5.4.1 Experimental considerations

Caution! For safety procedures see Appendix 2. For most optical measurements in DACs, it is advantageous to use the largest possible optical angular aperture (f-number) which is compatible with the maximum pressure required. Calculations of absorption coefficients from measurements of the transmittance of crystalline samples must include adequate corrections.

References

1. Elliott, R. J. *Phys. Rev.* **1957**, *108*, 1384.
2. Toyozawa, Y. *Progr. Theor. Phys.* **1958**, *20*, 53.
3. Dumke, W. P.; Lorenz, M. R.; Pettit, G. D. *Phys. Rev.* B **1972**, *5*, 2978.
4. Obriot, J.; Fondere, F.; Marteau, P. *Infrared Phys.* **1978**, *18*, 607.
5. Schmidt, S. C.; Moore, D. S.; Schiferl, D.; Shaner, J. W. *Phys. Rev. Lett.* **1983**, *50*, 661.
6. Tolles, W. M.; Nibler, J. W.; McDonald, J. R.; Harvey, A. B. *Appl. Spectros.* **1977**, *31*, 253.
7. Schmidt, S. C.; Moore, D. S.; Shaw, M. S. *Phys. Rev. Lett.* **1987**, *35*, 493.
8. Sandercock, J. R. In *Topics in Applied Physics*, Vol. 51; Light Scattering in Solids III, Cardona, M., Güntherodt, G. eds. Springer: New York, **1982**.
9. Grimsditch, M.; Polian, A. In *Simple Molecular Systems at Very High Density*; Polian, A.; Loubeyre, P., eds. NASI, series B, Vol. 186, Plenum Press: New York, **1989**, p. 237.
10. Shimizu, H.; Sasaki, S. **Science* **1992**, *257*, 514. **Jpn J. Appl. Phys.* **1993**, *32* (suppl. 32–1), 355.
11. Besson, J. M.; Calawa, A. R.; Paul, W. *Phys. rev.* **1968**, *173*, 4699.
12. Sherman, W. F.; Stadtmuller, A. A. *Experimental Techniques in High Pressure Research*. Wiley: New York, **1987**.
13. Paul, W.; de Meis, M.; Besson, J. M., *Rev. Sci. Instrum.*, **1968**, *39*, 928.
14. Besson, J. M.; Pinceaux, J. P.; Piotrzkowski, R. *High Temp. High Pressures* **1974**, *6*, 101.
15. Poulter, T. C. *Phys. Rev.* **1930**, *35*, 297.
    *Poulter, T. C.; Buckley, F. *Phys. Rev.* **1932**, *41*, 364.
16. Chervin, J. C.; Syfosse, G.; Besson, J. M. *Rev. Sci. Instrum.* **1994**, *65*, 2719.
17. Fournier, J.; Besson, J. M. *Revue Phys. Appl.* **1970**, *5*, 333.

18. Figuière, P.; Ghelfenstein, M.; Szwarc, H., *High Temp. High Press.* **1974**, *6*, 61.
19. Jodl, H. J.; Holzapfel, W. B. *Rev. Sci. Instrum.* **1979**, *50*, 340.
20. Cottle, A. C.; Lewis, W. F.; Batchelder, D. N. *J. Phys.* C, **1978**, *11*, 605.
21. Chervin, J. C.; Canny, B.; Besson, J. M.; Pruzan, Ph. *Rev. Sci. Instrum.* **1995**, *66*, 2595.
22. Johannsen, P. G. In *Simple Molecular Systems at High Density*; Polian, A., Loubeyre, P., eds. Plenum Press: New York, **1989**.
23. Ferraro, J. R. *Vibrational Spectroscopy at High External Pressures: The diamond anvil cell*; Academic Press: London, **1984**.

6. X-ray absorption

J. M. Besson

6.1 Principles of X-ray absorption spectroscopy (XAS)

X-ray absorption spectroscopy (XAS) is based on the absorption of photons by the excitation of electrons from the inner electronic shells of the atoms (K, L, etc., absorption edges). Various effects, which depend on the energy of the incident photon, due to interaction of the photoelectron with the neighbours of the parent atom can be observed and interpreted.[1]

The edge part (Fig. 3.35) is the onset of absorption, which occurs as soon as there is an empty state for the final state of the photoelectron that is allowed by symmetry. X-ray absorption near-edge spectroscopy (XANES) is used to observe energies immediately above the onset of absorption; the XANES energies are low enough for the mean free path of the photoelectron to be large and to extend over several (up to five) lattice parameters. At still higher energies, some 100 eV above the edge, oscillatory behaviour is observed in

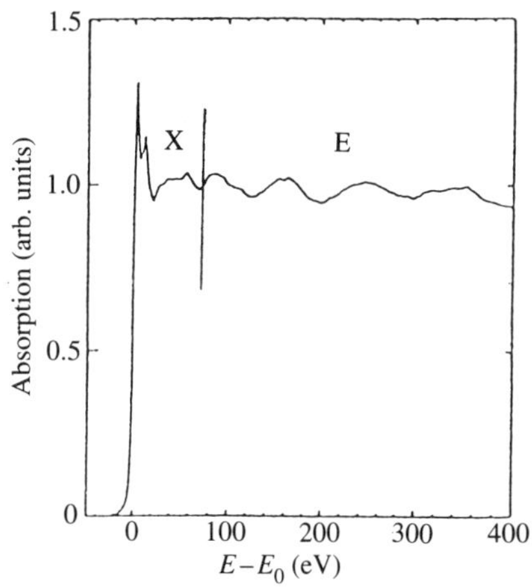

Fig. 3.35 An XAS spectrum of ZnSe under a pressure of 10 GPa at the Zn K-edge obtained with an energy-dispersive set-up: (X) the XANES structure, and (E) the EXAFS structure.

the absorption spectrum. This is due to retrodiffusion from the nearest neighbours, and this extended X-ray absorption fine-structure spectroscopy (EXAFS), which extends over several hundreds of electron volts, is readily interpretable, and thus it gives direct information on the local environment of a given atom, even in a disordered (amorphous or liquid) material. In EXAFS, the oscillatory part of the absorption coefficient, α, which depends on the wavenumber, K, of the photoelectron is given by

$$\alpha(K) = \sum_j \frac{N_j}{KR_j^2} A_j(K) \sin [KR_j + \Psi_j(K)] \exp\left(-\frac{2R_j}{\lambda}\right) \exp(-2\sigma_j^2 K^2), \quad (3.6.1)$$

where N_j is the number of jth neighbours at a distance R_j from the absorbing atom, $\Psi_j(K)$ is the phase shift function for the backscattering, which has an amplitude $A_j(K)$, λ is the photoelectron mean free path, and σ_j is the Debye–Waller factor, which includes dynamic (vibrational) and static (disorder) contributions. The wavenumber, K, of a photoelectron of mass m coming from a shell with a threshold energy of E_0 is related to the energy by

$$K = \hbar^{-1}[2m(E - E_0)]^{-\frac{1}{2}}. \quad (3.6.2)$$

When using XAS at high pressures, a number of simplifications have to be used to interpret EXAFS oscillations quantitatively. The variables Ψ_j, A_j, and λ depend only on the absorbing and backscattering atoms and are taken otherwise as being constant; the fit with experiment is made using the variables R, N, and σ, which are the structural parameters that vary with pressure. These quantities are determined under pressure by reference to the ambient pressure values which must be known beforehand.

Data analysis of EXAFS spectra involves double-Fourier-transform procedures whose precision is highly dependent on the accessible energy range. At high pressures, the contributions from the first shell of neighbours are readily accessible, and, less frequently, so are those from the second and third shells. A detailed review of XAS under high pressures can be found in reference 2 and the references therein.

6.2 Experimental procedure

In XAS experiments the optimum sample thickness is of the order of a few tens of micrometres, and they are therefore well adapted to the use of DACs. For this reason, almost all such measurements under high pressure have been performed with DACs. Since the diameter of the sample is, in this case, quite small, strong collimation of the incident beam occurs when using monochromatic radiation, which makes step-by-step measurements quite time-consuming. Currently, energy-dispersive XAS, using a focused beam of polychromatic radiation from a synchrotron source, is preferred for high-pressure measurements. The present section will deal with this type of arrangement.

Figure 3.36 gives a schematic layout of such a system.[3] The white beam from a synchrotron source is focused by a bent-crystal monochromator,

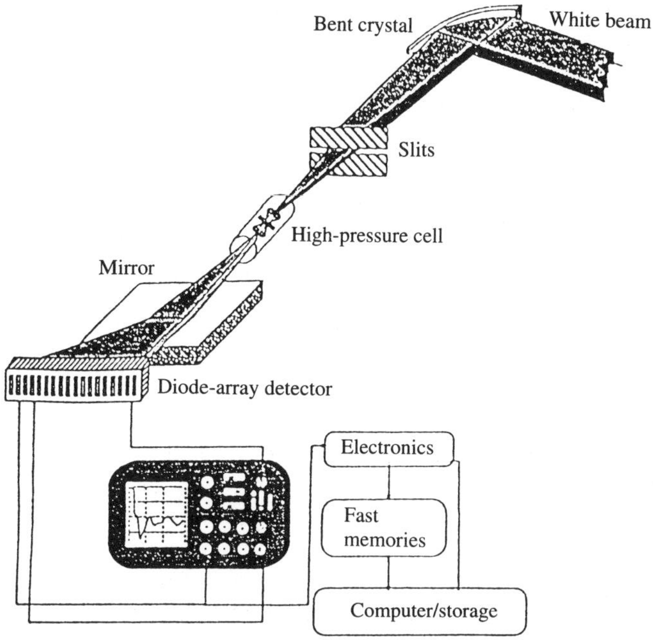

Fig. 3.36 A schematic layout of an energy-dispersive spectrometer for XAS under pressure. (From A. San Miguel, PhD Thesis, Université Paris 6 (1993).)

which separates the various wavelengths of the incident beam at the monochromatic focus point, where a diode array is used as a spectrographic detector. A total-reflection mirror eliminates the higher harmonics.

When using a DAC whose axis is in the direction of the X-ray beam, Bragg diffraction may occur within the energy range of interest. The anvils must be appropriately oriented to eliminate any such Bragg peaks from the energy range of interest. With the usual type of anvils used for high-pressure experiments, diamonds are transparent only to energies above 6–8 keV, depending on the thickness of the anvils. This means that no elements with an atomic number, Z, lower than ~ 28 can be studied in this configuration. To obtain spectra at lower energies, B_4C anvils[4] can be used, or the X-ray beam can be made to pass through beryllium gaskets[5], which are transparent down to energies of 2 keV.

With existing synchrotron sources, the size of the spot at the polychromatic focus point (the sample location) is several hundreds of micrometres, which is usually larger than the hole in the gasket of a DAC. Thus, powder samples in a suitable liquid medium (an ethanol–methanol mixture or silicone oil) are usually used to fill the hole entirely. Under such conditions, spectra such as those shown in Fig. 3.35 can be obtained in a matter of minutes. With the advent of the third-generation synchrotrons with high brilliance, the spot size will be reduced to less than 100 μm, allowing single-crystal samples to be studied in a quasihydrostatic (helium) environment.

6.3 Results of XAS under high pressures

Since X-ray absorption, especially EXAFS, is sensitive to the immediate environment of atoms, it can be used in ordered or disordered systems (amorphous solids or liquids) and, in particular, to study the disordered state which occurs at a first-order phase transition between two crystal structures under high pressure.

In a single phase, with known coordination number, N, both the Debye–Waller factor variation and the interatomic distance can be derived from fits to eqns (3.6.1) and (3.6.2). At the present time, EXAFS is about the only method which will measure the Debye–Waller factor variation under high pressures (Fig. 3.37). Interatomic distance variations can be determined in practice with an accuracy of 5×10^{-3} Å with respect to the ambient pressure value,[2] in both ordered and disordered systems.

When a phase transition occurs under pressure accompanied by a co-ordination change, XANES can be used[6] as a sensitive fingerprint of the symmetry around the absorbing atom (Fig. 3.38), especially in disordered (glassy) materials, where no other method is available. In fact, EXAFS is the most sensitive gauge for the occurrence of a new phase with a different coordination number. Figure 3.39 shows such a phase transition[7] on the upstrokes and downstrokes in GaAs, whose coordination number changes from 4 to 6 at 16 GPa with increasing pressure. This transition can be observed using EXAFS as soon as a sizeable (20 per cent) proportion of the

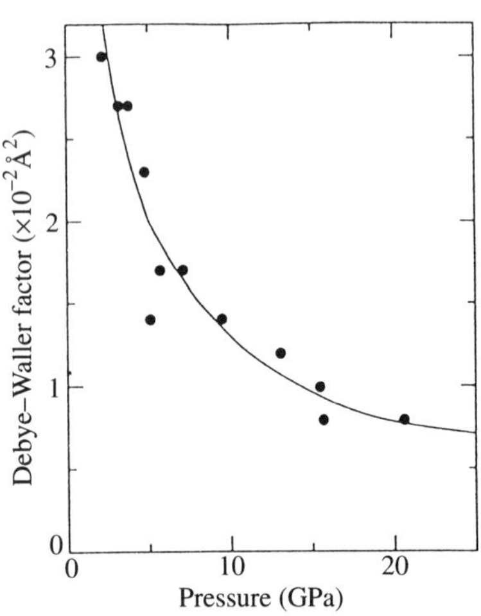

Fig. 3.37 Variation under pressure of the Debye–Waller factor of krypton[2]. (Reprinted with permission from Gordon & Breach Publishers.)

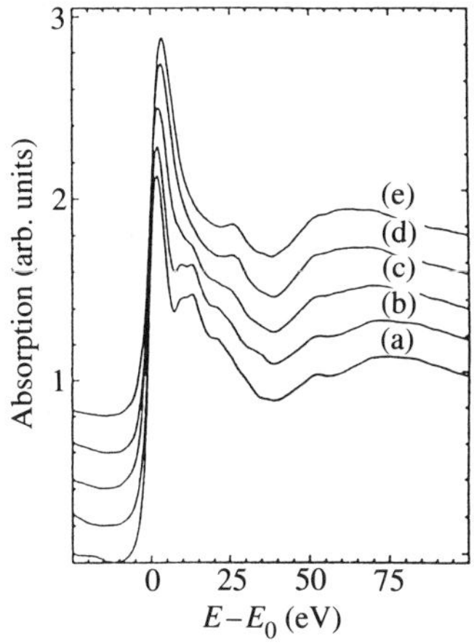

Fig. 3.38 XANES at the Ge K-edge for GeO_2 at various pressures between (a) 6.4 GPa and (e) 11.1 GPa[2]. (Reprinted with permission from Gordon & Breach Publishers.)

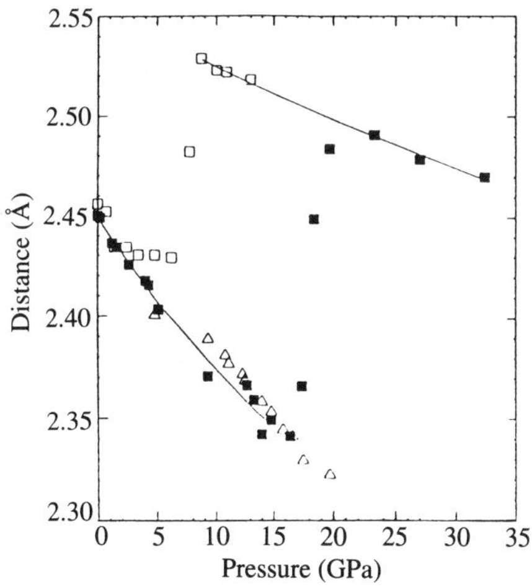

Fig. 3.39 Variation of the nearest-neighbour distance with pressure in GaAs: (■) EXAFS data for increasing pressure, (□) EXAFS data for decreasing pressure (from reference 7), and (△) X-ray-diffraction data. (Reprinted with permission from S. T. Weir *et al. Phys. Rev.* B **39**, 1280, (1989).)

sample changes its coordination number, although it will not yet be organized into the ordered high-pressure orthorhombic structure,[8] and thus it cannot yet be observed by X-ray diffraction.

6.4 Experimental considerations

Caution! For safety procedures see Appendix 2. X-ray absorption under high pressure can be best performed by energy-dispersive methods in DACs using synchrotron radiation, and it is particularly suited to the study of phase transitions and disordered materials such as amorphous systems or glasses.

References

1. Sayers, D. E.; Stern, E. A.; Lytle, F. W. *Phys. Rev. Lett.* **1971**, *27*, 1204.
2. Itié, J. P. *Phase Tans.* **1992**, *39*, 81.
3. Tolentino, H.; Baudelet, F.; Dartyge, E.; Fontaine, A.; Lena, A.: Tourillon, G. *Nucl. Instrum. Methods* A, **1990**, *289*, 307.
4. Hauser, B.; Alberdings, N.; Ingalls, R.; Crozier, E. D. *Phys. Rev.* B, **1988**, *37*, 6513.
5. Röhler, J.; Krill, G.; Kappler, J. P. *Nucl. Instrum. Methods* **1983**, *208*, 647.
6. Itié, J. P.; Polian, A.; Calas, G.; Petiau, J.; Fontaine, A.; Tolentino, H. *Phys. Rev. Lett.* **1989**, *63*, 398.
7. Polian, A.; Itié, J. P.; Jauberthie-Carillon, C.; Dartyge, E.; Fontaine, A.; Tolentino, H. *High Press. Res.* **1990**, *4*, 309.
8. Besson, J. M.; Itié, J. P.; Polian, A.; Weill, G.; Mansot, J. L.; Gonzalez, J. *Phys. Rev.* B **1991**, *44*, 4214.

7. Mössbauer spectroscopy

W. B. Holzapfel

Mössbauer's discovery[1] in 1958 of 'recoilless nuclear gamma resonance absorption' was rapidly applied in solid-state physics and chemistry because Mössbauer spectra provide detailed information on the local electron clouds around emitting and absorbing nuclei through the 'hyperfine' interactions, and on some lattice dynamical properties through the vibrational coupling of the nuclei to the lattice. Many textbooks,[2–5] together with the comprehensive bibliography by Stevens and Stevens,[6] give excellent introductions to the field. However, the information on high-pressure applications is more scattered in the literature.[7–36] High-pressure Mössbauer spectra to pressures of 10 GPa and above were already recorded by 1963 by Nicol and Jura[7] with the isotope ^{57}Fe and subsequently by Panyushkin and Voronov[8] with ^{119}Sn. Systematic studies with ^{57}Fe were also started at this time by Drickamer and his coworkers for many different applications at pressures up to 15 GPa.[9–16] Diamond anvil cells (DACs) were first used[17] with ^{119}Sn in 1965 and then in 1975 with ^{57}Fe.[18] Since 1986, when pressures in excess of 70 GPa were reached in Mössbauer measurements with DACs, this field has developed rather rapidly.[28–33]

Commonly, Mössbauer spectra are recorded by the use of a *source* containing *radioactive nuclei* with one specific Mössbauer isotope in its excited state, which decays by a characteristic sequence of transitions into the ground state, finally emitting a γ-ray photon in the energy range 5 keV $< E_\gamma <$ 100 keV. For the most commonly used Mössbauer isotope, ^{57}Fe, the decay

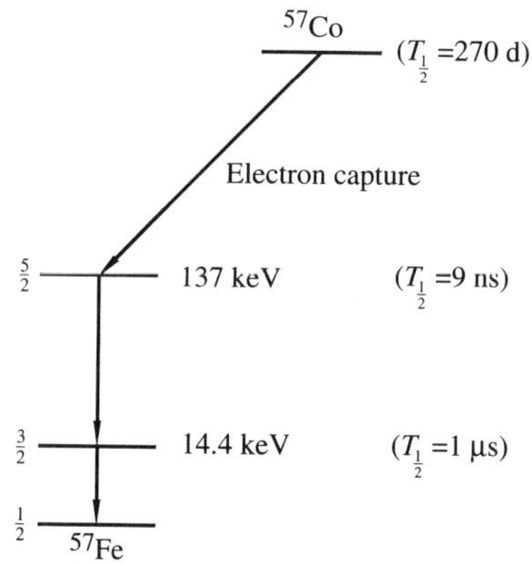

Fig. 3.40 The decay scheme of ^{57}Co for Mössbauer spectroscopy with ^{57}Fe[13]. (Reprinted with permission from ref. 13. Copyright 1974 Pion Limited, London.)

scheme is illustrated in Fig. 3.40. The radioactive parent isotope ^{57}Co decays by electron capture with a half-life $T_{1/2} = 270$ d into the higher excited state of the isotope ^{57}Fe. This state first decays into a lower excited state at 14.4 keV, which then decays with a half-life of 1 μs by the emission of a final γ-ray photon into the ground state of ^{57}Fe.

A nucleus of ^{57}Fe in the *absorber* can be excited into its first excited state by resonance absorption of such a γ-ray photon from the source, but only if this γ-ray photon has an energy E_γ precisely equal to the transition energy E_0 in the absorber. Neglecting all other effects, Heisenberg's uncertainty principle states that the natural line width Γ for the matching of these energies is given by $\Gamma = (\hbar \ln 2)/T_{1/2}$, which is a very stringent condition; for example, in the case of ^{57}Fe $\Gamma/E_\gamma \approx 10^{-12}$. By conservation of energy and momentum, a nucleus of mass M in the gas phase would recoil with an energy $E_R = E_\gamma/2Mc^2$ where $E_R/E_\gamma \gg \Gamma/E_\gamma$, and thus resonance absorption could occur only if the energy loss is compensated for by the Doppler shift $\Delta E = E_\gamma v/c$ caused by an emitting nucleus which is coincidentally moving towards the absorber with a velocity v. The velocity distributions in gaseous absorbers or sources would make this a very rare, and difficult to observe, event. For nuclei in solids there are, however, finite probabilities (given by the Debye–Waller factors, f) of events occurring in which no lattice vibrations are excited in either the source (given by f_s) or in the absorber (given by f_a), and in these cases the momentum is transferred to the whole crystal. Due to the large mass of the whole crystal ($M_{cryst} \approx 10^{23} \times M_{nuc}$) this recoil energy is practically zero (or, more precisely, much less than Γ). For these events, the resonance condition can be disturbed by differences in the *hyperfine* interactions for the nuclei in the source and the absorber. Well-defined Doppler shifts due to controlled motions of the source with respect to the absorber in a Mössbauer spectrometer can then be used to compensate

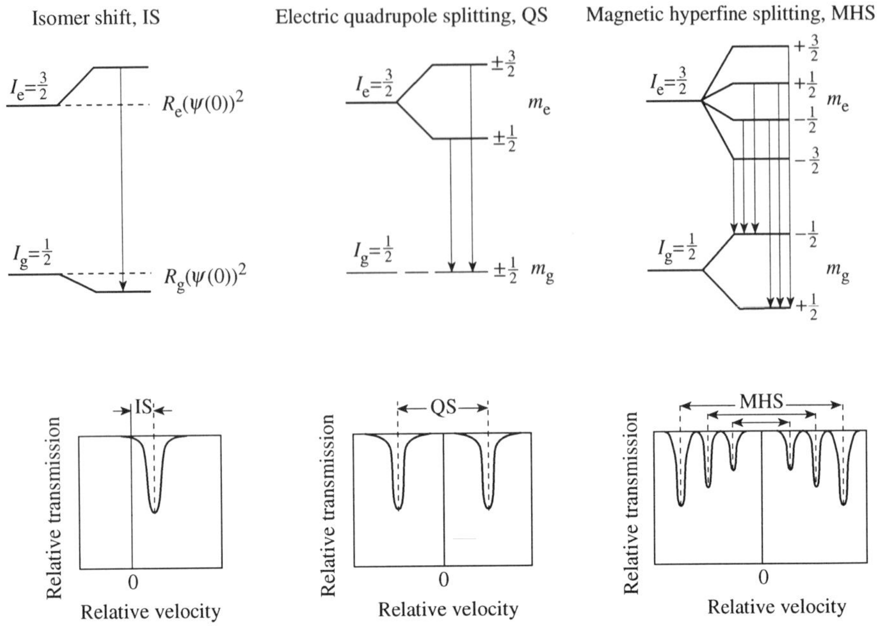

Fig. 3.41 A schematic representation of effects from hyperfine interactions in Mössbauer spectra[13]. (Reprinted with permission from ref. 13. Copyright 1974 Pion Limited, London.)

for these energy differences and to measure the resonance curves on an energy scale—usually calibrated in terms of the applied Doppler velocity. For the specific case of ^{57}Fe, the effects of hyperfine interactions on the nuclear energy levels are illustrated in Fig. 3.41 with the resulting spectra. Note that all the hyperfine interactions involve both nuclear and electronic parameters.

The *isomer shift*

$$\mathrm{IS} = \text{constant} \times (\Delta R)\Delta(|\psi(0)|^2)$$

is determined by the change ΔR in the nuclear radius in the transition and by the difference $\Delta|\psi(0)|^2$ between the electron density at the nuclei in the absorber $|\psi_A(0)|^2$ and that at the source $|\psi_S(0)|^2$, that is, by

$$\Delta(|\psi(0)|^2) = |\psi_A(0)|^2 - |\psi_S(0)|^2$$

The *electric quadrupole splitting*

$$QS = \text{constant} \times (eQ)\,(eq)$$

Table 3.1. List of spectral parameters

Effect	Spectral parameter	Solid-state information provided
Recoilless fraction	f_s, f_A	$<x^2>$, lattice forces
Quadratic Doppler shift	ΔIS	$<v^2>$, lattice energy
Isomer shift	IS	$\lvert\psi_A(0)\rvert^2 - \lvert\psi_S(0)\rvert^2$, electron densities at the nuclei
Electric quadrupole splitting	QS	eq_{zz}, electric field gradient at the nucleus
Magnetic hyperfine splitting	MHS	H_{eff}, effective magnetic field at nucleus
Line broadening	$\Delta\Gamma$	τ, relaxation times, diffusion
Superposition	c_1/c_2	K, chemical equilibria
Time scanning	t_i	reaction rates
Temperature scanning	T_c	transition temperatures
Pressure scanning	P_c	transition pressures

is proportional to the nuclear quadrupole moment, eQ, and to the electric field gradient at the nucleus, given by the factor eq.

The *magnetic hyperfine splitting* (MHS) depends on the nuclear spin quantum numbers I_e and I_g of the excited and ground state of the Mössbauer nucleus and on the effective magnetic field at the Mössbauer nucleus, H_{eff}, which includes contributions from the local electronic spin, from the orbital momentum, from dipole terms, and from external fields.

In addition, the area under the absorption curves can be used to extract information about the Debye–Waller factors and thus to extract information about changes in the average local vibrational amplitudes $\langle x^2 \rangle$, as indicated in Table 3.1.

In principle, changes in vibrational energy can change the line position by ΔIS; however, this effect is an almost negligible correction in high-pressure applications.

Diffusion and electronic relaxations can lead to broadening, $\Delta\Gamma$, of the line width, which would be difficult to resolve under pressure because of possible additional broadenings due to pressure gradients.

Superposition of various characteristic spectra give very useful fingerprints for the presence of different species, either due to different lattice sites or due to phase transitions or chemical reactions.

If, in a Mössbauer experiment, the relative transmissions at two velocities, v_+ and v_- (where $v_+ = -v_-$) are measured in a temperature scan characteristic thermal-scanning curves are obtained like those in Fig. 3.42, which can be very useful for the determination of magnetic-transition temperatures under pressure.[34,35]

Mössbauer isotopes are suitable for high-pressure studies if the nuclear parameters provide a high sensitivity for compressional changes and if the

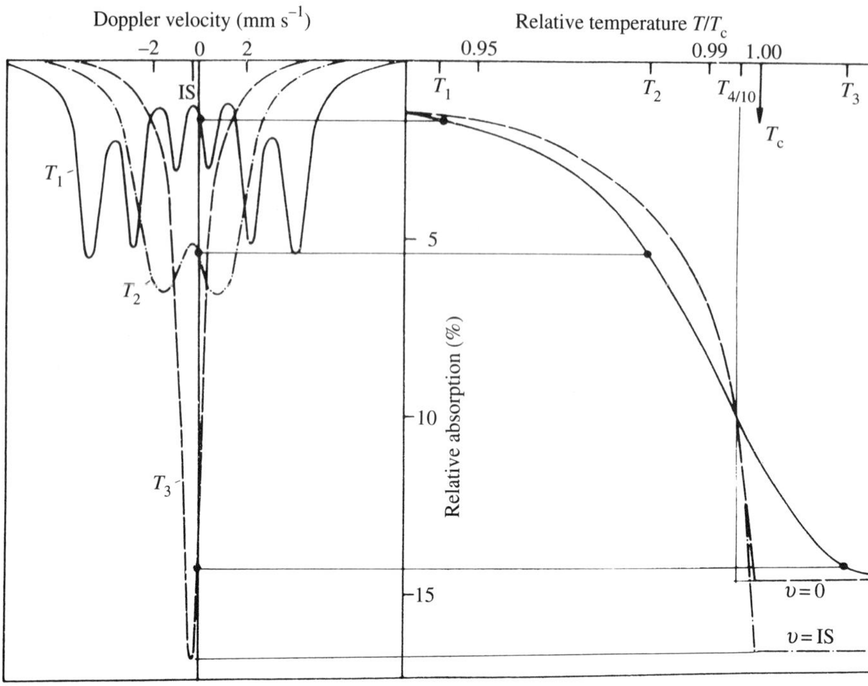

Fig. 3.42 A comparison of individual Mössbauer spectra at different temperatures with the form of thermal-scanning curves for two selected velocities.[34] (Reprinted with permission from ref. 34. Copyright 1974 Springer Verlag.)

energy of the γ-ray photons is high enough to pass through strong pressure windows. A first selection can thus be made by inspection of the transition energies included in Fig. 3.43, where all the elements with Mössbauer isotopes are untinted and transition energies are included for potential high-pressure candidates. With a transition energy of 6 keV, ^{181}Ta is a candidate with a rather low transition energy and ^{153}Eu with a transition energy of 103 keV is at the upper end, since larger transition energies lead to very small recoilless fractions, which are difficult to measure even at temperatures around 4 K. In Fig. 3.43, elements which have been used in high-pressure Mössbauer studies are framed in bold; detailed bibliographies are given in the literature.[13,16,20,33] In addition to the spectral sensitivity and transition energy, the preparations and the lifetimes of the sources are further experimental factors which must be considered in the application of a specific isotope. Experiments using absorbers under pressure are usually preferred to avoid tedious radiochemical work and to avoid the additional safety precautions which are necessary for studies of sources under pressure and for studies on the actinide isotopes.[24]

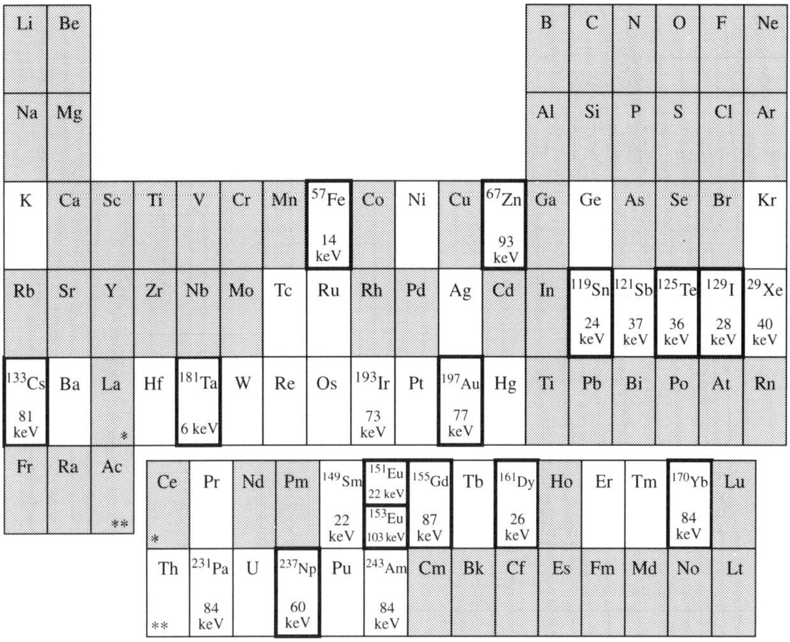

Fig. 3.43 Elements with Mössbauer isotopes are untinted. The energies are transition energies (for high-pressure candidates). Isotopes that have been used in high-pressure experiments are framed by bold lines.

A wide range of *experimental techniques* has thus been developed to meet the specific requirements of different isotopes, as illustrated in Fig. 3.44 for absorbers under pressure. The soft γ-rays (6.2 keV) of ^{181}Ta can pass only through thin beryllium windows, however, due to the very narrow natural line width, changes in the isomer shift can be observed very accurately for this tantalum isotope—even in the 500 MPa region—with conventional hydraulic systems.

For a long time, the Bridgman-anvil technique with carboloy anvils and γ-ray transmission through compacted boron gaskets (with LiH or an epoxy binder) was used predominantly for ^{57}Fe—which has a rather low transition energy of 14.4 keV—while direct transmission through B_4C anvils was more favourable for the higher-energy transitions given with heavy frames in Fig. 3.43. Since the study of many of these isotopes requires cooling to liquid-helium temperatures, a special clamp technique, (see Fig. 3.45) with an internal lead manometer was developed for these applications.[36]

To extend the pressure range and also because of the advantages of precise pressure calibration by the ruby-luminescence technique, gasketed DACs have also found wide application for high-pressure Mössbauer spectroscopy

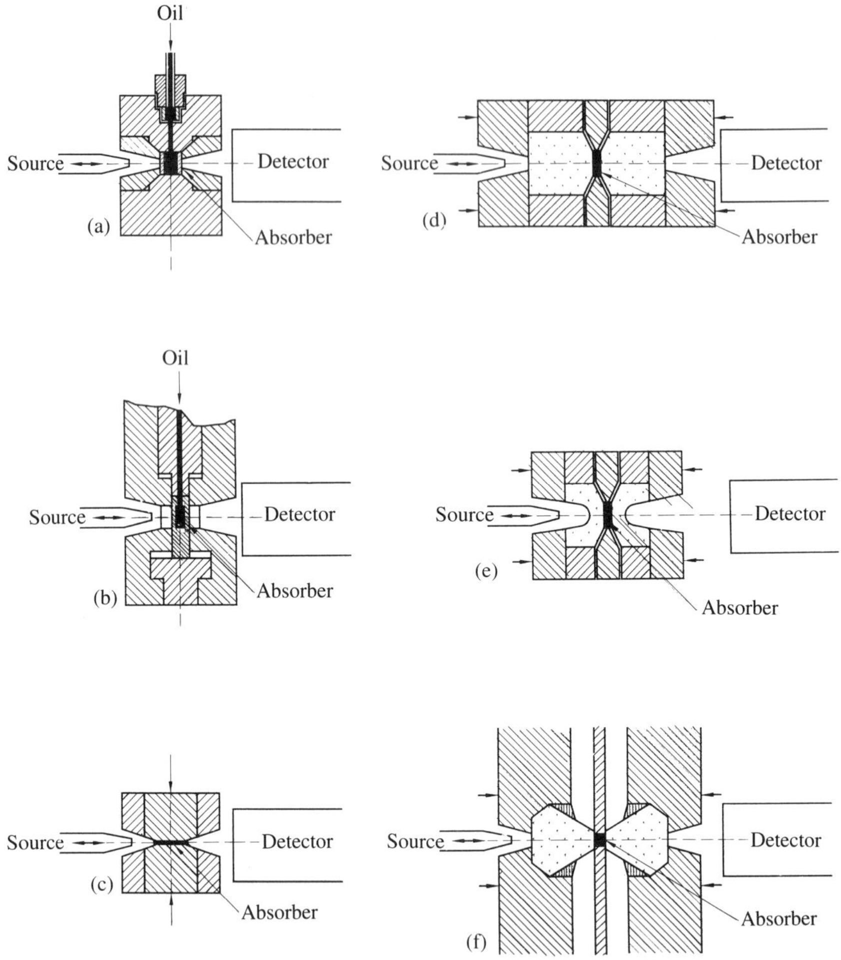

Fig. 3.44 Various high-pressure cells used for Mössbauer spectroscopy.[16] (a) hydraulic cell with beryllium windows for p<250 MPa, (b) the same for p<1GPa, (c) Simple Bridgman anvil cell for γ-ray transmission through the gasket, (d) as for (c), transmission through the anvils, (e) Belt-type cell with hollowed anvil, and (f) DAC. (Reprinted with permission from the publisher.)

in recent years;[29–33] strongly absorbing W–Re-gaskets are very useful in collimating radiation.

Two typical results from many measurements are illustrated in Figs 3.46 and 3.47. Figure 3.46 shows correlations between the isomer shift and the electric quadrupole splitting values for ^{57}Fe in various divalent and trivalent high-spin compounds and for a few low-spin compounds and pressure shifts. In Fig. 3.47, similar data are plotted for ^{197}Au in the Au(I) and Au(III) configurations. Both these figures illustrate, quite clearly, that characteristic

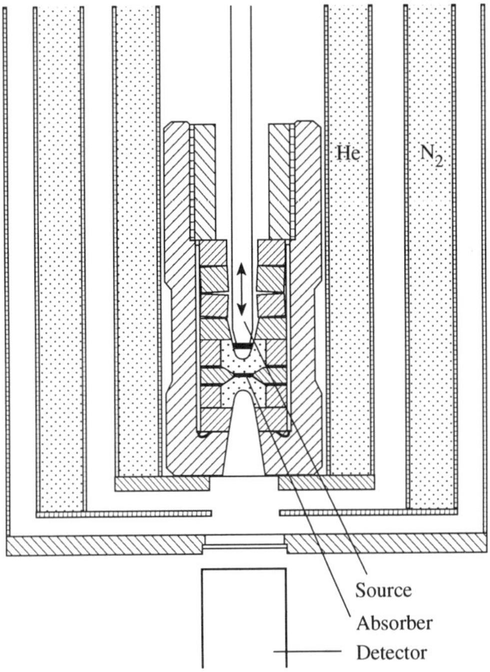

Fig. 3.45 A high-pressure clamp with B_4C anvils and a helium cryostat.[16,36] (Reprinted with permission of the publisher.)

changes in the electronic environments are easily observable under pressure in all these cases.

Recent developments in Mössbauer spectroscopy may also lead to interesting high-pressure applications. Many years ago it was proposed that the special properties of synchrotron radiation could be used to provide nuclear excitation without the use of radioactive sources, and recently progress with modern synchrotron-radiation sources could mean that such experiments could be feasible for ^{57}Fe. Due to the natural high collimation of the most favourable undulator radiation from synchrotron insertion devices, one can expect that high-pressure measurements will be one of the first applications of this technique, which will eventually be applied to isotopes for which no suitable radioactive sources exist.[37–41]

7.1 Experimental considerations

Caution! For safety procedures see Appendix 2. Since special hardware, software, experience, and special permission are required to work with radioactive sources for high-pressure Mössbauer spectrometry, any one interested in this subject is strongly advised to contact one of the groups already working in this field[19–33] before any experimental work is begun.

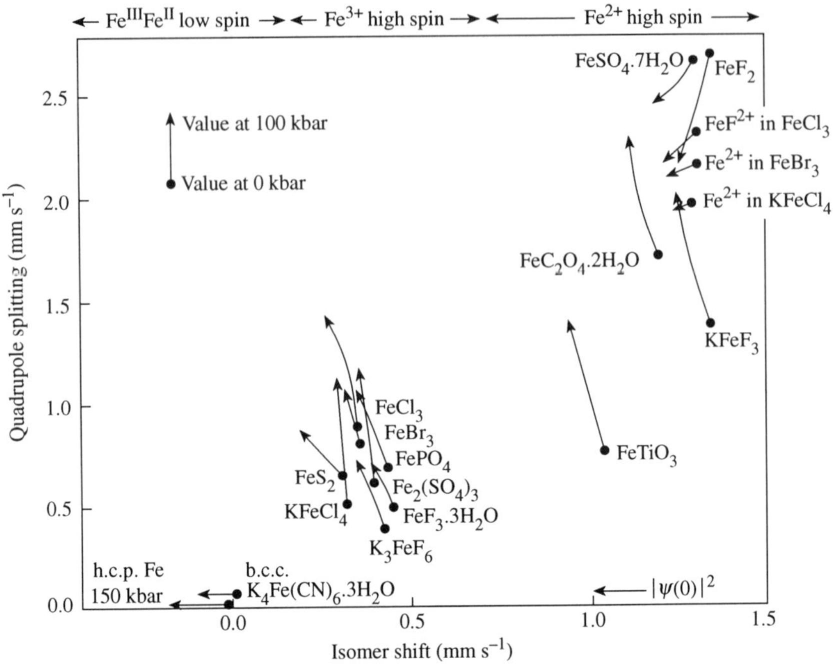

Fig. 3.46 Changes in the isomer shift and electric quadrupole splitting under pressure, where the regions of high-spin Fe^{2+} and Fe^{3+} and of low-spin Fe(II) and Fe(III) can clearly be distinguished.

References

1. Mössbauer, R. L. *Naturwissenschaften* **1958**, *45*, 538.
2. Frauenfelder, H. *The Mössbauer Effect*; Benjamin: New York, **1963**.
3. May, L. *An Introduction to Mössbauer Spectroscopy*; Plenum Press: New York, **1971**.
4. Greenwood, N. N.; Gibb, T. C. *Mössbauer Spectroscopy*; Chapman and Hall: London, **1971**.
5. Gütlich, Ph.; Link, R.; Trautwein, A. *Mössbauer Spectroscopy and Transition Metal Chemistry*; Springer-Verlag: Berlin, **1978**.
6. Stevens, J. G.; Stevens, V. E. *Mössbauer Effect Data Index*; Plenum Press: New York, **1969–1973**.
7. Nicol, M.; Jura, G. *Science* **1963**, *141*, 1035.
8. Panyushkin, V. N.; Voronov, F. F. *JETP Lett.* **1965**, *2*, 97.
9. Drickamer, H. G. In *Solid State Physics*; Seitz, F., Turnbull, D., eds.; Academic Press: New York, **1965**, Vol. 17.

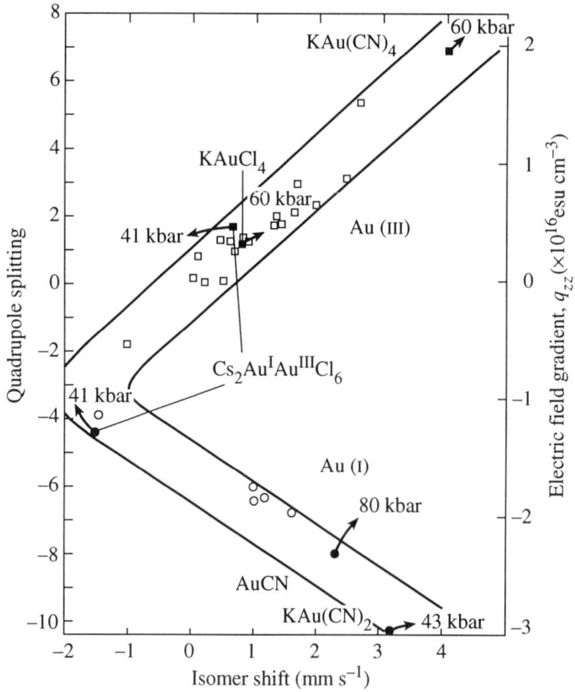

Fig. 3.47 Changes in the isomer shift and the electric quadrupole splitting for various gold compounds under pressure.[16] ○ = Au(I); □ = Au(III) (ambient pressure.) ● = Au(I); ■ = Au(III) arrows indicate the effect of high pressure. (Reprinted with permission of the publisher.)

10. Drickamer, H. G.; Ingalls, R. L.; Coston, C. J. In *Physics of Solids at High Pressure*; Tomizuka, C. T.; Emrick, R. M., eds. Academic Press: New York, **1965**.
11. Drickamer, H. G.; Fung, S. C.; Lewis, Jr, G. K. In *Advances High Pressure Research*, Bradley, R. S. ed.; Academic Press: New York, **1969**, Vol. 3.
12. Drickamer, H. G.; Bastron, V. C.; Fisher, D. C.; Grenoble, D. C. J. *Solid State Chem.* **1970**, *2*, 94.
13. Holzapfel, W. B. *High Temp.-High Press.* **1970**, *2*, 241.
14. Drickamer, H. G.; Frank, C. W. *Ann. Rev. Phys. Chem.* **1972**, *23*, 39.
15. Drickamer, H. G.; Frank, C. W. *Electronic Transitions and the High Pressure Chemistry and Physics of Solids*; Chapman and Hall: London, **1973**.
16. Holzapfel, W. B. *CRC Crit. Rev. Solid State Sci.* **1975**, **5**, *89*.
17. Herber, R.; Spijkerman, J. *J. Chem. Phys.* **1965**, *42*, 4312.
18. Huggins, F. E.; Mao, H. K.; Virgo, D. In *Carnegie Institute of Washington Year Book 74*; Carnegie Institute: Washington, DC, **1975**, p. 405.

19. Potzel, W.; Moser, J.; Potzel, U.; Litterst, F. J.; Kalvius, G. M. *Hyperfine Interactions* **1987**, *34*, 391.
20. Potzel, W. *Hyperfine Interactions* **1988**, *40*, 171.
21. Adlassnig, W.; Potzel, W.; Moser, J.; Schiessl W.; Schäfer, C.; Kalvius, G. M. *Hyperfine Interactions* **1988**, *41*, 531.
22. Adlassnig, W.; Potzel, W.; Moser, J.; Schäfer, C.; Steiner, M.; Kalvius, G. M. *Nuclear Instrum. Methods* A, **1989**, *277*, 485.
23. Adlassnig, W.; Potzel, W.; Moser, J.; Schiessl, W.; Potzel, U.; Schäfer, C.; Steiner, M.; Karzel, H.; Peter, M.; Kalvius, G. M. *Phys. Rev.* B, **1989**, *40*, 7469.
24. Potzel, W.; Adlassnig, W.; Moser, J.; Schäfer, C.; Steiner, M.; Kalvius, G. M. *Phys. Rev.* B, **1989**, *39*, 8236.
25. Potzel, U.; Moser, J.; Potzel, W.; Zwirner, S.; Schiessl, W.; Litterst, F. J.; Kalvius, G. M. *Hyperfine Interact.* **1989**, *47*, 399.
26. Potzel, W. *Hyperfine Interact.* **1992**, *71*, 1515.
27. Zwirner, S.; Potzel, W.; Spirlet, J. C.; Rebizant, J.; Gal, J., Kalvius, G. M. *Physica* B, **1993**, *190*, 107.
28. Nasu, S.; Kurimoto, K.; Nagatomo, S.; Endo S.; Fujita, F. E. *Hyperfine Interact.* **1986**, *29*, 1583.
29. Taylor, R. D.; Pasternak, M. P. *Hyperfine Interact.* **1990**, *53*, 159.
30. Taylor, R. D.; Chen, A.; Meade, Ch.; Falicov, L. M.; Giesekus, A.; Jeanloz, R.; Yu, P. Y. *Phys. Rev. Lett.* **1990**, *65*, 790.
31. Potzel, W.; Gleissner, A.; Moser, J.; Zwirner, S.; Potzel, U.; Gal, J.; Kalvius, G. M. *Physica* B, **1993**, *190*, 98.
32. Gleissner, A.; Potzel, W.; Moser, J.; Kalvius, G. M. *Phys. Rev. Lett.* **1993**, *70*, 2032.
33. Hesse, H.-J.; Wortmann, G. *Hyperfine Interact.* **1994**, *93*, 1499–1504.
34. Holzapfel, W. B. In *Les Proprietes Physiques des Solides sous Pression*; Editions du CNRS: Paris, **1970**, p. 355.
35. Christoe, C. W.; Forster, A.; Holzapfel, W. B. *Z. Angew. Physik.* **1971**, *31*, 263.
36. Schilling, J. S.; Klein, U. F.; Holzapfel, W. B. *Rev. Sci. Instrum.* **1974**, *45*, 1353.
37. Rüffer, R.; Gerdau, E.; Grote, M.; Hollatz, R.; Röhlsberger, R.; Rüter, H.-D.; Sturhahn, W. *Hyperfine Interact.* **1990**, *61*m, 1279.
38. Hastings, J. B.; Siddons, D. P.; van Bürck, U.; Hollatz, R.; Bergmann, U. *Phys. Rev. Lett.* **1991**, *66*, 770.
39. Sutrhahn, W.; Gerdau, E.; Börger, S.; Guse, W. *Hyperfine Interact.* **1990**, *58*, 2483.
40. Sturhahn, W.; Gerdau, E.; Hollatz, R.; Rüffer, R.; Rüter, H.-D.; Tolksdorf, W. *Europhys. Lett.* **1991**, *14*, 821.
41. Rüffer, R.; Chumakov, A. I.; Grünsteudel, H. F.; Grübel, G.; Leupold, O.; Metge, J.; Hesse, H.-J.; Wortmann, G.; Goodwin, H. A. *ESRF Newsletter* **1995**, *22*, 12.

8. Special techniques

W. B. Holzapfel

One less-well-known technique, which has many experimental aspects in common with Mössbauer spectroscopy, deserves special attention at this point, since it gives valuable information about the electric-field gradients and the magnetic hyperfine interactions of radioactive nuclei in solids at ambient conditions and under pressure. In this technique, two γ-rays with different energies from two different transitions of an individual nucleus in a radioactive-decay cascade are recorded consecutively. The spatial and temporal perturbation of the emission probability by the hyperfine fields is registered in the corresponding perturbed angular correlation (PAC) spectra.

In this case, the sample is a radioactive source which can be compressed for high-pressure experiments either in the gasket of a conventional Bridgman-anvil device[1–4] (as shown in Fig. 3.48) or with submillimetre-size samples in DAC devices.[5,6] While early applications of PAC spectra exclusively involved studies of the pressure dependence of electric-field gradients,[1–4] the subject of recent interest has been transferred magnetic hyperfine fields.[5,6]

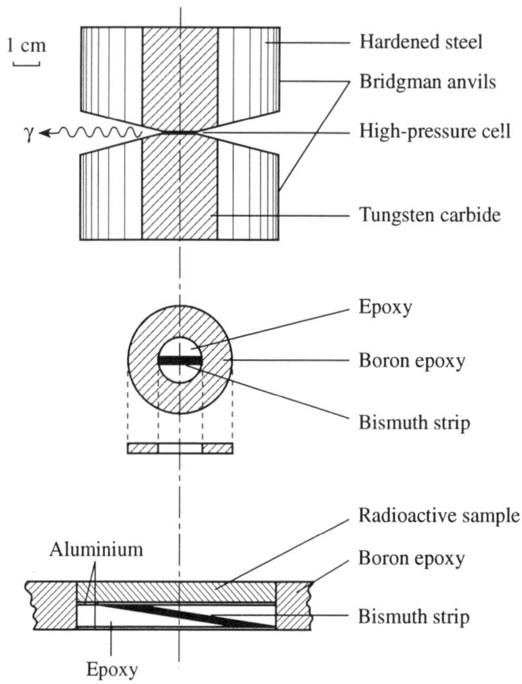

Fig. 3.48 A pressure cell for PAC studies.[3] (Reprinted with permission from Baltzer Sciences Publishers BV.)

The most favourable isotopes for this technique are ^{111}Cd and ^{181}Ta, and the most serious limitation is the fact that radioactive samples must be handled.

8.1 Experimental considerations

Caution! For safety procedures see Appendix 2. Experimental groups which have experience with PAC spectra at ambient pressure will find the development of new techniques for handling the tiny radioactive sources used in DACs in high-pressure measurements an interesting challenge.

References

1. Raghavan, P.; Raghavan, R. S.; Holzapfel, W. B. *Phys. Rev. Lett.* **1972**, *28*, 903–906.
2. Butz, T.; Wortmann, G.; Kalvius, G. M.; Holzapfel, W. B. *Phys. Lett.* A, **1974**, *50*, 127–128.
3. Butz, T.; Göbel, H.; Holzapfel, W. B.; Kalvius, G. M. *Hyperfine Interact.* **1975**, *1*, 1–14.
4. Butz, T. *Physica Scripta* **1978**, *17*, 87–93.
5. Lindgren, B.; Vijay, Y. K. *Hyperfine Interact.* **1981**, *9*, 379.
6. Chow, L.; Zhao, X.; Collins, G. S. *Phys. Rev.* B, **1992**, *45*, 4672.

9. High pressure viscometry

H. E. King Jr

9.1 Introduction

The study of fluid viscosity has occupied scientists for many years. Intellectual interest has been sparked by attempts to link the underlying molecular dynamics to the viscosity, especially the multiple-decade-viscosity change found in simple glass-forming fluids. The many practical consequences which arise from fluid-flow characteristics (from low-viscosity liquids to high-viscosity glasses) have provided ample incentive for applied studies. Variable-pressure data are valuable to both theoretical and practical investigations.

The theories of the viscosity of ordinary liquids are mainly scaling relationships; there is no first-principles theory for their viscosities. An important scaling relationship is that the viscosity is related to the ratio of the occupied volume to the free volume. The usefulness of variable-pressure studies lies in their ability to probe this directly. Such studies of low-density fluids (gases and supercritical fluids), interpreted through extensions to the kinetic gas theory,[1] have provided a quantitative understanding of their viscosities. However, these models fail as one approaches the densities of ordinary liquids, whose most important density-scaling law is the *free-volume theory* of Cohen and Turnbull.[2] However, variable-pressure data have only recently been obtained with high precision and over a sufficiently wide pressure range to test this theory adequately. The two devices described in this work provide data over many decades of viscosity, approaching the viscosity of the glass transition. Such data are necessary[3] to test stringently any model which describes the volume scaling of viscosity.

High-pressure viscosity measurements can also be of considerable practical importance. For example, in the field of lubrication it has been recognized for some time that the fluid in a lubricated contact can experience several gigapascals of pressure.[4] Theory predicts that the derivative of viscosity with respect to pressure is a critical parameter in determining the metal–metal contact and thus the wear in such a system.[5] However, this derivative is itself a strong function of pressure, and a more accurate estimate of viscosity effects is gained by experimentally determining the entire pressure–viscosity curve up to the actual pressures of operation.

9.2 Scope of the experimental problem

9.2.1 Pressure and viscosity range

The two most important qualities of a high-pressure viscometer are its pressure and viscosity range. The shear rate is of secondary interest. We can gain an appreciation of the ranges required by considering the pressures over which fluids exist and the viscosities they exhibit. Our focus in this work is on dense fluids at nearly ambient temperatures. Isothermal compression of such fluids lead to one of two states: crystallization, or glass formation. (This excludes complex fluids in which various mesophases may also form.)

Crystallization occurs for many common fluids, such as carbon tetrachloride, benzene, and cyclohexane, at pressures less than 200 MPa, thus their entire fluid range is rather limited.[6] The supercritical noble gases significantly extend this range, achieving a maximum crystallization pressure for helium[7] of $P = 11.6$ GPa. The viscosity increase prior to all such transitions is modest. The viscosity for a typical dense fluid is 1–100 mPa and this will increase by, at most, about three orders of magnitude. Experimentally, this viscosity and pressure regime is covered by many of the viscometers discussed below, and hard-sphere theory[1] can explain most of the viscosity increase.

Glass formation, the other common fate of a compressed fluid, however, gives a much larger viscosity increase. For mixtures and fluids with nonspherical molecules, it is common to produce a metastable glass at some pressure in excess of the equilibrium crystallization point. Methanol is one example. It can be easily superpressed past its crystallization pressure of 3.5 GPa at $T = 25°C$ towards its glass-transition[8] pressure of 11.4 GPa and beyond.[9] Being a kinetic property, the pressure at which this *transition* occurs is highly material dependent. For some lubricants,[10] this is only a few hundred megapascals, while it reaches 14 GPa for the commonly used methanol–ethanol–water pressure-transmitting media.[11] One might also suppose that the pressurization rate would influence the transition pressure, but this has not been systematically explored. The viscosity increase in the superpressed liquid is enormous: a value of 10^{12} Pa s is typical of the actual glass transition. The exact shape of the viscosity rise is of considerable importance in understanding transport in dense fluids, and it is clear that to define this shape a high-pressure viscometer must cover a multiple-decade viscosity range, and DAC viscometers (which have $\eta_{\max} \approx 10^7$ Pa s and $P_{\max} \approx 10$ GPa) are particularly well suited to the study of such glass formers.

An important area of theoretical study is exploration of density-mediated transport; the wide pressure range of DACs is an advantage. For the dense fluids of interest here, pressurization to 10 GPa can alter the density by about 50 per cent.

9.2.2 Shear rate

Experimental time scales must always be considered when discussing viscosity measurements. Viscosity can be defined through the proportionality between the strain rate, $\partial\varepsilon/\partial t$, of a liquid and the observed stress, σ:

$$\sigma = \eta \frac{\partial \varepsilon}{\partial t}. \tag{3.9.1}$$

At some elevated shear rate, above the Newtonian regime, this proportionality is lost and viscous behaviour is replaced by elastic behaviour. The full implications of this changeover are far outside the scope of this work.[12] For our purposes, it is sufficient to recognize whether a given technique measures a quantity associated with the viscous or the elastic response. These two regimes are roughly separated from one another by the product of the measurement frequency ω, and the relaxation time of the fluid, τ. The viscous-response regime applies for $\omega\tau \ll 1$. One can relate the values of τ for any given fluid to the viscosity through use of the Maxwell relaxation time, defined by $\tau_M = \eta/G$, where G is the shear modulus. Note that at the glass transition $\tau_M \approx 10^2$ and $\eta \approx 10^{12}$ Pa s and that these values can be used to obtain a universal estimate for G. Thus, for low-viscosity liquids the crossover to elastic behaviour is in the gigahertz time-scale range and it is in the megahertz range for more viscous fluids (10^4 Pa s). With a few notable exceptions, the shear rates of high-pressure viscometers are below 500 s^{-1}, well inside the viscous-response regime. (Complex fluids, that is, polymer solutions, often exhibit shear-rate-dependent effects even within these time scales, because this simple Maxwell-relaxation-time argument does not capture their long relaxation times.) As we discuss below, several devices have been built to address specifically the effects of high-shear-rate viscometry at high pressures. The shear rates for these devices can exceed 10^5 s^{-1}, and for high viscosities this is clearly within the viscoelastic regime. Finally, devices which employ an oscillatory strain often operate at frequencies which span both the viscous and the elastic regime, and both viscosity and shear moduli are measurable.

9.3 Summary of previous viscometric techniques

The methods used to measure viscosity at elevated pressures can be divided into three classes: (i) moving-object devices, in which the drag on an object moving within the fluid is used to measure viscosity; (ii) capillary-flow devices; and (iii) oscillatory-strain devices, in which the frequency shift and damping of a vibrating object is used to measure the viscosity. We review these devices here, keeping in mind the three experimental design criteria discussed in Section 9.3.2: pressure range, viscosity range, and limiting shear rate. Readers should also consult a recent review of the use of high-pressure viscometers in lubricant studies.[13]

9.3.1 Moving-object devices

9.3.1.1 Couette geometry. Couette geometry, in which a piston is rotated within a concentric cylinder, is probably the most quantitative method for measuring viscosity. The hydrodynamics are well defined, and through a knowledge of the speed and the piston-to-cylinder gap, the shear rates are variable and precisely determined. Typically, the inner piston is rotated at a fixed speed and the force transmitted to the outer cylinder is measured. This has proven to be a difficult device to adapt to high pressures, because it is difficult to make the transmission of the rotation through the pressure-chamber wall compatible with very high pressures and viscosities. A device capable of measuring viscosities in the range of $5 \leqslant \eta \leqslant 10^{6.5}$ Pa s at pressures up to 500 MPa has been described by Hutton and Phillips.[14] The maximum shear-rate is low, 10 s^{-1}, thus the effect of the shear rate on the viscosity is not measurable. Hutton and Phillips estimated that their accuracy was 5 per cent up to $\eta = 10^{3.5}$ Pa s and 10 per cent for larger viscosities. A device capable of much higher shear rates (10^5 s^{-1}) has been described recently.[15] Using this device up to its maximum pressure of 275 MPa, the authors have documented significant shear rate effects in lubricants, including a transition to plastic-flow behaviour. Such effects are the main focus of their study, and viscosity is only a by-product of their measurements. They note that their device gives fair agreement with falling-body viscosimetric data, but they do not give any estimates of the accuracies.

9.3.1.2 Falling plungers/swinging vane. Viscometers of the falling body type are among the most widely used for high pressures. The geometry is well suited to pressure vessels, since a cylindrical weight falls under the force of gravity within a concentric cylinder. The plunger is self-centring within the cylinder, producing well-defined hydrodynamics. The shear rate is low but well characterized, being proportional to the plunger-to-cylinder gap and the plunger velocity. Shear-rate effects cannot be explored because the plunger velocity varies with the viscosity. To measure a large pressure-induced viscosity increase often requires the use of several plungers. This can introduce errors since an empirical calibration coefficient is used for each plunger–cylinder combination. Variations of this calibration with pressure are also a concern. Furthermore, one must contend with exceedingly small velocities at the highest viscosities. The early studies by Bridgman[6] of pressures up to 1.2 GPa, which were later[16] extended to 3.0 GPa through the use of a swinging vane, demonstrated how this type of device could generate very accurate data over a wide pressure range. Bridgman's studies covered the viscosity range $10^{-3} \leqslant \eta \leqslant 10^{1}$ Pa s with an accuracy of about 0.2 per cent; later studies extended the viscosity range to $10^{3.5}$ Pa s, but there was a reduction in accuracy. Similar devices were later used in the study of lubricants.[17,18] Two major problems with these devices, which have been described by all the above authors are: (i) the entire pressure chamber has to be inverted to replicate plunger falls; and (ii) the time for the plunger to traverse the entire cylinder length is measured. This requires an empirical correction for the velocity changes of the plunger near the cylinder ends, and it results in very long fall-times at high viscosities. Later devices sought to overcome these limitations. Irving and Barlow[19] have described a device in which magnetic coils are used to lift the plunger and to detect its velocity. The pressure range is limited to pressures below 1.4 GPa, with an accuracy of 2 per cent within the measurable viscosity range, $10^{-3} \leqslant \eta \leqslant 10^{2.5}$ Pa s. In a series of papers, Bair and Weiner[20–22] have described two falling-slug viscometers. Both viscometers detected the slug magnetically, which further extends the measurable viscosities to 10^{7} Pa s at a maximum pressure of 1.1 GPa. These viscometers are used exclusively in the study of lubricants, consequently it is difficult to evaluate their accuracy. Measurements on reference fluids were not compared with previous work, and the authors did not report the accuracy of their data.

9.3.1.3 Falling/rolling-ball viscometers. The falling-ball/rolling-ball viscometer is among the simplest to envision. The distance of the sphere from the pressure-chamber walls (this is discussed further in Section 9.4) is an important parameter. If this distance is on the order of five ball diameters, then the Stokes' equation for an infinite fluid applies and the device is called a simple falling-ball viscometer. Because there is no self-centring force for a sphere, one cannot assume that the sphere will find its way into the centre; there must be some method to guide it there. For smaller separations, the measured drag includes a wall effect. This can be modelled empirically for a well-controlled sphere trajectory. A device meeting the latter requirements has been used in some unique high-pressure, high-temperature studies of oxide-melt viscosities.[23–25] For these studies, spheres of high-melting-point materials, such as platinum, were placed in the centre of an oxide powder in the pressure chamber of a piston-in-cylinder device. This assembly was then pressurized and heated to a temperature above the melting point of the oxide for a specific period of time. Following a temperature quench the resulting glass was removed from the chamber and the distance travelled by the sphere was measured. The range of accessible viscosities is $10^{2.7} \leqslant \eta \leqslant 10^{4.7}$ Pa s, for pressures of $0.5 \leqslant P \leqslant 3.0$ GPa. Temperatures were investigated in the range 1100–1500°C. At ambient pressure, good agreement was found with data from other viscometric techniques and the authors estimated that their maximum error was 10 per cent.

A more common device is the rolling-ball viscometer in which a close-fitting sphere rolls within a cylindrical tube. Operationally, this is similar to the falling-plunger devices discussed in Section 9.3.2; however, the hydrodynamics are not as well determined. The wall effect depends on the angle of tilt of the cylinder,[26] and due to the one-sided placement of the sphere within the cylinder, the shear rate is not uniform over the surface of the sphere. Typically, these effects are lumped into an empirical correction factor which characterizes a given sphere within a given cylinder. Using this approach, very accurate data can then be obtained; accuracies of better than one per cent have been reported.[26,27] Like falling-plunger devices, small velocities limit this technique, but the problem is more extreme. The small-mass sphere is driven with less than the full acceleration of gravity (due to the tilt angle), and the wall drag effect is enhanced by the small gap; hence, this type of viscometer is limited to low viscosities, that is $\eta \leqslant 10^2$ Pa s.

9.3.2 Capillary-flow viscometers

In capillary-flow viscometers the flow time for a liquid through a capillary is related to its viscosity. The capillary diameter and the fluid velocity determine the shear rate. At ambient pressure, gravity produces a pressure gradient as fluid flows out the bottom of a filled vessel into an open receptacle. Transforming this device for use at high pressures requires that a small pressure gradient between a filled and empty container is generated at some high overall pressure. Such a device, capable of measurements over a very wide pressure range (P_{max} = 6 GPa), was designed by Barnett and Bosco.[28] Although the design is unsuitable for low viscosities ($\eta < 10^4$ Pa s) data taken on several fluids demonstrate that quite high viscosities can be measured $\sim 10^9$ Pa s). However, as the authors themselves note, the purpose of this device was not to make high-precision, quantitative, viscosity measurements but to establish the limits of hydrostatic fluids for use in high-pressure devices. They estimate the uncertainties in their data to be approximately 50 per cent. A similar technique, operating to a maximum pressure of 0.65 GPa, has been used[29] to measure the viscosity of krypton to quite high accuracy—

about ±1.3 per cent. The effects of the shear rate are not considered in these two devices, but a viscometer employing variable shear rates of 10^2 to 10^5 s^{-1} has been demonstrated.[30] Pressures up to 200 MPa and viscosities in the range 10^{-1} to 10^3 Pa s are measurable.

9.3.3 Ultrasound frequency shift/attenuation

All of the techniques described in this section use the fact that both an oscillator's resonance frequency, f, and its resonance width, Δf, change when it is immersed in a liquid. To associate these effects with *viscous* damping, as discussed above, one must be well below a characteristic frequency, that is, $\omega\tau \ll 1$. Within this frequency range, and for low-viscosity fluids, these devices can be exceedingly precise. Extensive high-pressure viscosity measurements, up to 1.0 GPa, with an average precision of 2 per cent, have been made on the supercritical fluids methane and krypton using a vibrating wire oscillator.[31–33] Measurements with an accuracy of 0.5 per cent have been made on *n*-hexane[34] to pressures of 0.08 GPa. Furthermore, through an analysis of the various theoretical sources of error it has been suggested[35] that an accuracy of a few parts in a thousand is possible. It was recently shown[36] that a similar device can be used to measure even highly nonideal fluids, such as $^3He/^4He$ mixtures, at elevated pressures. Vibrating crystals have also been used to measure viscosity at high pressure. Using the alternating-current-driven torsional modes of a quartz crystal at frequencies of 40–60 kHz, the viscosities of several liquids have been measured[37,38] at pressures up to 0.7 GPa. The accuracy is estimated to be ~1 per cent, and the maximum viscosity is 1 Pa s.

These techniques all fail at some high value of the viscosity. This can be understood by considering the hydrodynamic equations which describe the force on a wire moving in the fluid.[35] It is assumed that the damping length in the fluid is short compared to any length of the device, particularly the radius of the wire. This is expressed through a dimensionless quantity, $m = a/2\lambda$, where a is the wire radius. The viscous penetration depth, λ, is given by $(\eta/2\pi f\rho)^{1/2}$. In theory, $m \geqslant 0.5$ is the applicable range;[39] although experiment shows that good results[35] can still be obtained for $m \geqslant 0.2$. However, the typical pressure-induced viscosity rise of several orders of magnitude will force m to become considerably smaller; for example, using vibrating-wire parameters of $a = 25$ μm, $f = 2$ kHz, and $\rho = 1000$ kg m^{-3}, $m \approx 10^{-2}$ for $\eta = 1$ Pa s. A compensating change in the measurement frequency, f, would extend the measurement range somewhat. But this frequency is difficult to change by more than one or two orders of magnitude, and eventually one reaches the viscoelastic limit. Similar arguments concerning the limited viscous-damping regime can be made for the quartz-crystal oscillators. For these devices, it is easier to extend the frequency range, and they are often used at higher frequencies (~30 MHz) to measure the viscoelastic properties such as the shear modulus, at elevated pressures.[40,41]

9.3.4 High-shear-rate studies

Consideration of the viscoelastic regime is outside the scope of this work, but some interesting pressure-induced behaviour will be mentioned. One puzzling effect is the limiting shear stress. Ordinarily a plot of the shear stress against the shear rate gives a line whose slope is the viscosity. Non-Newtonian behaviour transforms this line into a curve. The slope at any point on such a curve is the viscosity at that shear rate. A new phenomenon has been found at extreme shear rates. Through high pressure shock-wave studies,[42] and high-shear-rate viscometry[15] it has been found that the sheer stresses of some lubricants, for shear rates of $\sim 10^4\ s^{-1}$, reach maximum values and remain constant for further shear-rate increases. As the pressure increases, the shear rate at which this occurs moves to lower values, for example, at a few hundred megapascals the shear rate is $\sim 10^3\ s^{-1}$. This has important implications for lubrication studies where high shear rates accompany high-pressure conditions,[4] but the physics underlying this transition to plastic-like behaviour is not well understood. Another feature associated with pressure-induced changes in the relaxation times has recently been probed by Brillouin scattering.[43] Using a unique experimental technique which allows one to resolve the intrinsic line width of the longtitudinal acoustic modes of the scattering,[44] Oliver *et al.* were able to observe for the first time the line broadening associated with the $\omega\tau = 1$ crossover. They were then able to map the relaxation-time changes for pentane–isopentane mixtures from the liquid at ambient pressure up to the glassy state at $P_g = 7.4$ GPa.

9.4 The DAC rolling-ball and centrifugal-force viscometers

9.4.1 Introduction to DAC viscometry

9.4.1.1 Advantages. A DAC viscometer is a very useful device for a number of reasons. A wide range of fluids can be measured because the components of the diamond cell which contact the sample (Fig. 3.49) are chemically inert. The pressure is accurately determined because the manometer[45] (chromium-doped Al_2O_3) is actually immersed in the fluid. The sample and the falling ball are visible at high pressures, thus one can track the ball's trajectory and select that portion of the fall which is at the terminal velocity. (In contrast, the falling-mass techniques described above typically measure the entire fall time, and they use an empirical model to calculate the terminal velocity needed in the Stokes' equation.) Also, this ability to monitor the sample's appearance can reveal phase transitions. The pressure and temperature range over which such a device can operate is large. Finally, it is a significant experimental advantage to use the same sample and device to measure data over the entire accessible ten-decade range of viscosity.

9.4.1.2 History. Modern high-pressure physics is dominated by the use of DACs.[46] Their small size, simplicity of operation, and extreme pressure range make them ideal for many experiments. The first use of a DAC viscometer

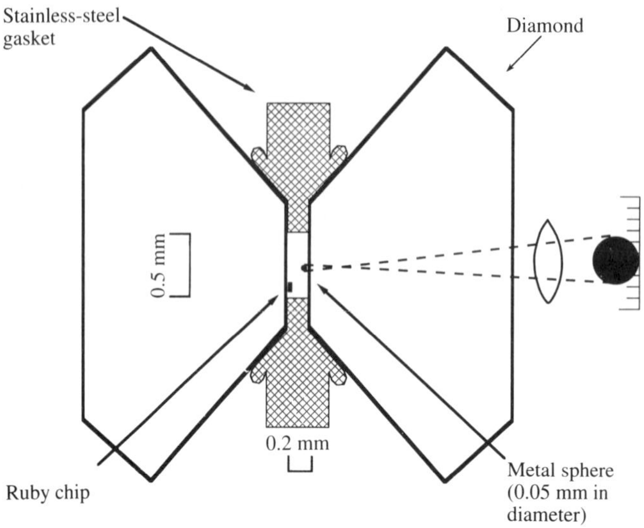

Fig. 3.49 All the components of a DAC viscometer are shown except for the triangular Merrill–Bassett-type clamp. The metal sphere rolls down the diamond surface under the influence of gravity, and its velocity is determined using a video system. The tilt angle (not shown) ensures a rolling-sphere trajectory: (from King *et al.*, reference 49)

was reported by Piermarini *et al.*[47] They presented results using their falling-ball viscometer on a 4 : 1 methanol:ethanol mixture at pressures up to 7 GPa, corresponding to a maximum viscosity of 10^4 Pa s. The uncertainties in their viscosity data ranged from 10 to 50 per cent. A major goal of their study was to demonstrate the fluid nature of the 4 : 1 methanol : ethanol mixture. The precision was adequate to achieve this goal, but it is unacceptably large for quantitative studies of high-pressure viscosity. This same device, with some changes in the optics and electronics used to measure the sphere position, has been further refined by Fujishiro *et al.*[48] They presented a graphical comparison of their data with that collected by Bridgman which allows the accuracy to be estimated as 20–30 per cent.

9.4.1.3 General hydrodynamic considerations. Our primary objective in altering the DAC viscometer was to transform it into a more accurate tool for the study of fluids. The main cause of poor precision is variability in the excess drag, that is, the drag in excess of the Stokes' drag, which is caused by the close proximity of the walls of the small pressure chamber. This drag is highly sensitive to changes in the sphere's position within the cell; consequently, uncontrolled trajectories can lead to large divisions in the measured velocity. We have found that this trajectory can be controlled by placing the sphere against the diamond surface, thus transforming the device into a rolling-ball viscometer.

One can write a modified Stokes' equation which accounts for both the viscous drag and the wall effect drag:

$$\eta = \gamma \left(\frac{2R_s^2 g(\rho_s - \rho_f) \cos \theta}{9v} \right) \tag{3.9.2}$$

where R_s is the sphere radius, g is acceleration due to gravity, ρ_s is the sphere density (7600 kg m^{-3}), ρ_f is the fluid density, θ is the tilt angle of the cell (0° = vertical), v is the velocity of the ball, and γ is the wall correction. Although eqn (3.9.2) can be used to calculate the viscosity directly, it is more accurate to use the relative velocities as follows:

$$\frac{\eta(P)}{\eta(0)} = \left(\frac{\rho_s - \rho_f(P)}{\rho_s - \rho_f(0)} \right) \frac{v(0)}{v(P)}. \tag{3.9.3}$$

Given the fluid density as a function of pressure, the relative viscosity can be directly determined by the ratio of settling velocities, independently of any determination of the sphere radius or the wall correction. The compressibility of the sphere is negligible over the pressure range used. Furthermore, if the fluid compressibility is unknown, only a relatively small error is made (about 5 per cent for a typical fluid compressibility) in neglecting the density term in eqn (3.9.3).

The utility of eqn (3.9.2) and the validity of eqn (3.9.3) both rest on the constancy of the wall correction, γ. It should not depend on parameters such as the viscosity or the ball velocity nor should it vary from one pressure to another. Hydrodynamic calculations[49] reveal two important features of the wall correction: (i) for the low-Reynolds-number flow experienced in DAC viscometers, it depends only on the instantaneous position of the ball relative to the walls of the cell, and (ii) its magnitude varies widely throughout the pressure chamber. Low-Reynolds-number flow implies, further, that there is no force tending to move the sphere along any one path in the pressure chamber, for example, into the centre of the two diamond culets, where the wall effect would be minimized. The starting point of the sphere determines its trajectory and, left uncontrolled, one can expect, as observed, a significant variation from run to run in the path traversed, and hence a wide variation in γ. Tilting the DAC, however, places the sphere near one of the diamond surfaces. This controls the trajectory, and experiment shows that the average error of the mean velocity is only about 4 per cent. Another important advantage of rolling the ball down an inclined surface is that the resistance due to the nearest surface dominates that due to the other walls of the cell. This effectively reduces the effects of both the gasket walls and the variable gap between the diamond culets (as the pressure is changed).

Transforming the cell to a rolling-ball viscometer makes γ constant, but if the device is to be used as a viscometer one must also show that eqn (3.9.2) remains valid even when nonhydrodynamic interactions with the walls, such as friction, are significant. The variation of γ with the tilt angle indicates that friction is present. By considering the force and torque balances for the sphere,[49] one can obtain results which show how friction contributes to γ. The important ratio is the ratio of the force on the sphere tangential to the diamond surface, F_T, to the normal forces, F_N. From eqn 9 in reference 49 a dependence on the tangent of the tilt angle can be predicted. The observed dependence verifies this functional form, except at the extremes of tilt. At the smallest tilt angles (a nearly vertical surface) γ is underestimated; that is, the measured deviation from the ideal is smaller than predicted. A similar effect has recently been reported by Smart *et al.*[50] who attribute it to nonuniform sliding motion of the sphere as it loses contact with the wall. For the DAC viscometer, deviations become important at angles below about 15°. At high angles (a nearly horizontal surface) purely rolling motion takes over, a condition not accommodated by eqn 9 in reference 49. Above the critical angle, which depends upon the friction coefficient, the magnitude of γ becomes fixed. This point is at angles near 80–90° for DAC viscometers, and data are not taken above 75°. We have found that the highest quality data are obtained for tilt angles near 45°, well away from both bounds.

The discussion above applies for gravity-driven spheres where the ratio, F_T/F_N, between the tangential force, which drives the sphere along the wall, and the normal force, which forces it into the wall, is entirely determined by the tilt angle. For the recommended 15–75° range of tilt angles, $3.7 \leqslant F_T/F_N \leqslant 0.3$. In centrifugal-force viscometers, the DAC is placed horizontally, thus F_N is proportional to g, the acceleration due to gravity. However, F_T varies significantly, and F_T/F_N ranges from 50 to 1700. Such ratios simulate those of exceedingly small tilt angles. However, unlike rolling-ball viscometers, this does not lead to variability in γ nor to magnitudes approaching unity. The magnitudes of γ for centrifugal-force viscometers are similar to those found at normal tilt angles (see below). Perhaps because $F_N = g$, the sphere maintains good contact with the wall.

To measure the absolute viscosity eqn (3.9.2) must be solved, and so a value of γ is needed. The discussion above indicates the conditions necessary to produce a constant wall correction, but there are unknowns, such as the sphere-to-wall separation, which make it impossible to calculate an actual theoretical value with sufficient accuracy. One must measure γ empirically by observing the sphere velocity under conditions for which the fluid viscosity is known. For the rolling-ball viscometer, comparison is made with ambient-pressure viscosities; whereas high-pressure data taken by the rolling-ball viscometer are used for comparison for the centrifugal-force viscometer.

The shear rates in any rolling-ball device are imprecisely defined (see above). However, one can estimate the maximum shear rate by taking the ratio of the sphere velocity to the dimension of the sphere-to-wall gap. For example, in a recent experiment[51] we found that a polymer solution with an apparent viscosity of 68 mPa s had a velocity of 330 μm s^{-1}, and the gap was estimated as 1 μm; the shear rate was 300 s^{-1}. From the known shear thinning of this fluid, we could then estimate that the true viscosity was 435 mPa s. This concept of shear in our experiment is a much simplified approximation to the actual shear effects, but for most fluids no such correction is needed. Further, as the viscosity increases and the velocity slows, the shear rate will decrease to small values.

9.4.2 The DAC rolling-ball viscometer

9.4.2.1 Principles of operation. A high-pressure viscometer is shown in Fig. 3.49. The DAC is a simple device. There are two $\frac{1}{4}$ carat diamonds with 1.2 mm diameter culets. These culets press into a stainless-steel foil into which is drilled a 0.5 mm hole. We typically use Inconel X750 stainless steel with an initial thickness of 0.25 mm. The pressure is applied by a mechanical clamp, which is not shown. Various clamping devices have been described.[46] We currently use a Merrill–Bassett-type clamp[52] which is modified for use in optical experiments by replacing the beryllium–diamond pushers in the original design by steel ones. There are 25° cone-opening optical-access ports in these pushers, and the ports are centred on the large diamond tables at the back of each diamond. Force is applied through these pusher plates by turning three screws in the clamp. We often use these cells at pressures up to 3.5 GPa. Using hardened-steel pushers and a technique in which the pressure is cycled up then down then up again, one can attain higher pressures (~ 10 GPa). A pressure measurement is made at each data point using a ruby-fluorescence manometer.

The pressure chamber in such a device is, of course, quite small; this is an advantage, because only a small quantity of fluid is needed. But, one must guard against sample contamination and develop a means of manipulating small objects. These procedures are discussed in Protocol 1.

Protocol 1.
Loading the DAC viscometer

Caution! For safety procedures see Appendix 2.

This procedure describes the cleaning of the diamond culets and the loading of the fluid, the ruby manometer, and the sphere for use in the two viscometers described in Protocols 2 and 3.

Equipment

- A DAC equipped as described in the text
- Lint-free cotton-tipped swabs (Puritan)
- Lint-free optical lens paper (Ross optical lens tissue)
- Syringe μL (Hamilton Model 1701N) or a Pasteur pipette
- Round toothpicks,
- Stereoscope ×140 equipped with incident and transmitted light (Bausch and Lomb stereozoom 7 with ×20 eyepieces and a calibrated reticule)
- DAC gasket, prepared as explained in text

Protocol 1. *Continued*

- Nickel sphere ~50 μm diameter (Hoeganaes Corporation)
- Ruby chips ~40 μm square, 0.5 wt % chromium doped
- Sewing needles
- Glass fibres ~75 μm diameter (pulled by hand from flame-heated glass rods)
- Razor blade
- Tweezers, #4 size (General 423-4)
- Allen wrench for the DAC

Materials

- Methanol cleaning fluid, spec grade
- Mounting wax (Cenco softseal tackiwax)
- The liquid whose viscosity is to be measured

1. Rough clean the diamond surfaces to remove large dirt particles and any fluid remaining from previous measurements. Moisten a cotton swab with methanol or with a solvent appropriate for the fluid to be removed. Lightly rub the surface of the diamond to remove the majority of the contaminates. Use methanol and a clean swab for the final cleaning (the swab may leave residual fibres).
2. Limit fine cleaning of the lower diamond to the culet face itself. Wrap a small strip of optical lens tissue over the tip of the toothpick, and moisten the tip with methanol—use only a tiny amount. Wipe dust particles and grease from the culet by lightly swabbing the surface (do this under the microscope). Do not touch the epoxy diamond mount since this tends to dislodge grease and other dirt. Replace the tissue when it starts to leave residual particles.
3. Fine clean the upper diamond, including the girdle (diamond sides) and the culet. Push upwards, with a toothpick wrapped as above, on the girdle facets to remove the contaminants and grease on the culet. Clean the culet as above.
4. Clean the gasket in two steps. Clean, firstly, all gaskets in bulk with an ultrasonic cleaner. Clean, for approximately five minutes each, the gaskets in solutions of water containing soap, clean water, and methanol. Clean, secondly, each gasket individually before use. Rub both surfaces with a swab moistened with methanol, then inspect the inside of the gasket hole under the microscope—turn the incident lighting off to aid viewing of the hole interior. Remove any dirt particles or metal flakes within the hole with the clean tip of a needle (dirt from the surface and/or hole of the gasket is the main source of contamination in the cell). Blow pressurized gas through the hole to dislodge any residual dirt.
5. Hold the gasket in place during the final loading with small pillars of tackiwax. Prepare and mount these now. Remove any existing pillars, leaving a layer of wax on the lower diamond mounting plate to serve as a base. Roll out a string of ~2 mm diameter wax between two microscope slides. Cut three ~5 mm long pieces with a razor blade. Install these around the

diamond at positions which will allow the gasket to be supported above the diamond. Press the bottom edge of each pillar into the base plate to anchor it, then flatten the top. (Each pillar should be $\sim \frac{1}{2}$ mm above the culet of the diamond.)

6. Select a ruby chip, about $20 \times 40 \times 40$ μm^3. Clean the chip (leaving no small adhering particles) and place it near the centre of the culet using the end of a very clean needle. Select a nickel sphere that is spherical, that has no other particles adhering to it, and that has a highly reflective surface (indicating a smooth surface)—typically a diameter of ~ 50 μm is useful, but for very high pressures ($P > 5$ GPa) the gap between the diamonds can become small and a smaller diameter of ~ 25 μm is used. Use a clean needle to place the sphere at the centre of the culet to within ~ 15 μm. Avoid direct contact between the sphere and ruby chip.

7. Now install and seat the gasket. Position the cell so that the centre of the culet is near the centre of the field of view. Look through the microscope as you lower the gasket onto the wax pillars: as the gasket approaches the level of the diamond you will see the sphere and ruby through the gasket hole. Sight on the sphere and rest the gasket on top of the wax pillars using tweezers. Move the gasket around until you can see the sphere approximately centred in the gasket hole (autocollimation of the incident illumination indicates the correct gasket tilt). Use clean tweezer tips to push the gasket down until it contacts the diamond, always keeping the sphere in view. Now seat the gasket using the DAC. Place the two halves of the cell together after re-examining the top diamond to check for cleanliness, and lightly tighten the three pressurization screws. Use the sphere's position as a guide to verify that the gasket hole is centred on the culet. Make adjustments by gently pushing on the edge of the gasket, between the two halves of the cell. Use care, as you can easily push the gasket off the diamond culet thus destroying the mount. Gently tighten each screw incrementally until the gasket is seated onto the lower diamond. The gasket hole diameter should decrease by $\sim$1–5 μm.

8. Now we load the fluid (this can be difficult because everything must be done quickly to avoid evaporation and contamination). Disassemble the cell and reclean the upper diamond. Remove any particles from the surface of the gasket using a clean needle. Check the edges of the gasket for wax if you are loading a volatile fluid (the fluid will be wicked down beneath the gasket if any wax contacts the upper gasket surface or edge so clean it away with a needle point). Lay out the following for immediate use:

- the top diamond-cell half
- the allen-head screws
- a glass fibre, diameter ~ 75 μm
- a syringe or Pasteur pipette

Protocol 1. *Continued*

- the sample fluid
- the allen-head wrench for the DAC

9. Centre the lower diamond-cell half in the microscope field of view. Extract several drops of the sample liquid and place the fluid over the gasket hole (if you are working with a volatile fluid, put on enough fluid to cover the entire gasket)—surface tension will produce a meniscus, preventing it from flowing over the sides of the gasket. Examine the gasket hole to see that is filled with fluid and that there is no bubble trapped within it (high-surface tension liquids can have difficulty entering the gasket hole. Placing the glass fibre through the liquid and into the gasket hole can help. Similarly, a trapped bubble can be removed using the glass fibre. Also, the ruby chip or sphere can sometimes become dislodged and the glass fibre is then used to place them back into the hole). Now, immediately put on the upper half of the DAC and press down on it firmly with your fingers—do not let go until the screws are snug. Check for bubbles using the microscope (sometimes bubbles can be removed by slightly releasing the screws—be careful of admitting dirt in this way. If the bubble persists, you must remove the top and re-install it with as little delay as possible. There will be no time to reclean it). Tighten the screws sufficiently to place the liquid under a few megapascals of pressure, thus ensuring a leak-free cell. (See *Protocol 2* for a description of attaining lower pressures for data collection.)

The choice of the sphere is itself important. A reasonable density contrast with the fluid, uniform roundness, and a smooth surface finish are all important characteristics. Among the several commercially available metals we examined, the nickel–boron spheres described in *Protocol 1* best fulfil these requirements. We determined their density to be 7600 kg m^{-3}, which gives a good density contrast with hydrocarbon fluids whose densities are in the range 800–1500 kg m^{-3}, even at high pressures. Because we select an individual sphere by hand it is easy to choose a sphere with good sphericity and surface finish. Recently, it has been reported (I. Fujishiro, personal communication) that gold spheres (nominal density 19300 kg m^{-3}) can be produced by passing an electric current through a small gold wire.

The viscosity is measured by tracking the velocity of the ball as it rolls down the surface of the diamond. Tracking the velocity of such small objects requires some care, both because the velocity range can be quite large ($0.5–10^{-7}$ mm s^{-1}) and because vibrations can cause problems in longer (12 h) experiments. The experimental set-up is shown schematically in Fig. 3.50, and the procedure is described in detail in *Protocol 2*. The optics are mounted on an optical rail along with the DAC. Combined with ×10 video magnification, they magnify the image of the 0.5 mm gasket hole so that it nearly fills the video screen. The resultant image is either recorded or analysed directly. For

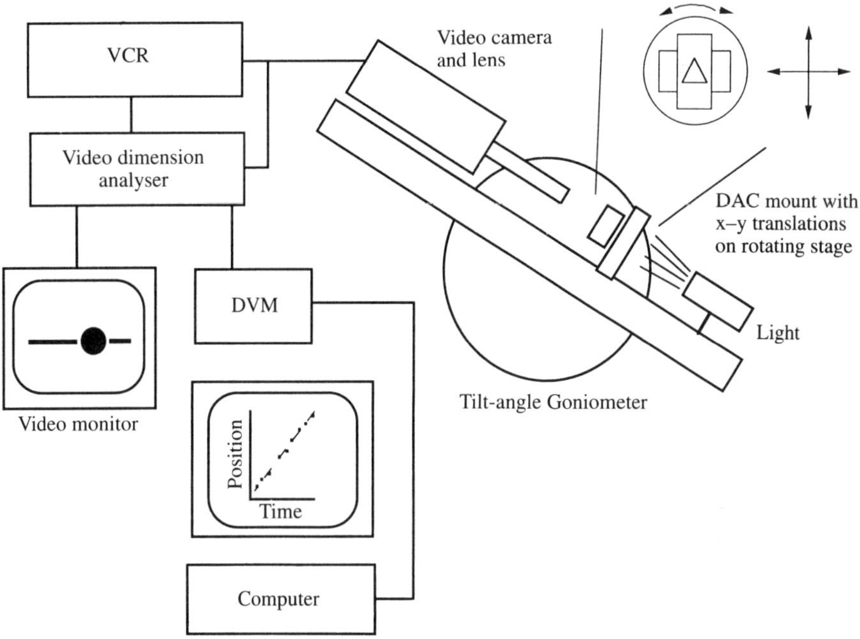

Fig. 3.50 The viscometer optics and electronics. The tilt-angle goniometer is mounted on an optical table. The optics, video camera, and the DAC are attached to an optical rail mounted on the goniometer. The velocity of the sphere is analysed electronically from a video image. The video dimension analyser is triggered by the image of the sphere (moving to *the right* in this case), and it produces a voltage proportional to the distance along a linear sampling bar. Using a computer and digital voltmeter (DVM), this voltage is converted to a relative position and then correlated with time; this results in a linear plot whose slope is the velocity. (From King *et al.*, reference 49)

a sphere moving at high velocity, we would record the image and then use the calibrated, 1/10, playback speed on the video cassette recorder (VCR) for analysis. An electronic tracking device, the video dimension analyser (VDA), records the instantaneous position of the sphere. The voltage from this device, which is proportional to the linear position of the sphere (the sphere moves to the right in Fig. 3.50), is periodically sampled by a computer, and these positions and times, provided by the computer clock, are placed in a file for later analysis. After the ball reaches the end of its fall the cell is rotated 180° around an axis parallel to the optical axis, and the ball is allowed to traverse the cell again. We would measure 6–12 traverses for $\eta < 10^3$ Pa s (see *Protocol 2*). For analysis, it is important to select data in the central two-thirds of the gasket hole, and to then examine the data for linearity. If the data are visually linear, we would use linear regression to determine the velocity. Typical fits give correlation coefficients greater than 0.98. At each pressure we would calculate an average velocity and the standard deviation. The standard deviation averages 4 per cent.

Protocol 2.
Measuring viscosity with the rolling-ball DAC viscometer

Caution! For safety procedures see Appendix 2.

This procedure describes the use of a DAC rolling-ball viscometer which is useful in the range $0.5 \leqslant \eta \leqslant 10^7$ m Pa s. As explained in the text, the procedures described here for producing well-defined hydrodynamic conditions are crucial for attaining high precision. Loading of the DAC with a ruby manometer chip, a nickel sphere, and the experimental fluid is described in *Protocol 1*.

Equipment

- Optical table
- Vertical goniometer, 90° rotation, equipped with an optical rail containing a video-camera mount and a rotating stage mounted on an *x*–*y*–*z* translation (Fig. 3.50)
- Rotating stage equipped with a split-ring mount on *x*–*y* translators
- Video camera (Dage model 66) equipped with a ×7 lens, ×2 magnifier, and a ×3 amplifier (Bausch and Lomb, models 31-29-77, 31-29-89, and 31-29-96)
- Video cassette recorder (VCR) equipped with a calibrated, 1/10 speed playback (Panasonic model NV-8950).
- Video monitor (Panasonic model CT-1330M)
- Video dimension analyser (VDA, model 303, Instrumentation for Physiology and Medicine, San Diego, CA)
- Digital voltmeter with IEEE-488 computer control (Kiethley 195 DMM)
- Microcomputer (IBMAT)
- Spreadsheet program (Lotus 1-2-3)

Method

1. Select the tilt angle. As explained in the text, the constant wall correction that is required is maintained only by fixing this angle throughout the experiment. Using eqn (3.9.2) and assuming $\gamma = 0.2$, calculate the sphere's velocity for ambient-pressure viscosity. This is the largest velocity to be encountered, and under the constraint that 15° ⩽ tilt angle ⩽ 75°, the tilt angle should be set to give $v \leqslant 0.5$ mm s^{-1}.
2. Align the cell which is mounted on the rotating stage via a split-ring, so that the gasket-hole image will not wander as the stage is rotated. Work from low to high magnification and use the split-ring *x*–*y* translators to bring the gasket hole and the rotation axis of the stage into coincidence; then use the *x*–*y*–*z* stage translations to place the image in the centre of the field of view. Translate along the optical axis, if necessary, to maintain focus.
3. Make a video recording of the sphere for low-viscosity liquids, $0.5 \leqslant \eta \leqslant$ 100 m Pa s. Position the DAC so that the sphere is at the bottom, then rotate the cell 180° and allow the sphere to roll down the diamond face. Repeat this more than thirty times for a complete data collection (the sphere and ruby chip must be separated by more than a sphere diameter for a meaningful measurement; typically the ruby chip remains on the lower diamond surface and the sphere rolls on the upper one). Rotate the cell by 90° if the ruby chip begins to move, and let the ruby fall to the side of the gasket hole

and then rotate by another 90° and resume the 180° rotations for data collection (the vast difference in drag between the sphere and the ruby will keep the ruby to the side of the gasket hole).

4. Measure the velocity by analysing the video images. The camera is orientated so that the sphere trajectory is from left to right in the video image, along the raster lines. The VDA measures the position of an object along a given set of raster lines and produces a voltage proportional to that position; the raster lines to be analysed are indicated on the video screen by a darkened bar whose position is adjustable by front-panel controls. Place this measurement bar on the beginning sphere position for each sphere traverse of the gasket hole. Observe the constantly varying voltage as the sphere moves. Sample this voltage at fixed times using a BASIC language program running on an IBM AT computer.[53] Analyse the data in 1/10 speed playback mode for the highest velocities. Play the tape at normal speed to locate those traverses in which the sphere goes nearly through the centre of the gasket hole and in which it is well clear of the ruby chip. Note the location on the tape of each such sequence and play them back at 1/10 speed, setting the VDA measurement bar and starting the computer sampling at the beginning of each fall (the resulting data are automatically saved in a file for later processing; typically each data set consists of a selection of 6–12 traverses of the gasket hole).

5. Make a direct analysis of the video image for high-viscosity liquids, $\eta >$ 100 mPa s. Use a 180° alteration of the DAC rotation as described above. Position the VDA measurement bar at the sphere position at the beginning of each sphere traverse, then start the sampling program. Apply the criteria of acceptability mentioned above for each traverse, but, in addition, do not sample positions within one ball diameter of the gasket hole edge (such data have an additional wall correction originating from the gasket; these data can be eliminated from the analysis later, but significant time savings at high viscosities will result from maintaining the sphere in the centre of the gasket hole). Reverse the DAC position after the sphere travels $\sim$100 μm. Be content with six measurements, except at the highest viscosities, $\eta \approx 10^3$ Pa s, where such long distances would consume an enormous time; only use three 25 μm traverses, which take $\sim$ 20 h each, for the highest viscosities.

6. Analyse the voltage-versus-time file generated by the computer program described above using a spreadsheet program such as Lotus 1-2-3. Use an experimentally determined proportionality constant to convert the observed voltage to position, and then plot the resulting position versus time. Use the linear equation $y = a + bx$ as a least-squares fit to the linear region of the data. Average the velocities b which result from the analysis of all the traverses at a given pressure, and their calculated standard deviation. Use macros to perform these procedures automatically in Lotus 1-2-3.

Protocol 2. *Continued*

(For viscosities up to 10^3 Pa s, one should obtain approximately a 4 per cent precision; for higher viscosities the errors can reach 8 per cent.)

7. Apply eqn (3.9.3) to the above data to yield the relative viscosities directly. (Measurement of the absolute viscosities requires a measurement of the velocity at ambient pressure (compare eqns (3.9.2) and (3.9.3)). This is accomplished by lowering the pressure in the cell to the vapour–liquid coexistence point.) Observe the DAC under the microscope, using incident lighting, as the three pressurizing screws are loosened: as the pressure approaches the ambient value, a set of concentric rings should be seen at the junction between the gasket and the upper diamond culet (when the screws are sufficiently loose, these rings will merge with the gasket hole and a vapour bubble will appear; this is the desired pressure for the velocity measurement, but such a setting is difficult to maintain for long periods). Place the DAC back onto the viscometer and immediately measure the velocity. Note the temperature also. Use a conventional viscometer operating at this temperature to measure the viscosity of the same fluid. Obtain the sphere diameter using a calibrated reticule in the optical microscope. Calculate a value for the wall correction, γ, using eqn (3.9.2) and apply this to each high-pressure data point, that is, to η/γ. (See the text for a discussion of typical magnitudes and the origin of the wall correction.)

9.4.2.2 Sample data. We first tested the ambient-pressure accuracy of the viscometer through measurements of standard viscometer-calibration fluids with various viscosities. These data showed that the wall correction is essentially constant at a given tilt angle, and that it does not depend upon the viscosity of the fluid nor on its chemical make-up. To test the high-pressure accuracy of our viscometer we made measurements on several fluids,[49] which were selected because: (i) they had a range of chemical characteristics, (ii) high-precision data was available in the literature, and (iii) they remained fluid over a wide pressure range. These measurements demonstrate the accuracy of the DAC viscometers over a wide viscosity and pressure range. Variation in the wall-correction factor is the greatest potential source of error. Our data are typically within one estimated standard deviation of previous measurements; thus the assumption of a constant correction appears to be valid.

Two representative fluids, methanol and 1,2-propanediol, are described here to illustrate the capabilities of rolling-ball and centrifugal-force viscometers. Both our samples were from Aldrich (with stated purity of greater than 99 per cent). The methanol was dried over a molecular sieve to remove water. Each fluid was loaded with a ruby chip and a metal sphere into a diamond cell. For methanol, a high tilt angle (65°) was necessary because of

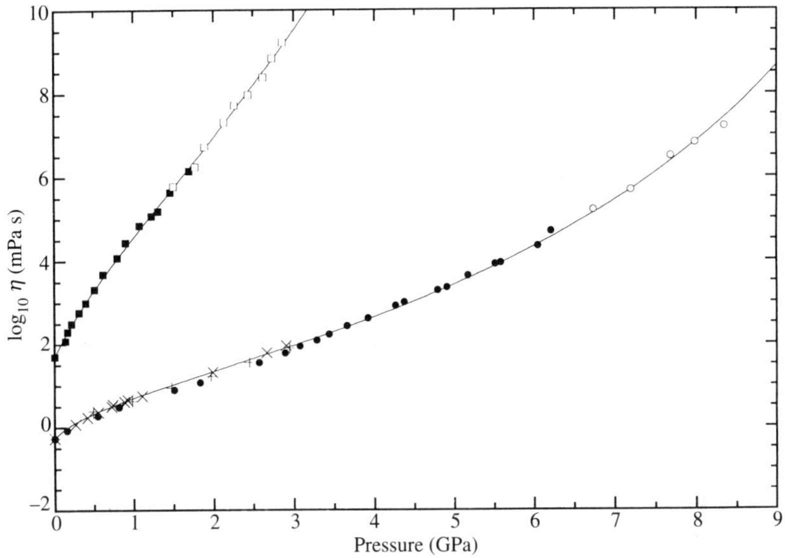

Fig. 3.51 The high-pressure viscosity of (●, ○) methanol and (■, □) 1,2-propanediol. The filled data points were obtained with a rolling-ball viscometer and the open data points were obtained with a centrifugal-force viscometer, the curves are a free-volume fit (Fig. 3.53). Methanol data, for pressures up to 3 GPa obtained by (×) Bridgman[16] and (+) Herbst *et al.*[54] are in excellent agreement with our results. Crystallization of the methanol sample limited measurements to 8.3 GPa while the maximum viscosity for 1,2-propanediol limited the experimental pressure to 2.9 GPa.

its low ambient-pressure viscosity (0.55 mPa s at 22.5° C). Then the sphere's velocity at ambient pressure was measured for each liquid, and was used to calculate η/γ from eqn (3.9.2). (This low viscosity in the diamond cell arises from the low pressure at which the liquid and vapour coexist; see *Protocol 2*.) Then, at the pressures shown in Fig. 3.51 the sphere's velocity was measured and $\eta(P)/\eta(0)$ was determined from eqn (3.9.3). This requires a calculation of the buoyancy correction, $\rho_s - \rho_f$. We fitted the pressure–density data to the Tait equation of state[8] and calculated the density from the values of the bulk modulus and its pressure derivative. The change in the sphere density was insignificant—no correction was applied. Using the absolute viscosity, previously determined in a Couette-geometry viscometer, the wall correction value was then calculated (Table 3.2) and this value was then used to obtain the absolute viscosities. The resulting values, plotted in Fig. 3.51 along with data on methanol from the literature,[16,54] are in excellent agreement.

The DAC viscometer described here is a useful device for measuring viscosities over a wide range (values from 0.5–10^7 mPa s have been measured). By taking a ratio of the sphere's settling velocity at high pressure to the velocity measured in the cell at ambient pressure (and taking account of the

Table 3.2. The wall-correction parameter, γ, at a given tilt angle, θ, for the rolling-ball method eqn (3.9.2) and γ_{cent} for the centrifugal-force method eqn. (3.9.4)

Compound	θ	γ	γ_{cent}
Methanol	65°	0.11 ± 0.02	0.24 ± 0.02
1,2-propanediol	20°	0.23 ± 0.02	0.24 ± 0.02

changes in the buoyancy as the pressure is increased), the relative viscosity, $\eta(P)/\eta(P=0)$, can be obtained independently of the sphere radius and the wall correction. Although DAC rolling-ball viscometers span a wide viscosity range, the approximately exponential dependence of the viscosity on the pressure still limits their use. This is both a restriction on their use as practical devices and a limitation of our exploration of density scaling of the viscosity. In these devices the only workable means of increasing the viscosity range is to increase the force on the sphere by several orders of magnitude. In the next section we describe how centrifugal-force viscometers achieve this.

9.4.3 The DAC centrifugal-force viscometer

In this section we describe a device which extends the range of continuous high-pressure viscosity measurements above 10^6 Pa s at pressures which can exceed 10 GPa.

9.4.3.1 Introduction. Rolling-ball DAC viscometers, although sensitive over many decades, are still restricted in their viscosity range. For example, variable-temperature viscometric studies often exceed 10^{10} Pa s, and such work demonstrates that, over limited viscosity ranges, many viscosity models can appear similar. To differentiate between these models, data spanning many decades are essential.[3,8,55] Thus, extending the high-pressure experimental viscosity range makes it possible to test density-mediated transport theories adequately. Also, it is now recognized that lubrication conditions can easily generate high viscosities,[4] and a quantification of the rate of the viscosity rise with pressure could be significant in defining the friction and wear in lubricated contacts.[5]

To measure higher viscosities, we have introduced a modification of the rolling-ball viscometers. The viscosity range of the rolling-ball technique is limited by the speed of the rolling ball. For example, at a viscosity of 10^4 Pa s, a 50 μm diameter ball rolls at a rate of only 2 μm per day; this displacement is too small to be accurately measured over any reasonable length of time. To overcome this limitation, we have developed a novel method in which we use centrifugal force instead of gravity to drive the ball. By applying a maximum acceleration of 17 000 m s^{-2} to the ball we are able to measure viscosities up to 4×10^6 Pa s and to maintain an acceptable uncertainty of ~ 8 per cent.

9.4.3.2 Principles of operation. A schematic diagram of a centrifugal-force viscometer is given in Fig. 3.52. A bench-top centrifuge was fitted with a specially designed rotor to accommodate the DAC. The hole in the rotor which accepts the triangular DAC contains a flat surface perpendicular to the rotor's radius; thus the cell may be placed so that force is in one of three cell directions, and it is positioned so that the ball will always travel near the centre of the cell to avoid unwanted wall effects. The ball will travel in a straight line along the diameter of the rotor because the low velocity of the ball results in a negligible Coriolis force. At the beginning of a measurement, the position of the ball is measured by the video set-up used for the rolling-ball viscometer. The cell is then placed in the centrifuge and accelerated for a specified time. The ball position is measured again after the run to determine the total displacement. During the course of a run, the rotor is initially accelerated to its maximum angular velocity, ω, which is determined by the capability of the motor. The centrifuge is later turned off and the rotor's angular velocity slowly coasts to zero at the end of the run. Since the force applied to the sphere is proportional to ω^2, it is critical to know ω as a function of time. This

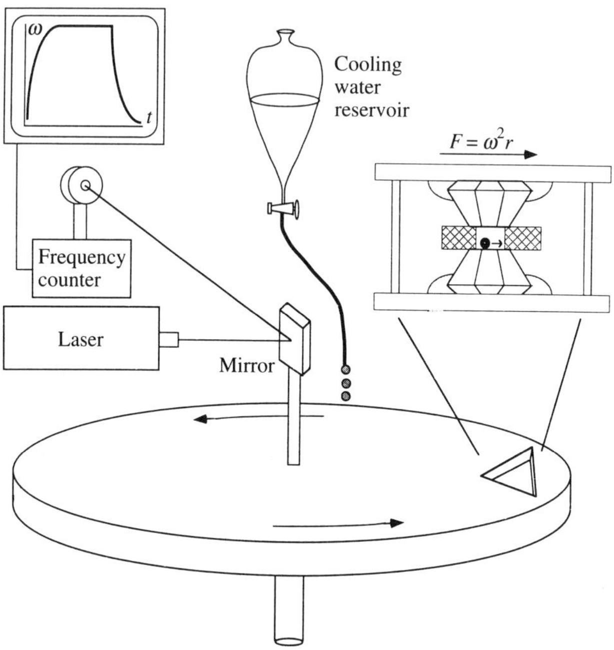

Fig. 3.52 A centrifugal-force viscometer. The ~50 μm diameter nickel ball is driven along the horizontal diamond surface by centrifugal force. The rotation of a mirror on the shaft of the rotor deflects a laser beam across a detector to produce a signal whose frequency gives the speed of the rotor. Our current centrifuge is able to generate an acceleration of 17 000 m s^{-2}. (Reprinted with permission from *J. Phys. Chem.* **1993**, *97*, 2355. Copyright 1993. American Chemical Society.

is measured by reflecting a laser beam from a mirror attached to the shaft of the rotor and onto a photodiode. This produces electrical pulses that are read by a frequency counter which then tracks ω. A computer records the measured frequency and the total time of the measurement and integrates over $\omega^2 t$. This quantity is used in addition to the total sphere displacement (as discussed above) to calculate the viscosity using the equation

$$\eta = \gamma_{\mathrm{cent}} \left(\frac{2R_{\mathrm{s}}^2 (\rho_{\mathrm{s}} - \rho_{\mathrm{f}}) r\omega^2 t}{9(x_{\mathrm{f}} - x_{\mathrm{i}})} \right), \tag{3.9.4}$$

where x_{i} is the initial position of the ball, x_{f} is the final position of the ball, r is the distance of the DAC from the centre of rotation (= 73.7 mm), and γ_{cent} is the wall correction for the centrifugal-force measurements. The Stokes' drag model of eqn (3.9.2) can be applied because—throughout the entire run—the sphere is at its terminal velocity with respect to the instantaneous value of $\omega^2 r$. We can determine γ_{cent} in a manner similar to that described for the rolling-ball viscometer; however, the viscosity at some elevated pressure (which is obtained from a rolling-sphere experiment) is used as the reference value. Typically a value of $\eta \sim 10^2$ Pa s is used because of the accuracy and limitations of rolling-ball viscometers. Although the rolling-ball technique is capable of measuring viscosities up to 10^4 Pa s, the extremely low sphere velocities at such a viscosity can result in an experimental error of $\sim$ 10 per cent; therefore, we begin by using the centrifugal technique at viscosities near 10^2 Pa s, where the error of the rolling-ball technique is $\sim$ 4 per cent. The centrifugal wall correction is then determined by equating the viscosity from eqn (3.9.2) to the right-hand side of eqn (3.9.4).

As for the rolling-ball experiments, a constant value for γ_{cent} throughout the experiment is essential for high-precision data. Viscosity measurements throughout the overlap range for the rolling-ball and centrifugal-force techniques, 10^2 to 10^4 Pa s, show that γ_{cent} is indeed constant to within the experimental error of the technique. Also, we have found that γ_{cent} is not dependent upon the rotor velocity or on the fluid composition. As discussed above, it is predicted that the magnitude of γ_{cent} is similar to that at low-tilt angles in rolling-ball viscometers; and this is indeed the case, as can be seen in Table 3.2. Thus, to within experimental error, γ_{cent} is constant. Taking averages over the entire range of experimental conditions we find that our data are reproducible to within 8 per cent.

Runs on the centrifuge varied from 3 min at a speed of 800 r.p.m. (with an acceleration of 500 m s^{-2}) for $\eta \approx 10^2$ Pa s to 1 h at 4600 r.p.m. (with an acceleration of 17 000 m s^{-2}) for $\eta \geqslant 10^6$ Pa s. The most critical experimental parameter in making accurate viscosity measurements is a constant temperature. Heating of the bench-top centrifuge on high-speed runs for longer than 5 min would cause a significant decline in the observed viscosity. To combat this problem, we cool the centrifuge by dripping water on the rotor for measurements of viscosity greater than 10^5 Pa s where long runs are required. This evaporative cooling is able to maintain the temperature to within 2°C of room temperature (that is 22.5 ± 2.0°C).

Protocol 3.
Measuring viscosity with the centrifugal-force DAC viscometer

Caution! For safety procedures see Appendix 2.

This procedure describes the use of DAC centrifugal-force viscometers which are useful in the range $10^2 \leqslant \eta \leqslant 10^7$ Pa s. Viscosities from DAC rolling-ball viscometers (see *Protocol 2*) are used here to convert the measured relative viscosities into absolute values. Loading of the DAC with a ruby manometer chip, a nickel sphere, and the experimental fluid is described in *Protocol 1*.

Equipment

- Optics, electronics, and the computer from *Protocol 2*
- A computer-controlled frequency counter (Hewlett-Packard universal counter model 5316A)
- An optical detector (Newport Corporation model 815 power meter)
- A 5000 r.p.m. centrifuge (International Equipment Company, centrifuge model 428)
- A modified rotor for the centrifuge above (American Design Corporation)
- A water cooling device
- A low-power laser

Method

1. Ensure that the sphere is well separated from the ruby manometer chip and that it is approximately centred on the diamond culet. Move the sphere to this position before increasing the viscosity through application of pressure. Measure the position of the sphere from the gasket edge using the optics of the rolling-ball viscometer (this measurement is made along a line perpendicular to one of the flat sides of the triangular-shaped DAC because the centrifugal force will be applied along this direction).
2. Mount the DAC in the modified rotor so that the DAC's flat side selected in step 1 is perpendicular to the rotor's diameter (it is possible to run two DACs at one time by placing a second cell into the holder 180° away; alternatively, an empty cell is placed there to balance the rotor). Fasten the hold-down straps to keep the cells in place.
3. Use a cooling water spray on the rotor (because heating of the cell by the centrifuge motor can be a serious source of error) for runs longer than ~ 5 min ($\eta > 10^5$ Pa s), and allow the water to drain from the bottom of the centrifuge. Prepare a water-drip cell and receptacle (Fig. 3.52). Record the temperature of the cell and rotor before and after each run (these values must remain within 2°C of room temperature).
4. Monitor the speed of the centrifuge rotor with an optical probe. Ensure that the laser beam is properly reflected from the mirror (attached to the rotor)

Protocol 3. *Continued*

and into the optical detector (Fig. 3.52). Amplify the detector pulses recorded by the frequency counter with the power meter listed above.

5. Select the centrifuge speed and time to keep the sphere at least one ball diameter away from the gasket walls (it is best to keep the sphere near the centre of the diamond culet throughout). Use eqn (3.9.4) to estimate the rotor speed and total time required to achieve a sphere movement of approximately 100 μm. Use $\gamma_{cent} = 0.2$ for this calculation (for high viscosities ($\eta \approx 10^5$ Pa s) this will indicate times greater than 1 h). Reduce the displacement in the calculation to the minimum, 10 μm, for such conditions.
6. Measure the rotation speed as a function of time using a BASIC program running on the IBM AT computer (this is essential, since the force is very sensitive to variations in the speed). Start the centrifuge and this program at the same time (the total force applied to the sphere is proportional to the product $\omega^2 t$, and the program samples the speed at fixed time intervals and performs an integration). Turn off the centrifuge at approximately the time calculated in step 5 (the program will continue to sample the speed until the rotor slows to ~50 r.p.m.; the integrand is displayed continuously). Record the final value and the temperature of the cell and rotor.
7. Place the DAC into the optics of the rolling-ball viscometer and record the position of the sphere relative to the gasket hole's edge. Remember, that the direction for this must be along the direction selected in step 1. Use the difference between the starting and final sphere positions in eqn (3.9.4) to calculate η/γ_{cent}. Use the procedure described in the text to calculate γ_{cent}.
8. Repeat the above procedure several times at each pressure. Average the results from six runs for low viscosities (a few minutes per run is sufficient). Make three runs at the highest viscosities. Make each successive run at an angle of 120° from the previous one, due to the triangular shape of the cell. Rotate the DAC, with the cell already mounted on the optics, as in step 7, so that a new flat is in position, and measure the sphere–gasket hole separation as in step 1 and repeat steps 2–7. (The precision of the data averages 8 per cent.)

9.4.3.3 Sample data and free-volume analysis. The data for two examples are plotted in Fig. 3.51. For 1,2-propanediol, the maximum pressure at which measurements could be made was limited by the upper viscosity limit of the centrifugal viscometer. The methanol crystallized at 8.5 GPa (well above its equilibrium freezing point of 3.5 GPa) thus limiting the measurements. The centrifugal-force data smoothly overlay those from the rolling-ball viscometer. For methanol a compression of 44 per cent was obtained at the highest measurement pressure of 8.3 GPa. For 1,2-propanediol the highest pressure data of 2.9 GPa corresponds to a compression of 26 per cent. The canonical

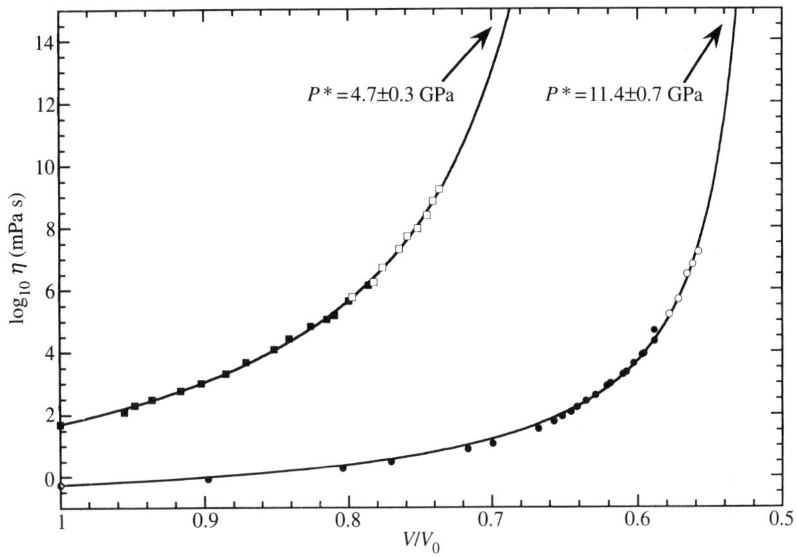

Fig. 3.53 Viscosity data from Fig. 3.51 plotted against the volume (normalized by the ambient-pressure volume). The curves are fits obtained by the free-volume theory (eqn 3.9.5), the resulting parameters being: for methanol, $B = 2.22 \pm 0.05$ and $V_\infty = 33.8 \pm 0.1$ Å^3; for 1,2-propanediol, $B = 6.7 \pm 0.3$ and $V_\infty = 71.9 \pm 0.5$ Å^3. Extrapolation of these fits to $\eta = 10^{12}$ Pa s gives the indicated glass-transition pressures, P^*. The strong curvature of the methanol data indicates the more fragile behavior of this fluid, which is reflected in the fragility parameters (calculated from equation 5 given by Cook *et al.*).[57] For methanol $m_P = 289 \pm 6$ and for 1,2-propanediol $m_P = 123 \pm 5$.

shape of all pressure–viscosity data is exhibited by both fluids.[16] At low pressures a plot of log η against pressure is concave towards the pressure axis, and at high pressures it is convex. This high-pressure upturn comes from the glass transition, while the inflection point can be shown to be proportional to the elastic modulus and the density at the glass transition point.[55]

The complex shape of the viscosity–pressure behaviour gives way to simple universal scaling when volume replaces pressure (Fig. 3.53). The physics underlying this scaling can be expressed through the free volume equation.[2] The normalized form of this equation is

$$\ln(\eta) = \frac{BV_\infty}{(V - V_\infty)} - \frac{BV_\infty}{(V_0 - V_\infty)} + \ln(\eta_0), \qquad (3.9.5)$$

where η_0 and V_0 are the ambient-pressure viscosity and volume and V_∞ and B and fitting parameters. Through the use of rolling-ball and centrifugal-force data, we have shown that this equation successfully models the density-mediated viscosity rise in many materials.[53,56] Several features of this modelling are noteworthy. Extrapolation to the glassy-state viscosity (10^{12} Pa s) gives glass-transition pressures, P^* (see Fig. 3.53), which are in good agreement with other measurements. Also, the resulting values of V_∞ closely match the volumes of effective hard spheres. Finally, it has recently[57] been shown that the parameter B can be related to the 'fragility' of the fluid. This is a powerful scaling concept, which has proven useful in understanding glassy behaviour in variable-temperature studies.[58] Its application to our high-pressure data suggests that the parameter B is a measure of correlated motion within the fluid. Previous volume-scaling theories have not explicitly considered such correlations. The experimental signature of this behaviour is seen in Fig. 3.53, where the methanol data exhibit considerably more curvature. Using the definition of fragility given by Cook *et al.*[57] in their equation 5, methanol has fragility twice that of 1,2-propanediol. By analogy with temperature fragility, this implies that the molecular motion in methanol is relatively uncorrelated. Thus high-pressure viscosity data may help in redefining our understanding of the general problem of transport in dense fluid.

References

1. King, H. E., Jr. In *High Pressure Science and Technology*; American Institute of Physics: Colorado Springs, **1994**; pp. 1055–58.
2. Cohen, M. H.; Turnbull, D. *J. Chem. Phys.* **1959**, *31*, 164–1169.
3. Cohen, M. H.; Grest, G. S. *Phys. Rev.* B, **1979**, *20*, 1077–1098.
4. Bair, S.; Winer, W. O. *Trans. ASME-J. Tribol.* **1992**, *114*, 1–13.
5. Booser, R. E. (ed.) *CRC Handbook of Lubrication*; CRC Press: Boca Raton, FL, **1984.**
6. Bridgman, P. W. *Proc. Amer. Acad. Arts Sc.* **1926**, *61*, 57–99.
7. Le Toullec, R.; Loubeyre, P.; Pinceaux, J.-P. *Phys Rev* B, **1989**, *40*, 2368–2378.
8. Cook, R. L.; Herbst, C. A.; King, H. E., Jr. *J. Phys. Chem.* **1993**, *97*, 2355–61.
9. Zaug, J. M.; Slutsky, L. J.; Brown, J. M. *J. Phys. Chem.* **1994**, *98*, 6008–6016.
10. Alsaad, M.; Bair, S.; Sanborn, D. M.; Winer, W. O. *Trans. Amer. Soc. Mech. Eng.—J. Lubric. Technol.* **1978**, *100*, 404–417.
11. Fujishiro, I.; Piermarini, G. J.; Block, S.; Munro, R. G. In *High Pressure in Research and Industry, Eighth AIRAPT Conference*; Uppsala, Sweden, 1982; pp. 608–11.
12. Ferry, J. D. *Viscoelastic Properties of Polymers*, 3rd edn. Wiley: New York, **1980.**
13. Berthe, D.; Vergne, P. *J. Rheol.* **1990**, *34*, 1387–1414.
14. Hutton, J. F.; Phillips, M. C. *Nature* **1973**, *245*, 15–16.
15. Bair, S.; Winer, W. O. *Trans. Amer. Soc. Mech. Eng. J. Tribol.* **1990**, *112*, 246–253.
16. Bridgman, P. W. *Proc. Amer. Acad. Arts Sci.* **1949**, *77*, 117–128.
17. ASME. *ASME Pressure Viscosity Report*; American Society of Mechanical Engineers, New York, **1953.**
18. Abbot, L. H.; Newhall, D. H.; Zilberstein, V. A.; Dill, J. F. In *ASLE/ASME Lubrication Conference*; ASLE Preprint: Dayton, HO, **1979;** pp. 1–7.
19. Irving, J. B.; Barlow, A. J. *J. Phys. E*, **1971**, *4*, 232–236.

20. Bair, S.; Winer, W. O. *Trans. Amer. Soc. Mech. Eng.—J. Lubric. Technol.* **1980,** *102*, 229–235.
21. Bair, S.; Winer, W. O. *Trans. Amer. Soc. Mech. Eng.—J. Lubric. Technol.* **1982,** *104*, 357–364.
22. Bair, S.; Winer, W. O. *Tribol. Trans.* **1988**, *31*, 317–324.
23. Kushiro, I. *J. Geophys. Res.* **1976**, *81*, 6347–6350.
24. Kushiro, I.; Yoder, H. S.; Jr., Mysen, B. O. *J. Geophys. Res.* **1976**, *81*, 6351–6356.
25. Sharma, S. K.; Virgo, D.; Kushiro, I. *J. Non-Cryst. Solids* **1979**, *33*, 235–248.
26. Nishibata, K.; Izuchi, M. *Physica* **1986**, *139*, 903–906.
27. Jonas, J.; Hasha, D.; Huang, S. J. *J. Chem. Phys.* **1979**, *71*, 3996–4000.
28. Barnett, J. D., Bosco, C. D. *J. Appl. Phys.* **1969**, *40*, 3144–3150.
29. Vermesse, J.; Provansal, M.; Brielles, *J. Physica* A, **1978**, *92*, 282–294.
30. Galvin, G. D.; Hutton, J. F.; Jones, B. *J. Non-Newt. Fluid Mech.* **1981**, *8*, 11–28.
31. Trappeniers, N. J.; Van Der Gulik, P. S., Van Der Hooff, H. *Chem. Phys. Lett.* **1980**, *70*, 438–443.
32. Van Der Gulik, P. S.; Trappeniers, N. J. *Physica* A, **1986**, *135*, 1–20.
33. Van Der Gulik, P. S.; Mostert, R.; Van Den Berg, H. R. *Physica* A, **1988**, *151*, 153–166.
34. Assael, M. J.; Papadaki, M.; Dix, M.; Richardson, S. M.; Wakeham, W. A. *Int. J. Thermophys.* **1991**, *12*, 231–244.
35. Bruschi, L.; Santini, M. *Rev. Sci. Instrum.* **1975**, *46*, 1560–1568.
36. König, R.; Pobell, F. *Phys. Rev. Lett.* **1993**, *71*, 2761–2764.
37. Collings, A. F.; McLaughlin, E. *Trans. Faraday Soc.* **1971**, *67*, 340–352.
38. Rein, J. R. G.; Charng, T. T.; Sliepcevich, C. M.; Ewbank, W. J. *ASLE Trans.* **1972**, *18*, 123–129.
39. Tough, J. T.; McCormick, W. D.; Dash, J. G. *Rev. Sci. Instrum.* **1964**, 35, 1345–1348.
40. Barlow, A. J.; Harrison, G.; Irving, J. B.; Kim, M. G.; Lamb, G.; Pursley, W. C. *Proc. R. Soc. London* A, **1972**, *327*, 403–412.
41. Hutton, J. F.; Phillips, M. C. *Nature* **1972**, *238*, 141–142.
42. Ramesh, K. T.; Clifton, R. J. *Trans. ASME J. Trib.* **1987**, *109*, 215–222.
43. Oliver, W. F.; Herbst, C. A.; Lindsay, S. M.; Wolf, G. H. *Phys. Rev. Lett.* **1991**, *67*, 2795–2798.
44. Oliver, W. F.; Herbst, C. A.; Lindsay, S. M.; Wolf, G. H. *Rev. Sci. Instrum.* **1992**, *63*, 1884–1895.
45. King, H. E.; Jr., Prewitt, C. T. *Rev. Sci. Instrum.* **1980**, *51*, 1037–9.
46. Jayaraman, A. *Rev. Mod. Phys.* **1983**, *55*, 65–108.
47. Piermarini, G. J.; Forman, R. A.; Block, S. *Rev. Sci. Instrum.* **1978**, *49*, 1061–1066.
48. Fujishiro, I.; Nakamura, Y.; Matsuhiro, S. *Bull, J. Soc. Mech. Eng.* **1986**, *29*, 1280–1286.
49. King, H. E.; Jr., Herbolzheimer, E.; Cook, R. L. *J. App. Phys.* **1992**, *71*, 2071–2081.
50. Smart, J. R.; Beimfohr, S.; Leighton, J., D. T. *Phys. Fluids* A, **1993**, *5*, 13–24.
51. Cook, R. L.; King, H. E., Jr.; Peiffer, D. G. *Macromolecules* **1992**, *25*, 629–635.
52. Merrill, L.; Bassett, W. A. *Rev. Sci. Instrum.* **1974**, *45*, 290.
53. Cook, R. L. Ph.D. thesis, Harvard, 1993.
54. Herbst, C.; King, H. E., Jr.; Gao, Z.; Ou-Yang, H. D. *J. Appl. Phys.* **1992**, *72*, 838–844.

55. Herbst, C. A.; Cook, R. L.; King, H. E., Jr. *Nature* **1993**, *361*, 518–520.
56. Herbst, C. A.; Cook, R. L.; King, H. E., Jr. *J. Non-Cryst. Solids* **1994**, *172–174*, 265–271.
57. Cook, R. L.; King, H. E., Jr.; Herbst, C. A.; Herschbach, D. R. *J. Chem. Phys.* **1994**, *100*, 5178–5189.
58. Angell, C. A. *J. Non-Crystal. Solids* **1991**, *131–133*, 13–31.

4

Techniques for chemical studies

1. Vibrational spectroscopy in dense fluid phases

M. Buback and C. Hinton

1.1 Introduction

Chemical reactions in fluid phases have several interesting aspects, particularly in the supercritical fluid state where both the temperature, T, and the pressure, P, are above the critical values: (i) the availability of an extended range of pressures and temperatures offers opportunities to run a process under conditions which optimize the kinetics and selectivity; (ii) since the solvent properties can be significantly varied in the supercritical fluid state, homogeneity of the reaction mixture may be easily obtained and phase separation for product recovery may be achieved after the reaction without major problems; (iii) additional advantages of carrying out chemical processes in supercritical phases accrue from the excellent heat- and mass-transfer properties in these states; (iv) free-radical polymerizations in extended ranges of pressure and temperature are particularly important since the kinetics of the process and the properties of the polymeric product are tied together. Thus, by varying the pressure and temperature over a wide range, polymers with widely differing properties may be obtained from the same monomer.

A large number of chemical reactions have already been investigated at high pressures at subcritical temperatures,[1] where the main interest has been directed at elucidating the reaction mechanism in some detail and also at improving the chemoselectivity, the regioselectivity, and the stereoselectivity of chemical reactions. The availability of spectroscopic techniques for online measurements on fluid systems at high pressures and temperatures is very important for the detailed study of reactivity under pressure. Vibrational spectroscopy is particularly useful in this respect, since almost all materials have characteristic, narrow vibrational components. Absorption techniques in the infrared (i.r.) and the near-infrared (n.i.r.) ranges have been extensively used for quantitative analysis of fluids at high temperatures and pressures,[2] typically up to 5 kbar and 500 °C[2] and (in some cases) up to 7 kbar and to 700 °C[3]. The simultaneous measurement of the i.r. and the n.i.r. range

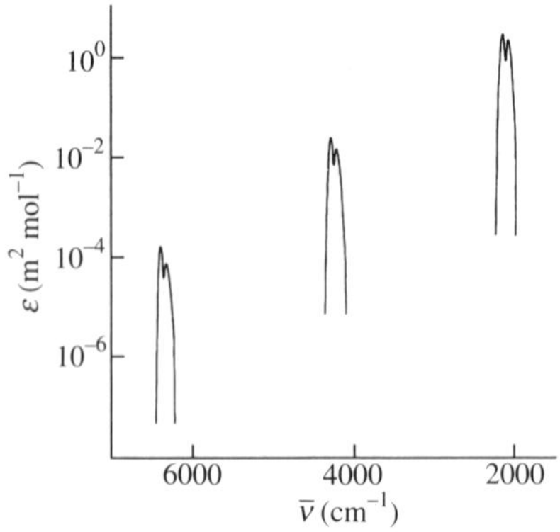

Fig. 4.1 The molar absorption coefficient, ϵ, for pure carbon monoxide (at 127 °C and 193 bar) in the fundamental, first-overtone, and second-overtone regions.

(for example, from 2000 cm^{-1} up to 8000 or even 10 000 cm^{-1}), which is routinely performed on modern Fourier-transform interferometers in less than a second, allows the dynamic range for quantitative analysis via vibrational absorption spectroscopy to be enormously increased. This situation is illustrated in Fig. 4.1 where the absorption of pure carbon monoxide at 127 °C and 193 bar is taken as an example. The three absorption components occurring slightly above 2000 cm^{-1}, 4000 cm^{-1}, and 6000 cm^{-1} are due to the fundamental, the first-overtone mode, and the second-overtone mode, respectively. The most striking feature of Fig. 4.1 is the large decrease in the molar absorption coefficient, ε, toward the higher overtones (ε is defined by $\log \varepsilon = A/lc$, where A is the (decadic) absorbance, l is the optical path length, and c is the concentration (or density) of the material). The change in the vibrational intensity necessitates the use of a logarithmic scale to present the three absorption bands in one coordinate system. One way to take advantage of the significant change in ε is to use the low wavenumber (fundamental) range for the detection of small CO concentrations, and to detect high concentrations from the overtones, for example, at around 6000 cm^{-1}. For CO in the wavenumber range in Fig. 4.1, an absorbance of $A = 1$ in the band maximum, at an arbitrarily chosen optical path length of 250 mm, is brought about by CO concentrations of 3.9×10^{-5} g cm^{-3} (fundamental), 4.7×10^{-3} (first overtone), and 0.67 g cm^{-3} (second overtone). Similar changes of ε between the fundamental and the overtone modes are found for almost all kinds of materials. The great potential for using i.r. spectroscopy in conjunction with n.i.r. spectroscopy for the online measurement of concentrations which may vary by orders of magnitude during the reaction is particularly important for high-pressure processes. Thus high-pressure polymerizations may be monitored from the initiation of the reaction, when the monomer is the sole species, up to almost complete conversions, when the monomer is present only in trace amounts. It should be noted that the derivation of concentrations from high-pressure, high-temperature, absorption spectra can be quite a complex task, especially in situations where, as a consequence of strong intermolecular interactions, the Lambert–Beer law does not hold and the vibrational intensity is not proportional to the concentration. These aspects will be briefly addressed in Section 1.3 of this chapter. The cases in which the Lambert–Beer law fails, especially in polar and hydrogen-bonded materials, are far from being unimportant, since the vibrational spectra provide much information about the structure and dynamics in the fluid state. Again, spectroscopic experiments in the supercritical-fluid range are highly desirable because the density of a substance can be altered continuously between the gaseous and liquid-like states. Thus the fluid properties can be systematically studied over a wide range.

Absorption studies of this kind are nicely supplemented by high-pressure, high-temperature, Raman experiments. While Raman spectroscopy is less suitable for quantitative work, it is invaluable for measuring speciation and for determining structures and dynamics. In particular, high-pressure investigations of low-lying vibrational modes should be performed using Raman spectroscopy. The wavenumber range below 2000 cm^{-1}, and especially below 1200 cm^{-1}, is not as easily accessible by absorption (i.r.) spectroscopy because of the limited availability of suitable high-pressure window materials. For i.r. experiments, the window material (and the other components transmitting the i.r. radiation) must necessarily be transparent in the wavenumber range under investigation. On the other hand, vibrational Raman modes are recorded as wavenumber differences with respect to the laser line, which is in the visible (and partly in the n.i.r.) range. Raman high-pressure cells may thus be equipped with the type of windows used for absorption spectroscopy in the visible and n.i.r. ranges. Sapphire and quartz are suitable window materials for this purpose.

In Section 1.2 of this chapter, optical cells used for vibrational spectroscopy on fluid systems under conditions of both high pressures (up to 7 kbar) and high temperatures (up to 400 °C) are described. Four protocols are provided which outline in some detail the procedures for mounting the optical windows and for putting together some of the optical cells. In Section 1.3 the potential of high-pressure vibrational spectroscopy for the study of polar and nonpolar materials in an extended fluid range, and for the quantitative study of chemical transformations at high pressures and temperatures, will be outlined.

1.2 Optical cells

A survey of high-pressure optical cells has been given in the books by Isaacs[4] and by Sherman and Stadtmuller.[5] However, only a few of the cells described in these books are suitable for vibrational spectroscopy studies under conditions of simultaneous high pressure and high temperature. Cells which meet these requirements and which can be used up to 7 kbar and 700 °C (where these maximum values of pressure and temperature are not simultaneously reached) have been developed by Franck and his co-workers,[6] and they may be traced back to ideas from Franck's Karlsruhe group. Several of these devices have already been presented by one of the authors.[7] In the present article a few of the optical cells which are used for quantitative vibrational spectroscopy (i.r., n.i.r., Raman) on fluids under pressures up to several kilobar and temperatures up to several hundred degrees Celsius will be described, together with specific information, contained in Protocols 1–4, on the assembly of these cells.

A crucial factor in high-pressure, high-temperature, spectroscopy is the proper choice and the effective sealing of the optical windows. In addition to high pressures and temperatures, the windows often have to withstand considerable chemical attack. As has been mentioned previously, the choice of suitable window materials is less difficult for high-pressure Raman spectroscopy, where quartz and sapphire can be used. For absorption (i.r.) measurements below 2000 cm^{-1} only a very limited number of window materials are available. Polycrystalline silicon may be used down to about 1450 cm^{-1}, and windows of single-crystal CaF_2 may be used down to approximately 1000 cm^{-1}. With these materials, however, the accessible pressure range is restricted to 3 kbar and 2 kbar, respectively. At even lower wavenumbers, windows made out of diamond (type IIa) or of quartz and silicon may be used.

Above 2000 cm^{-1}, and up to 50 000 cm^{-1}, with the maximum accessible wavenumber depending on the quality of the specific single-crystalline material, sapphire is unrivalled as a high-pressure, high-temperature, window material. Moreover sapphire can withstand appreciable chemical attack so long as the medium remains neutral or acidic. Additional and more detailed information about high-pressure window materials has been given by Ferraro and Basile[8] and by Sherman and Stadtmuller.[5] The optical cells described subsequently are usually equipped with sapphire windows. They are thus sufficiently transparent from 2000 to 50 000 cm^{-1}. When fitted with windows of polycrystalline silicon, most of these cells have also been used in an extended i.r. range down to 1450 cm^{-1}.

A very efficient method of sealing optical windows is provided by the Poulter principle.[9] The arrangement is shown in Fig. 4.2. The optical window rests on an optically flat polished surface of a steel ram which has a cylindrical bore. The pressure at the seating face exceeds the internal high pressure (HP) by the factor $d_1^2/(d_1^2 - d_2^2)$, which is the ratio of the total surface area of the window to the supported area. The arrangement is thus self-sealing at this supported interface. To compensate for minor imperfections on the two surfaces in the sealing area, a spacer of gold foil of about 10 μm thickness is positioned between the window and the steel ram. In experiments of temperatures below 250 °C, spacers of poly(tetrafluoroethylene) (PTFE) of approximately the same thickness can be used. Sealing may become quite difficult in experiments at high temperatures and low pressures, where the cap may not

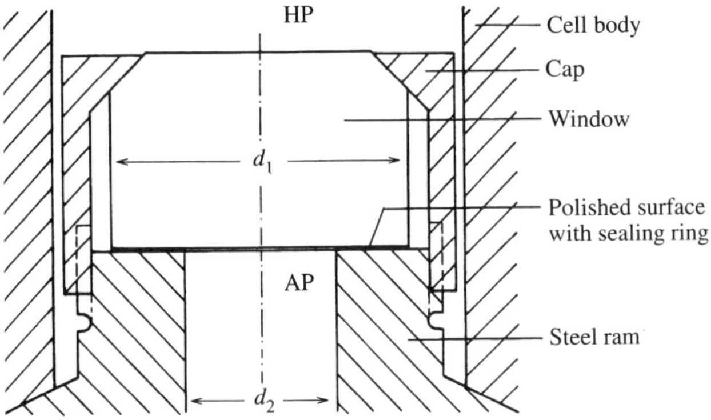

Fig. 4.2 The sealing arrangement of the high-pressure optical window: (HP) high pressure; (AP) ambient pressure; d_1, the window diameter; and d_2 the diameter of the area available for the passage of beam.

adequately hold the window under compression if the coefficient of thermal expansion of the cap material exceeds that of the window material. Hence care must be taken so that the thermal expansions of both components match each other. Stainless steel is a suitable cap material at temperatures up to about 400 °C. At higher temperatures, niobium should be used as the cap material in conjunction with a sapphire window. A facet is cut on the internal side of the window where the cap acts on the window. This arrangement allows rather small optical path lengths to be used in cells with two windows (see Fig 4.3), since the minimum distance between the windows is not affected by the geometry of the cap. Some comments on window size are given below in connection with Fig. 4.4. The individual steps of sealing the window on to the steel ram are given in Protocol 1. The steel ram, with the optical high-pressure window, is sealed conically against the cell body of the autoclave, as is detailed in subsequent figures and in Protocol 2.

Protocol 1. Sealing of a high-pressure optical window

Caution! For safety procedures see Appendix 2.

Equipment

- A high-pressure optical window
- A steel ram of the chosen length and bore
- A sealing cap fitting the ram
- A hand-held circle borer
- Fine-point tweezers
- A small sheet of thick cardboard
- An adjustable clamp for cap tightening (optional)

Materials

- A strip of PTFE film of ~ 10 μm thickness
- Lens paper

Protocol 1. *Continued*

Method

1. Invert the steel cap on a clean, flat surface with a sheet of lens paper under it.
2. Wipe the polished surfaces of the optical window carefully with lens paper, holding it by the sides.
3. Hold the window, faceted face down, at the top of the upturned cap, and check that it fits comfortably into the cap (the window should slide into the cap with ease, and under no circumstances should it be forced (into the cap); if the window does not fit, a new window and/or cap must be found). Leave the fitted window in the cap.
4. Inspect the polished surface of the steel ram (see Fig. 4.2) to ensure that it has no visible imperfections (slight grooves or marks). If the surface of the steel ram contains any marks it must be polished (this procedure is not included here), and if large grooves are present it must be machined. Wipe the surface briefly with lens paper if no imperfections are apparent and place the ram beside the cap.
5. Take the thin PTFE film and wipe it with lens paper to free it of dust and dirt, and then place it on the small sheet of cardboard.
6. Select a hand-held circle borer with a diameter matching that of the optical window (d_1, see Fig. 4.2).
7. Hold the borer just above and perpendicular to the PTFE film and then press down on it. Turn the borer slightly clockwise and then anticlockwise several times until the PTFE film has been cut through.
8. Lift the borer to leave the PTFE circle on the cardboard. Remove the circle carefully with the tweezers if it remains attached to the cutter.
9. Using the tweezers, lift the PTFE circle on to the upturned window sitting in its cap. Press down on the PTFE gently with the tweezer tips until it sits flat against the window surface.
10. Hold the cap between the thumb and forefinger of one hand and with the other hand screw the steel ram into the cap (this must be done slowly and carefully). Screw the ram right in, and finally lift the assembled unit and turn the cap clockwise with full strength.
11. Support the unit in a vice at the ram base, place the special adjustable clamp around the cap, and turn clockwise for maximum tightening of the cap and thus optimal sealing.
12. Look down the ram bore and inspect the sealing surface carefully: if there is any fold or crinkle in the fitted PTFE circle, or if the circle appears to have slipped at all, the unit must be disassembled and a new PTFE circle cut and fitted.

13. The visible circle (of width d_2; see Fig. 4.2) of the PTFE film is now 'cut out' using the fine point of the tweezers. Rest one tweezer tip against the ram bore and trace around the circumference of the visible circle, moving around the circle slowly and pressing down slightly. Use both tweezer tips to pull the PTFE film gently away from the ram, working gradually around the circle circumference until the excess film comes right away.

Figure 4.3 shows a simple transmission-type cell for absorption experiments up to 3500 bar and 350 °C. The probing light penetrates the cell along the cylindrical axis. Each window is mounted on a steel ram according to the sealing principle given in Fig. 4.2 and in Protocol 1. Each ram is pressed against the cell body by means of a flange which is secured by six bolts. The optical path length is determined by the distance between the internal surfaces of the high-pressure windows. The sample material is introduced through holes which are drilled perpendicularly to the axis of the cylindrical cell. The system under investigation is connected to the pressure-generating unit, where the pressure is also measured, by these borings. The borings also house sheathed thermocouples which are positioned in such a way that they probe the temperature inside the fluid close to the sample volume irradiated by the probing light. The autoclave is heated electrically from the outside. The resistance-heating wire is mounted on a brass support in order to simplify installation and removal of the heating wire.

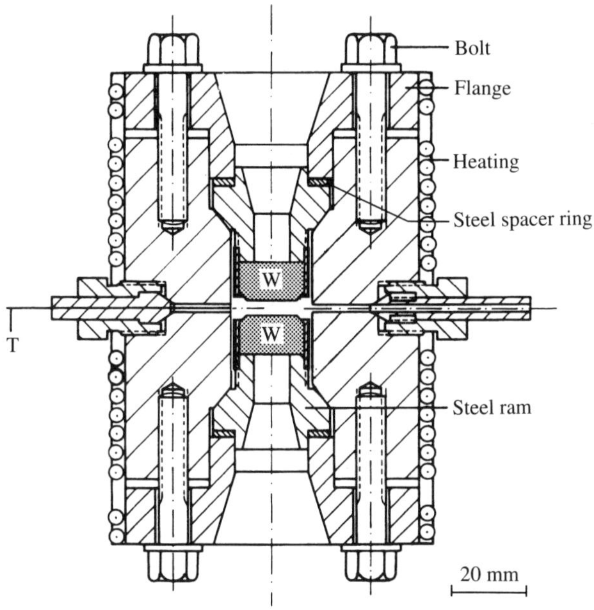

Fig. 4.3 A high-pressure optical cell: (W) window, (T) a sheathed thermocouple.

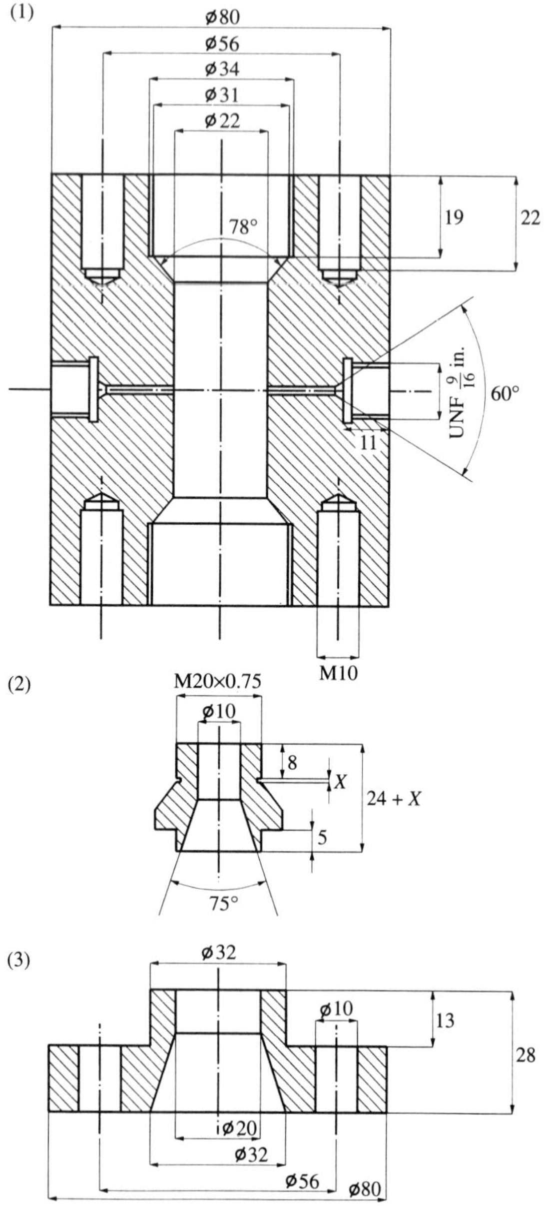

Fig. 4.4 A technical drawing of (1) the cell body, (2) the steel ram, and (3) the flange for an optical cell similar to that shown in Fig. 4.3.

Figure 4.4 is a technical drawing of a transmission-type cell similar to that depicted in Fig. 4.3. The length of the cylindrical cell body (1) is chosen to ensure that a suitable optical path length is obtained. The path length is also changed by using a single cell body (1) in combination with several steel rams (2) which have different lengths, X (see Fig. 4.4). It should be noted that the maximum conditions of pressure and temperature which are attainable depend crucially on the materials from which the cell body, flanges, and bolts are made. It should also be realized that any steel parts which are forced against each other to make a conical seal, such as the ram and the cell body in Figs 4.2 and 4.3, should not both be made from the same material, so that welding is prevented. This argument also applies to the selection of materials for the bolts and for the cell body. The stainless-steel qualities which were used for the construction of a cell with the dimensions shown in Fig. 4.4—for operation up to 3500 bar and 350 °C—are W.St.–Nr. 2.4668 for the cell body and flange and W.St.–Nr. 2.4969 for the steel ram and bolts, where the numbers refer to the German standard.[10] These materials contain a large quantity of nickel, as well as cobalt and molybdenum. In addition to the steel parts, the dimensions of the optical window and the aperture, which is given by the size of the boring in the steel ram, have to be carefully considered. For operation up to 3500 bar and 350 °C, high-quality sapphire windows with a diameter of $d_1 = 18$ mm and a height of ~ 10 mm are used with a boring in the steel ram with a diameter of $d_2 = 10$ mm (see Fig. 4.2). Windows of the same size made of polycrystalline silicon and also with $d_2 = 10$ mm have been used up to 3000 bar and 200 °C. It should go without saying that work at these pressure and temperature conditions requires that any risk of chemical attack is negligible.

Protocol 2.
Fitting of the steel ram with a high-pressure optical window to the high-pressure cell body

Caution! For safety procedures see Appendix 2.

Equipment

- High-pressure-cell components (see Fig. 4.3):
  - (i) cell body
  - (ii) flange
  - (iii) six sealing bolts
  - (iv) steel ram of chosen length with a mounted optical high-pressure window (see Protocol 1)
  - (v) steel spacer rings (preselected)
- tools for assembling the high-pressure cell:
  - (i) vice
  - (ii) torque wrench (up to 100 N m)
  - (iii) T-spanner
  - (iv) socket wrench fitting the cell bolts
  - (v) metal spacers (0.1 mm gauge) ranging from 0.3 to 1 mm
- Tweezers
- A toothbrush
- A scalpel (not always needed)

Materials

- Lens paper
- Tissues
- Acetone (Technical grade)
- Antiseize lubricant (OKS 240)

Protocol 2. *Continued*

Method

1. Support the cell body in a vice in the upright position (with respect to the cylindrical axis, as in Fig. 4.3). Clamp the cell a little more than half-way up the body. Tighten the vice so that no movement of the cell is possible.
2. Use a tissue to wipe the cell cavity removing any loose dirt and dust.
3. Inspect the conical surface of the cell bore to ensure that it is free of any residue and that it contains no visible imperfections. Remove any residue with acetone and a tissue, and if this fails then scrape away the stubborn residue very carefully with a scalpel blade. Machine the cone if any imperfections (marks or grooves) are apparent.
4. Inspect the corresponding cone section of the steel ram/window unit in the same way. Remove any accumulated dirt and machine if marks are visible on the conical surface.
5. Insert the clean steel ram, window first, into the top of the upright cell body; the ram will sit snugly into the corresponding clean conical seat of the cell cavity (as depicted in Fig. 4.3).
6. Place the steel spacer rings on the base of the inserted ram (these rings will have been prechosen to ensure that when the end is sealed there is a separation between the flange and the cell body in the range 0.3–1 mm).
7. Use the toothbrush to put a small amount of lubricant on each of the six sealing bolts. Lubricate not only the bolt thread, but also the smooth area of the bolt to prevent welding of the bolts to the cell, and also lubricate under the bolt cap in order to ensure there is no friction between the bolt and the flange.
8. Place the steel flange on a tissue on a flat surface and locate the marker on the outer edge which indicates where bolt number 1 must be inserted.
9. Now insert the numbered bolts sequentially into the flange going clockwise around the flange (each of the six bolts has a number engraved on its cap, and the flange has a mark to indicate the position for bolt number 1 which is inserted first).
10. Place the flange on top of the ram and cell body, lining up each bolt with its corresponding cavity (each bolt fits into the cavity of the same number: each cavity is specific to a particular bolt).
11. Use the appropriately sized socket, fitted to the T-spanner, to screw the bolts loosely into their cavities. First screw one bolt down, and then screw down the bolt diagonally opposite it across the flange (screwing the bolts down in this way, as diagonal pairs, ensures that the flange sits in place more evenly). Do this for all bolt pairs, screwing until the bolt just touches the flange.

12. Remove the socket and fit it to the torque wrench, and set the torque wrench to 10 N m, once this rough bolting is finished.

13. Use the torque wrench, again working in diagonal pairs (as outlined in step 11), to screw the bolts in the distance allowed by the 10 N m setting.

14. Use the metal spacers to ensure that the distance between the flange and cell body is uniform, when all bolts have been screwed in to the 10 N m limit, by:

(a) Taking the 1 mm spacer and attempting to insert it in the gap between the flange and the cell body at every point around the circumference of the cell (run the spacer around the separation); if the 1 mm spacer does not fit, try the remaining spacers in sequence (0.9 mm, then 0.8 mm and so on) to find the one which first fits at any point around the circumference of the gap. Take the spacer which first fits at any point as the standard (the separation is now made uniform to the distance of this spacer).

(b) Run this spacer around the gap again. Unscrew the bolts (1, 2, or 3) where it 'catches' (that is, where it cannot pass through the gap unhindered) because the separation is too narrow (set the catch on the torque wrench to allow unscrewing). First rescrew the bolts diagonally opposite this area (adjacent bolts are screwed one after the other in this case) with the 10 N m limit restored on the wrench. Rescrew the bolts so that they are all adjusted to 10 N m.

(c) Remeasure the separation. Unscrew the bolts in the offending area if the spacer again 'catches', and rescrew their diagonal pairs, then rescrew all the remaining screws, as before.

(d) Repeat until the spacer passes right around the circumference unhindered.

15. Set the wrench to 20 N m and repeat steps 13 and 14 for this new setting, beginning with the uniform-distance spacer determined for the 10 N m bolting.

16. Repeat for 30 N m (with the flange uniformly secured to 30 N m, the high-pressure cell is now sealed at one end).

17. Wrap a piece of lens paper around the end of a pair of tweezers and extend the paper down the bore of the steel ram/window unit and carefully wipe the optical window to remove any loose dirt and dust.

In studies of fluids at high pressures and temperatures it is frequently desirable, if not mandatory, to separate the system under investigation from the pressure-generating unit. Separation may be achieved with a bellows, using a device such as that shown in Fig. 4.5. The bellows—which is usually made of stainless steel but can also be made from noble metals, or even from PTFE—allows materials which are potentially corrosive at high pressures and

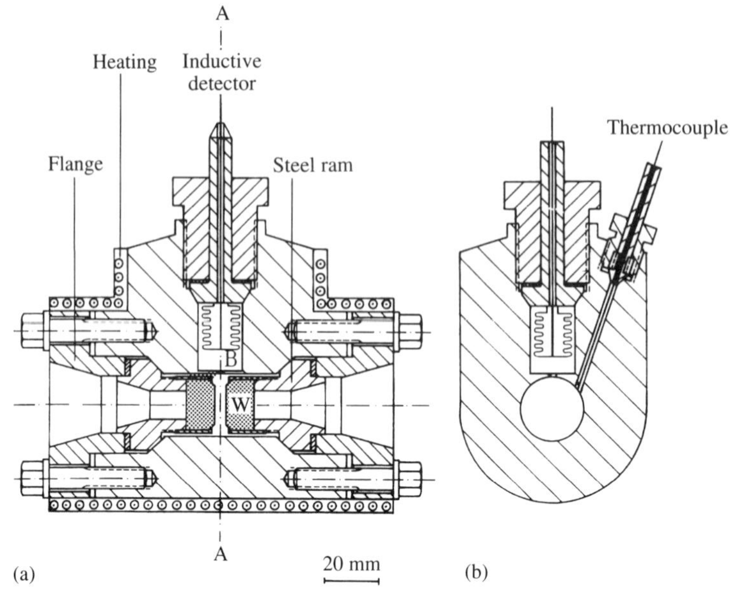

Fig. 4.5 (a) An optical high-pressure cell with an internal bellows: (W) window and (B) bellows. (b) A cross section through A–A.

temperatures to be isolated from sensitive parts in the pressure-generating or pressure-detecting units. It also prevents precipitation of any solid species from the reaction system in the capillaries, gauges, or pressure generators at ambient temperature. The position of the bellows is detected from the outside by induction, using a coil system which is positioned on a high-pressure capillary tube connected to the optical cell. With the exception of the bellows and the compartment in which it is contained, all parts of the cell correspond exactly to the components already shown in Figs 4.3 and 4.4. However, it should be noted that, in this case, the cylindrical bore is positioned off the axis A–A, as can be seen in Fig. 4.5(b). The machining of the cell body depicted in Fig. 4.5 is consequently much more complicated than the making of the devices shown in Figs 4.3 and 4.4.

Nickel, cobalt, and molybdenum, which are the main constituents of autoclave materials, may induce considerable catalytic activity in chemical reactions at high pressures and temperatures. This provides a strong motivation for isolating the fluid under investigation from the steel parts of a high-pressure vessel. This problem may be solved by applying a procedure developed by le Noble and Schlott,[11] where the system under investigation was contained within a cylindrical inner cell of quartz with a movable quartz piston which was ground to give a close fit. Two alternative devices for preventing any metal contact in high-pressure investigations are illustrated in Figs 4.6 and 4.7, and the procedures for their use are described in Protocols 3 and 4, respectively.

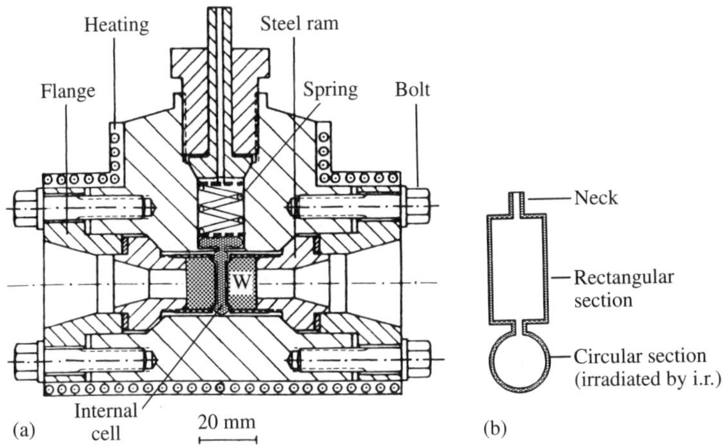

Fig. 4.6 (a) An optical high-pressure cell with an internal PTFE cell: (W) windows. (b) A side view of the internal cell, before its rectangular section is squeezed with the spring.

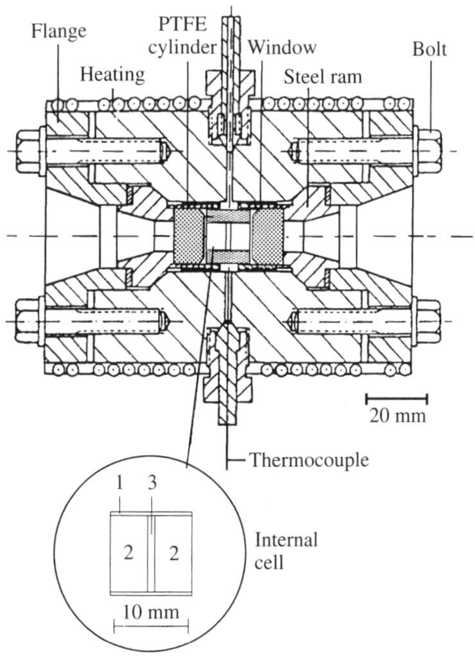

Fig. 4.7 An optical high-pressure cell with an internal cell that consists of: (1) a PTFE tube, (2) two CaF_2 windows, and (3) the sample volume.

One of the techniques (Fig. 4.6) uses an internal cell (Fig 4.6(b)) which consists exclusively of PTFE. After inserting the fluid, by means of a Pasteur pipette, into the internal volume, the PTFE cell is sealed by welding it (see Protocol 3). The internal cell is fitted into the high-pressure autoclave in such a way that the thin walls of the PTFE container are in direct contact with the inner surfaces of the high-pressure windows. This is easily achieved by using a metal spring to compress slightly the inner PTFE cell. The appropriate high-pressure cell (Fig. 4.6) is identical to the cell shown in Fig. 4.5 except that the metal spring now sits where the bellows was in Fig. 4.5. The pressure-transmitting fluid has free access to the autoclave except for the interior of the PTFE bag. This is different to the situation in Fig. 4.5, where the pressure-transmitting fluid was confined to the inner volume of the bellows.

The essential idea behind the technique shown in Fig. 4.6 is that the sample which is isolated in the PTFE bag can be studied spectroscopically in those i.r. and n.i.r. regions where the thin walls of the PTFE container are sufficiently transparent to the probing light. These regions are fairly extended, and high-pressure absorption spectroscopy is possible over a wide wavenumber range. Another important advantage of this technique lies in the ease with which the sample material can be removed from the autoclave. The sample, which is conveniently isolated in the PTFE bag, may be subsequently subjected to further analytical studies, such as nuclear magnetic resonance (NMR) or chromatographic methods. With the exception of rare cases in which the PTFE bag is destroyed during operation, cleaning of the cell is also unnecessary. Some details of the manufacture of the PTFE inner cells and of the filling and sealing procedures are presented in Protocol 3.

Protocol 3. The making and filling of a PTFE inner cell bag and its insertion into the high-pressure cell

Caution! For safety procedures see Appendix 2.

Equipment

- A welder with a raised outline (template) of the bag (the desired shape is depicted in Fig. 4.6)
- A welder whose jaws meet at a straight, fine edge
- A strip (of arbitrary length) of poly(tetrafluoroethylene) (PTFE) ~ 0.1 mm thick and ~18 cm wide
- Scissors
- Round nose tweezers
- A small beaker (~ 50 mL)
- A bunsen burner
- Pasteur pipettes
- A high-pressure cell (of the type shown in Fig. 4.6)
- Tools for assembling a high-pressure cell, as listed in Protocol 2, and an additional torque wrench rated above 150 N m

Materials

- Tissues
- A small quantity of acetone (technical grade)
- A sample of the solution for analysis

Method

Take several (at least three) Pasteur pipettes and quickly pass their tips through a bunsen-burner flame in order to smooth the glass (all the Pasteur pipettes referred to in the ensuing method are presmoothed).

Part A: Making the bag

1. Switch on the template welder and allow it to heat up to the operating temperature of 470–490 °C (this will take approximately 45 min).
2. Take the PTFE strip, place it on a clean, flat surface, and remove any dust and dirt by wiping with a tissue and a small amount of acetone. Allow it to dry.
3. Fold the PTFE strip in half lengthways, so that it is a 9 cm wide double-sided strip.
4. Open the jaws of the welder with one hand once it is fully heated. Hold the PTFE strip with the other hand (a safe distance along the strip), and place a section of the strip between the welder jaws, feeding the strip in width-ways (with the 9 cm edge leading), and taking care that the double strip remains flat. Allow the jaws to close briefly. Open the welder and remove the end of the PTFE strip, which now has the bag outline welded into it.
5. Use the scissors to cut off the welded bag. (Normally steps 4 and 5 are repeated many times until the end of the strip is reached, or until enough bags have been pressed. The welder should be turned off when finished with.)
6. Take the rectangular piece of doubled PTFE film now imprinted with the bag outline and carefully cut around the edge of the bag. Cut as close as possible to the edge, but do not cut through the welded outline.
7. Turn on the second, straight-edged PTFE welder and allow it to heat up (this takes ~25 min).
8. Use the second welder to trace around the already roughly welded bag in order to seal the edges properly (except of course at the top of the neck, which is left open for filling). Hold the cut-out PTFE bag in one hand with a pair of firmly gripped tweezers, place the edge of the bag between the welder jaws, and work right around the edge of the bag, opening and closing the welder with the other hand.
9. Squeeze open the neck of the bag, fit a Pasteur pipette to the end of an air hose, and introduce a fine, slow stream of air to the bag via the neck. Check to see if any air escapes from the edges of the bag (if air escapes at any point that section must be welded again, as in step 8).
10. Test the seal with acetone once the bag has passed the 'air check'. Inflate the bag a little with the air stream, so that the neck and body sit open. Use a Pasteur pipette to introduce a small amount of acetone to the bag. Use your thumb and forefinger to squeeze gently the acetone around the in-

Protocol 3. *Continued*

terior of the bag, looking closely for any leakage. Reweld any section which leaks.

11. Turn off the welder when satisfied that the bag is sealed (unless filling is to be performed immediately).

Part B: Filling and sealing the bag

(In principle a degassed sample can be used and the filling steps which are subsequently described can be carried out in a dry box. In this case, all equipment, including the straight-edged welder, should be introduced to the dry box as outlined in steps 2–5 of Protocol 4. The bag filling should then be performed as outlined below, but with the bag being opened manually and not with the stream of air used in step 2.)

1. Turn the straight-edged welder on and allow it to heat up to ~480 °C.
2. Take the PTFE bag and squeeze the neck open. Use a fine stream of air to slightly inflate the bag (as described in step 9 of Part A) so that the walls sit apart. Stand the partially inflated bag upright in a small beaker.
3. Draw an aliquot of the solution for analysis into a Pasteur pipette, so that it half fills the pipette (for a cell with the dimensions depicted in Fig. 4.6 this volume is typically 1.3–1.5 mL).
4. Hold the neck of the PTFE bag between the thumb and forefinger of one hand, push the tip of the filled pipette down through the neck of the bag, and slowly introduce the sample into the bag cavity. Take care not to overfill the bag, which would result in failure of the PTFE bag to fit into the high-pressure cell and/or probably cause the bag to burst in the cell during analysis (the solution should come to about half-way up the rectangular section of the bag body; see Fig. 4.6).
5. Remove the filled bag from the beaker and gently squeeze around the edges to remove any air bubbles. Push the solution within the bag upwards gradually towards the neck, so as to remove all the air from the bag. Squeeze until the solution nearly comes out of the neck. Clamp the base of the neck firmly with the tweezers (no air bubble should be present within the bag).
6. Use the hot welder to seal the neck of the clamped bag, first at the base where the tweezers are holding it, and then at several points stepwise up the neck (due to the heat of the welder, a bubble is usually formed within the PTFE bag during sealing. Do not panic—the gas bubble in the solvent will disappear with cooling, and the bag should be left free of any bubbles; however, with more volatile solvents such as dichloromethane, it is often very difficult to avoid solvent loss and so a small bubble may remain within the PTFE bag).
7. Turn the welder off; the sample is now ready for analysis.

Part C: Inserting the bag into the high-pressure cell

1. Support the high-pressure cell (of the type shown in Fig. 4.6) in a vice as outlined in Protocol 2. The cell should be closed only at one window side, without the second steel ram and the spring with retaining screw (see Fig. 4.6) being installed. The lengths of the steel rams should have been pre-chosen to give the desired path length in the assembled cell: the ram lengths define the optical path length.
2. Hold the entire PTFE bag lengthways with the tweezers and push it down into the spring cavity, perpendicular to the axis of the windows.
3. Push the circular section of the bag with a little force through the small slit which allows access to the windows (see Fig. 4.6).
4. Look through the open-window side of the cell and check to see if the circular section of the bag has been pushed completely through the slit and is sitting flat against the window which is in place. Use the tweezers to adjust the bag as necessary, gaining access from the spring cavity, until the circular cross section of the bag lies neatly in front of the window.
5. Put the second steel ram in place so that the window sits on top of the circular cross section of the bag. Seal this side of the cell as described in Protocol 2, but only tighten the eight flange bolts to 20 N m.
6. Turn the cell body in the vice so that the spring/retaining-screw cavity is facing upwards.
7. Push down the protruding rectangular cross section/neck of the PTFE bag with a finger, and place the spring and retaining screw on top of it. Tighten the retaining screw, first manually, and then to 150 N m (*not* to the 20 N m tightness for the flange bolts) using a torque wrench.

Another type of internal cell which avoids metal contact is shown in Fig. 4.7. The cell is used in combination with a high-pressure optical cell as in Fig. 4.3. The dimensions of the autoclave are chosen to ensure that an internal cell which is about 12 mm in length can fit between the high-pressure windows. The internal cell consists of a PTFE tube (with an internal diameter of 9 mm, an outer diameter of 10 mm, and a length of about 12 mm) and two cylindrical CaF_2 windows (with a diameter of 10 mm and a thickness of 5 mm). The PTFE tube is widened on either side before inserting the CaF_2 windows so that a tight fit is gained, leaving the fluid inside the cell effectively isolated from the metal walls of the autoclave. The procedure for joining the PTFE tube with the CaF_2 windows and of filling this cell, preferably under an inert-gas atmosphere, is described in Protocol 4.

The internal cell is positioned in a PTFE cylinder which fits into the bore of the autoclave. The PTFE cell is surrounded by the pressure-transmitting fluid. In contrast to the internal cell shown in Fig. 4.6, the whole (internal) sample volume can be irradiated by the incident light. However, the PTFE bag (Fig. 4.6) allows the use of larger sample volumes because the size of the cell is not limited by the necessity of a path length adequate for quantitative optical detection. In particular, for i.r. studies on compressed fluids the path length has to be fairly small, usually around or below 1 mm, even in experiments involving rather dilute systems. The two techniques illustrated by Figs 4.6 and 4.7 have been used extensively in kinetic studies on fluids up to high pressures. A few examples from these investigations will be presented in Section 1.3 of this chapter.

Protocol 4.
The assembly and filling of a tubular PTFE inner cell under an inert gas atmosphere

Caution! For safety procedures see Appendix 2.

Equipment

- Poly(tetrafluoroethylene) (PTFE) tubing with an internal diameter of 9 mm, an external diameter of 10 mm, and an arbitrary length
- Two calcium-fluoride (CaF_2) windows with diameters of 10 mm and thicknesses of ~5 mm (see Fig. 4.7)
- One solid cylindrical piece of aluminium (or another inert metal) with a diameter of 10 mm and a length of ~10 cm (this is the *small aluminium tool*)
- One solid cylindrical piece of aluminium with a diameter of 15 mm and a length of ~10 cm, with a cone-shaped nose at one end (this is the *large aluminium tool*)
- Vernier calipers
- One cylindrical piece of PTFE with an internal diameter of 10 mm, an external diameter of ~21 mm, and a length of ~2 cm; this slides over the small tool
- A micrometer screw gauge
- Flat nose tweezers (preferably of inert material)
- A scalpel with very sharp blade
- Pasteur pipettes
- A dry box with an external evacuating chamber
- A vessel fitting into the dry box evacuating chamber
- A cork ring

Materials

- Lens paper
- Tissues
- Paper towelling
- Acetone (technical grade)
- A sample of the solution to be analysed (which should already have been degassed by freeze–pump–thaw cycles; this procedure is not included here)

Method

Part A: Preparing the inner cell

1. Take the PTFE tubing in one hand, holding it close to the end, and in the other hand take the large aluminium tool. Widen the end of the tubing by pressing the nose of the tool into the tube and at the same time rotating it briefly.
2. Take the smaller aluminium tool and push it into the now widened end of the PTFE tubing. With some effort, and again with simultaneous rotation, push the tool in about 16 mm.

3. Cut off the tubing at the end of the tool with a sharp scalpel, leaving a piece ~16 mm long fitted over the aluminium. Trim the end of the tubing with the scalpel blade, using the end of the tool as a guide for the blade, to make it as uniform as possible.
4. Take the large cylindrical piece of PTFE, fit it over the other end of the small tool, and use it to push the thin PTFE tube off the end of the aluminium.
5. Refit the PTFE tube to the small tool with your thumb and forefinger, placing it in such a way that the opposite (nontrimmed) end is at the end of the aluminium.
6. Trim this end of the tubing, and then remove the tube again, as in step 4.
7. Use the vernier calipers to measure the length of the piece of PTFE tubing, which should now be closer to 12 mm in length.
8. Refit the tube to the small tool, use the large cylindrical piece of PTFE to push a fraction of it off the tool, and again trim the end. Push the whole piece off the aluminium and remeasure it.
9. Continue trimming the ends alternately in this manner until the tube reaches a length of 12 mm.
10. Rinse the PTFE tube once the desired length has been reached with a little acetone, and place it on a piece of paper towelling to dry.
11. Take both of the CaF_2 windows and inspect them to make sure that they both have polished edges (this facilitates better sealing) and that the edges are not significantly chipped or split. Replace one or both of the windows, if splitting or excessive chipping is apparent (the new window(s) should already have been polished; this procedure is not included here).
12. Rinse both CaF_2 windows with acetone, allow them to dry, and then wipe them with lens paper and place them with the PTFE tube.
13. Rinse the aluminium tools, the PTFE cylinder, and a pair of flat nosed tweezers with acetone and allow them to dry.

Part B: Filling the inner cell

1. Place together the CaF_2 windows (wrapped loosely in a piece of lens paper), the PTFE tube, a Pasteur pipette, the tweezers, the aluminium tools and PTFE cylinder, the degassed sample of the solution to be analysed (usually in a round-bottomed flask), and the cork ring on a piece of paper towelling in the vessel which fits into the evacuating chamber of the dry box.
2. Place the loaded vessel into the evacuating chamber, seal the chamber, and evacuate it for at least 10 min (this time will of course vary with the dry box being used, and with the quality of its vacuum pump).
3. Stop the pump after 10 min, and slowly flood the chamber with the inert gas.

Protocol 4. *Continued*

4. Remove any hand jewellery and carefully insert both hands into the gloves of the dry box.
5. Introduce the vessel to the chamber of the dry box and assemble the equipment on the paper towelling on a flat surface in the dry box (the base). Support the sample solution on the cork ring and open it to the inert atmosphere.
6. Push the PTFE tube on to the small tool with your gloved thumb and forefinger. Fit the PTFE cylinder over the other end of the aluminium tool, and use it to push 3–4 mm of the thin PTFE tube off the end of the aluminium. Use the nose of the large tool to widen the end of the tube as in step 1 of part A. Pick up one of the CaF_2 windows carefully and insert it into the widened end of the tube (do not drop the window—it chips!). Press down with one forefinger on the inserted window, and use the PTFE cylinder (which should still be around the small aluminium tool) to push the PTFE tube up to the edge of the CaF_2 window, so that the window sits inside the tube (see Fig. 4.7).
7. Push the tube right off the aluminium and place it, window side down, on the paper towelling.
8. Widen the upturned end of the tube with the large aluminium tool using one hand, as described before. Take care not to insert the nose too far to avoid scratching the window. Use your other hand to draw an aliquot of the sample carefully into the Pasteur pipette.
9. Remove the nose from the PTFE tube and fill it with the sample so that the sample almost spills over the edge of the PTFE (the solution will have a distinctly convex surface).
10. Hold the second CaF_2 window in the tweezers and slowly and carefully place it centrally on top of the filled tube. (If a lot of solution is spilt from the tube in this step—for instance by the accidental pushing sideways of the tube—the second window should be removed, the solution refilled, and the fitting of the window repeated because bubbles will form inside the cell if there is insufficient solution.)
11. Push the second window into the tube with the flat end of the large aluminium tool, easing the window very gently from side to side to aid insertion, and to help prevent the formation of bubbles. Push the window in so that it lies almost flush with the end of the PTFE tube.
12. Hold the assembled internal cell between the tweezers and inspect it carefully (if it contains any bubbles or if the second window is at a significant angle the cell must be dismantled, wiped clean and the assembly reperformed). If no bubbles are present and the windows lie neatly in place, wrap the cell in the lens paper and remove it and all the tools from the dry box. Recap the sample and leave it in the box.

13. Place a piece of lens paper each side of the cell (at each window) outside the dry box, and place the cell in the barrel of the micrometer screw gauge so that it is held quite loosely. Rotate the cell a little and at the same time wind the gauge in a fraction in order to ensure that the windows lie straight.
14. Wind the gauge in carefully until the desired length is almost reached. Remove the lens paper from the windows, wipe the windows and sides of the micrometer barrel with a tissue, and place the cell back in the micrometer.
15. Adjust the micrometer to the desired length (usually 12 mm) very slowly, so as not to scratch the windows.
16. The tubular inner cell should now be as depicted in Fig. 4.7 and it is ready for insertion into the high-pressure cell.

As has already been mentioned, optical path lengths should be rather small in high-pressure absorption experiments on compressed fluids. Path lengths down to several micrometres are required in i.r. studies on pure dense materials, in order to avoid intolerably large absorbances which render quantitative spectroscopic analysis difficult or even impossible. In transmission-type cells, the changes of path length induced by applying pressure and increasing the temperature are also of micrometre size, which may give rise to considerable uncertainty in high-pressure quantitative measurements on strongly absorbing materials. Moreover, it is extremely difficult to join the parts of a cell shown in Fig. 4.3 in such a way that path lengths below 100 μm are achievable. These problems may be circumvented by using the device shown in Fig. 4.8, which allows the optical path length to be changed from outside the cell while it is under pressure.[12]

The cell in Fig. 4.8 is designed for operation up to 2500 bar and 250 °C. The probing light is transmitted through the cell along its cylindrical axis. One of the windows (W_1) is sealed to a fixed conical steel ram, which is pressed against the cell body by means of a central screw. A steel ring is inserted between the steel ram and the screw to prevent almost all transfer of rotational motion from the screw to the steel ram. The second window (W_2) is attached to a movable piston which is sealed at a packing by means of the screw A. The screw B, situated between the movable piston and screw A, has different threads on the inner and outer diameter. In the cell depicted in Fig. 4.8, the pitches of the screws at the inner and outer side of screw B are 1.25 and 1.50 mm, respectively. Given this difference, turning screw B, while screw A and the movable piston (with auxiliary equipment not shown in Fig. 4.8) are fixed against rotation, moves the piston in the direction of the cylindrical axis. For this particular device, a full turn of screw B corresponds to a displacement of the piston by 250 μm, corresponding to the difference between the pitch sizes of 1.50 and 1.25 mm. The cell is charged through cross

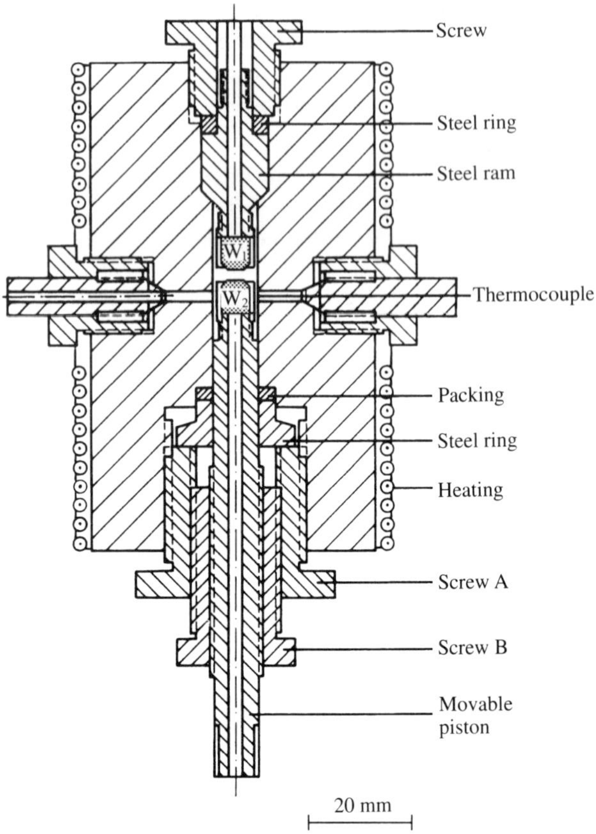

Fig. 4.8 A high-pressure cell with a variable optical path length: (W_1, W_2) windows.

bores perpendicular to the optical axis, and a sheathed thermocouple is introduced. The cell is heated electrically from the outside.

The procedure for obtaining absorbance spectra is as follows. The cell is heated to the experimental temperature and pressurized. By turning screw B, the path length, which is initially rather large (around 0.5 mm), is reduced until the absorbance in the wavenumber region of interest is below a value of $A = 1.5$. A spectrum is taken at this position and the optical path length is then further decreased by turning screw B through a precisely measured angle which corresponds to an accurately known path length difference Δl. A second spectrum is now recorded in this new position. When the concentration (or density) of the species c is known, the molar absorption coefficient, $\varepsilon = A/(cl)$, can be immediately obtained from the precisely measured path length, $l = \Delta l$, and from the absorbance, A, which is calculated from the two spectra using the first spectrum as the sample and the second spectrum (taken after changing the path length by Δl) as the reference spectrum. In addition to enabling high-pressure, high-temperature spectra to be measured for strongly absorbing materials with very small sample layers, this technique, where the sample and the reference for the system are taken at identical pressure and temperature, avoids uncertainties in the measured absorbances due to bending of the high-pressure windows or due to a difference between refractive indices of the sample and the (usually chosen) reference state of the empty optical cell at the experimental temperature, T. The device shown in Fig. 4.8 is, however, not without disadvantages: with ordinary sample light geometries there is an appreciable loss of radiation intensity because of the long and narrow boring. Increasing the aperture and the window size makes turning of screw B under pressure more laborious.

Using the construction principles applied in the design of cells for absorption measurements, high-pressure, high-temperature Raman cells have been devised, chiefly by Franck and his co-workers.[13] A Raman cell for operation[14] up to 7 kbar and 250 °C is shown in Fig. 4.9. The laser beam passes through two windows (W_1) which are sealed against steel rams fitted into opposite cross bores of the cylindrical autoclave. The Raman-scattered light is observed through a larger window (W_2) at right angles to the laser path. The fluid under investigation is separated from the pressurizing medium by means

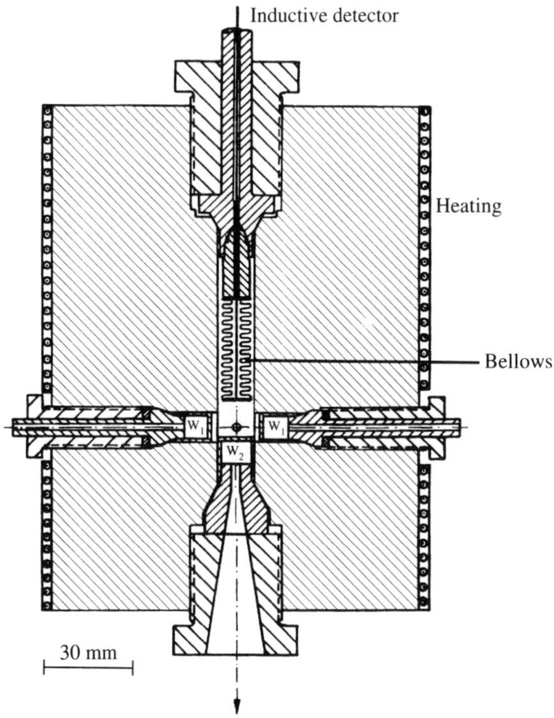

Fig. 4.9 A high-pressure Raman cell: (W_1) windows for the passage of a laser beam, and (W_2) a window for the detection of Raman-scattered light.

of a bellows in the cylindrical bore of the high-pressure cell. The cell is charged and a sheathed thermocouple may also be introduced into the fluid through additional small capillary borings perpendicular to both the laser beam and the cylindrical axis of the autoclave.

The optical cells described so far have mainly been used as batch-type reactors to study discontinuous fluid-phase reactions. Continuously operated high-pressure processes in tubular reactors are easily monitored by transmission-type cells, like that shown in Fig. 4.3, which are introduced into the product stream and perhaps also into the feed flux. The spectroscopic analysis may be performed on the main stream or by choosing some suitable by-pass arrangement.

A continuously operated stirred tank reactor (CSTR) for use up to 3 kbar and 300 °C is shown in Fig. 4.10. The reactor is equipped with a fairly large sapphire window (d_1 = 34 mm) which allows visual observation and spectroscopic analysis.[15] Spectroscopic studies may be conducted using a reflectance technique developed by Franck and Roth:[6] probing light enters the cell, passes through a sample layer with a precisely known thickness, and is reflected from a mirror which is positioned inside the fluid under investiga-

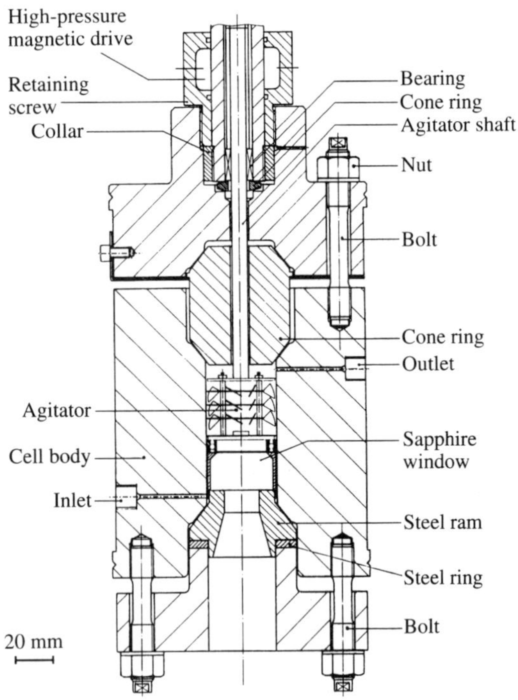

Fig. 4.10 A CSTR for operation at pressures up to 3 kbar and temperatures up to 300 °C, equipped with a window for visual observation or for spectroscopic analysis.

tion. After passing the sample layer a second time, the radiation leaves the cell through the same high-pressure window. The components required for this technique are not shown in Fig. 4.10.

The stirrer is magnetically driven from the outside. The geometry of the stirring device is chosen so that the centre of the internal boring is not intersected by the stirrer blades. Thus, in addition to affording visual observation, this reactor allows photochemically induced high-pressure processes to be studied. It is beyond the scope of this article to describe the reactor in any detail, and a full description is given elsewhere.[15] The concentration inside the CSTR is best obtained from high-pressure spectra taken in a simple transmission-type cell which is positioned directly after the reactor. Heating the capillary between the CSTR and the spectroscopic cell ensures that no precipitation of solid material and no fluid-phase separation take place. In addition, the residence time between the CSTR and the optical cell should be small to minimize additional conversion.

An interesting aspect of continuously operated devices is the residence-time distribution of the reactor. Quantitative absorption spectroscopy offers a convenient means to determine this distribution. A short tracer pulse may be added to the feed flux just before it enters the reactor (tubular or stirred), and the residence-time distribution can be found by measuring the tracer concentration immediately behind the reactor outlet (again in a separate optical cell). Carbon dioxide, with its stretching mode at about 2350 cm^{-1}, appears to be a very suitable tracer (unless CO_2 is a component of the system under investigation), since the CO_2 concentrations required are very small and since, in general, no influence of CO_2 on the course of the reaction is to be expected given the low tracer levers. It should be realized that the absorption lines of gaseous CO_2 in air do not interfere because they are easily distinguished from the broad band of CO_2 under pressure. In order to illustrate this method, Fig. 4.11 gives the CO_2 absorbance spectra around 2350 cm^{-1} measured as a function of time, t, after a CO_2 tracer pulse is injected into an ethylene stream (3 kg h^{-1}) at 2000 bar and 50 °C. The area under the absorption bands is proportional to the CO_2 concentration. By integrating over the wavenumber range 2200–2500 cm^{-1}, the time dependence of CO_2 concentration is obtained, which in turn yields the residence-time distribution. Detailed analysis also necessitates measurement of the equipment function, which is determined by the method of pulse injection, and by the geometry of the high-pressure connections to the reactor and between the reactor and the spectroscopic cell. Several applications of this technique to the measurement of residence-time distributions in reactors under conditions of both high pressure and high temperature can be foreseen. The method has already been used in high-pressure ethylene polymerizations in a tubular reactor and in a CSTR.[16]

By a suitable combination of the construction principles in Figs 4.3–4.10, a large number of optical cells may be constructed for the study of fluids up to

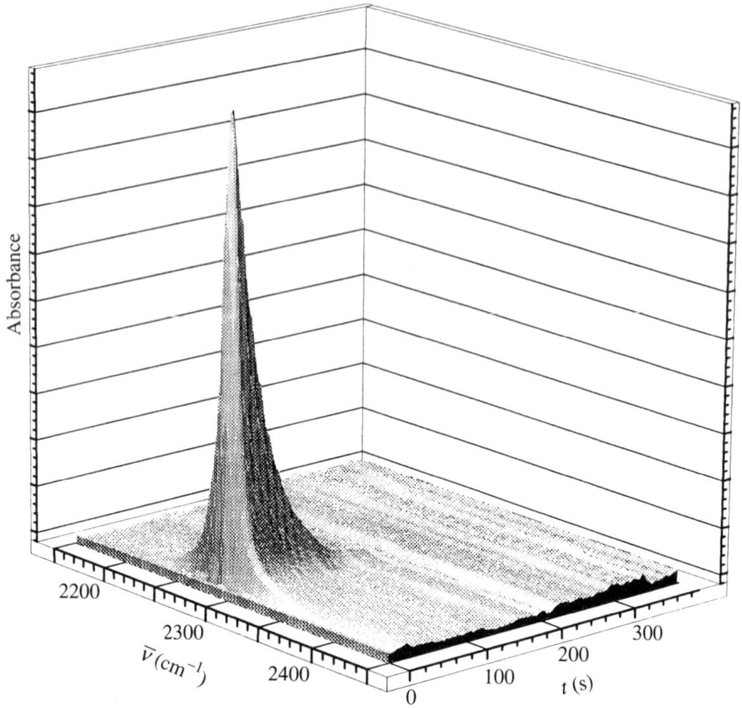

Fig. 4.11 The absorbance spectra of CO_2 measured at the outlet of a CSTR, operated with ethylene at 2000 bar, 50°C, and a mass flow of 3 kg h^{-1}, after injecting (at $t = 0$) a tracer of CO_2 into the ethylene flow just before entry to the CSTR.

high pressures and temperatures. High pressures up to 7 kbar, and perhaps up to 10 kbar, can be reached, even at higher temperatures in devices with a reduced internal diameter. Additional information on optical cells used for vibrational spectroscopy is provided in the books by Isaacs[4] and by Sherman and Stadtmuller.[5]

1.3 Examples of vibrational-spectroscopy studies in fluids

The main emphasis of this section is on a detailed description of the experimental devices and procedures used for measuring vibrational (i.r., n.i.r., and Raman) spectra in fluids up to high pressures and temperatures. Therefore, only a few examples of the results which can be obtained with these devices will be given. Only absorption work (i.r. and n.i.r.) is considered because this is the area in which the authors have the greatest personal experience. Examples on Raman fluid phase studies are presented elsewhere.[7,13,14]

In i.r. and n.i.r. experiments performed within a wide range of pressures and temperatures, two limiting types of vibrational modes will be found: type

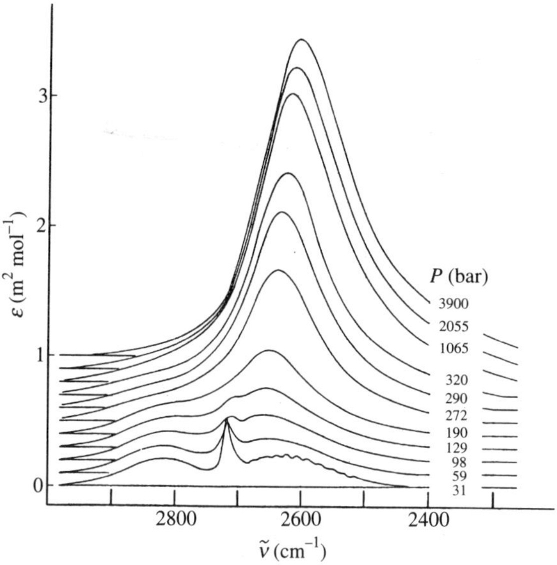

Fig. 4.12 The molar absorption coefficient, ϵ, for the O–D stretching mode of HDO diluted in H_2O at 400 °C at various pressures (data from Franck and Roth).[6]

I, modes which are strongly affected by intermolecular interactions and which show significant changes in the vibrational intensity, and also in the band-maximum position between gaseous and liquid-like states; and type II, modes without any change in the vibrational intensity when pressure and temperature are varied. Materials showing type-II behaviour are perfectly suited for quantitative analysis via online absorption spectroscopy.

Type-I behaviour is illustrated in Fig. 4.12, where the i.r. spectra of supercritical water, measured by Franck and Roth,[6] is taken as an example; the molar absorption coefficient ε for the O–D stretching mode of HDO (9.5 mol % of HDO in H_2O) is plotted for various pressures at 400 °C (T_c = 374 °C). The densities at the lowest and at the highest pressure are 0.010 and 0.900 g cm^{-3}, respectively. At 31 bar, a pressure-broadened spectrum with envelopes of the P-, Q-, and R-branches is found. Toward higher pressures the spectrum changes into a broad band, which, at the highest densities, shows no clear indication of any shoulder. The vibrational intensity, which is given by the area under each ε versus wavenumber $\bar{\nu}$ curve, increases enormously with pressure, by about a factor of 3 in the pressure range covered by the spectra in Fig. 4.12. (It should be noted that the baselines of the curves have been slightly shifted, which makes the changes appear even larger.) The substantial difference in the vibrational intensity between gaseous and liquid-like states makes quantitative analysis in the fluid range via the oxygen–hydrogen

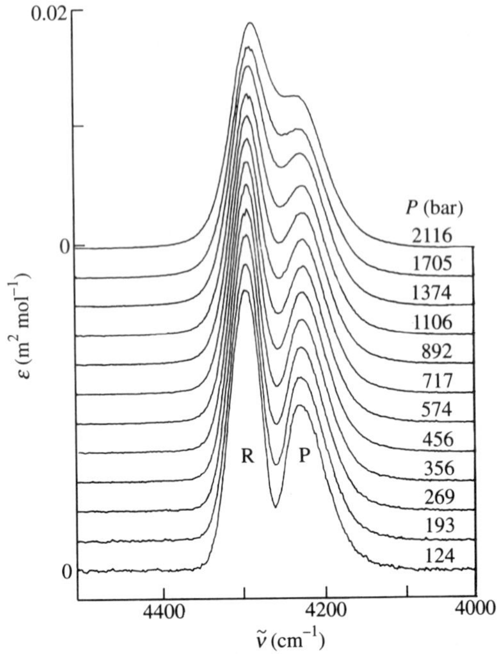

Fig. 4.13 The molar absorption coefficient, ϵ, of the stretching first overtone of pure carbon monoxide at a temperature of 127 °C for various pressures. (The curves have been displaced for clarity.)

stretching mode extremely difficult, if not impossible. On the other hand, the large changes in band shape, band-maximum position, and vibrational intensity may provide important information about the structure and dynamics in polar fluids.

It should be expected that type-II behaviour will be found more frequently with less polar and with nonpolar materials. Figure 4.13 shows the molar absorption coefficient, ε, of the stretching first overtone in pure carbon monoxide at 127 °C ($T_c = -140\,°C$).[17] The pressure varies from 124 to 2116 bar, corresponding to densities ranging from 0.10 to 0.65 g cm^{-3}. The baselines of the individual spectra have been appreciably shifted to avoid intersection of the curves. The appearance of the spectra when measured over an extended region of states is rather similar; a P-branch and an R-branch contour is observed up to the highest pressures, where the gap between the two branches does, however, become less pronounced. Integration of the ε versus $\tilde{\upsilon}$ curves gives almost identical values for the vibrational intensity, *B*; *B* decreases by only about 2 per cent over the full range from gaseous to liquid-like densities.[17] Thus favourable conditions are met for the quantitative spectroscopic determination of CO over a wide range of fluid states. In addition, by taking advantage of the implications of Fig. 4.1, quantitative analysis of CO up to high pressures and temperatures may be performed at widely differing concentrations via quantitative i.r./n.i.r. spectroscopy. The validity of the Lambert–Beer law, in the sense of the vibrational intensity, B, being independent of the pressure and temperature, has also been tested for several other species, including carbon dioxide[18] and ethylene,[19] in extended wavenumber ranges. Many vibrational modes have been found, even in polar molecules, where B does not, or does not significantly, change with temperature and with density. Whether a particular mode shows such type-II behaviour and is therefore suitable for quantitative spectroscopic analysis has to be determined via calibration spectra taken in the pressure and temperature range of interest. This laborious procedure will remain unavoidable until the theory of the vibrational intensity in dense phases is significantly advanced.

Among the technical fluid-phase processes, the high-pressure, free radical polymerization of ethylene is of outstanding importance. Careful calibration work on this system, in particular on ethylene,[19] has shown that the Lambert–Beer law holds for several vibrational modes. The polymerization can be accurately monitored via online high-pressure, high-temperature i.r. and n.i.r. spectroscopy.[2,20] Figure 4.14 shows a series of n.i.r. spectra in the wavenumber region of the first C–H stretching overtones taken during the polymerization of pure ethylene at 200 °C and at 2.5 kbar. The arrows in Fig. 4.14 indicate the direction of spectral changes with time. The polymerization kinetics can be quantitatively followed either by the decrease of the ethylene (E) bands, with maxima above 5900 cm^{-1}, or by the increase of the polyethylene (PE), with maxima below 5900 cm^{-1}.

The course of an ethylene polymerization can also be precisely measured

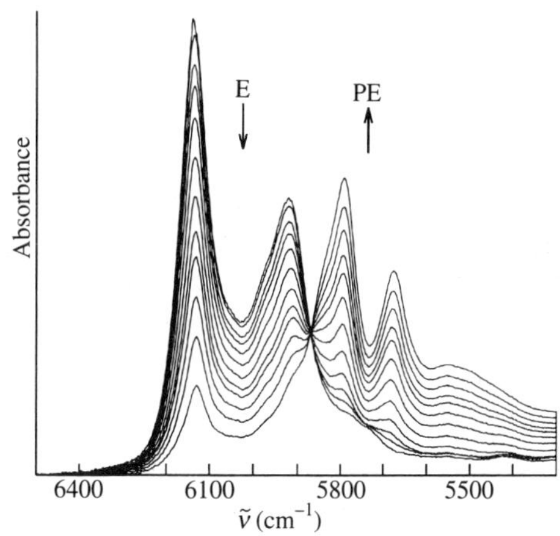

Fig. 4.14 Near-infrared absorbance spectra recorded in the first C–H stretching overtone region during an ethylene polymerization at 200 °C and 2.5 kbar: (E) ethylene bands, and (PE) polyethylene bands.

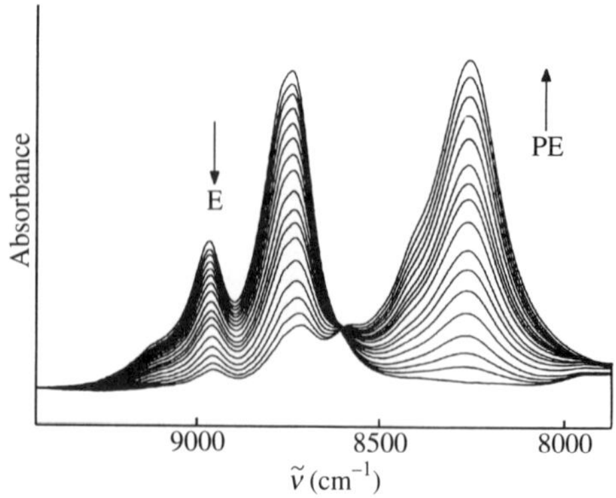

Fig. 4.15 Near-infrared absorbance spectra recorded in the second C–H stretching overtone region during an ethylene polymerization at 190 °C and 2.63 kbar: (E) ethylene bands, and (PE) polyethylene bands.

via n.i.r. spectra recorded in the region of the second C–H stretching overtones. A series of spectra taken in this range for a pure-ethylene polymerization at 190 °C and 2.63 kbar is given in Fig. 4.15. The ethylene bands around 8730 and 8960 cm^{-1} decrease during the reaction, whereas the polyethylene band at 8260 cm^{-1} increases.

For the two polymerization reactions depicted in Figs 4.14 and 4.15, which both proceed to high conversions (above 75 per cent), no time scale has been indicated. This is because the polymerization rate is determined by the choice of initiation process, with free radicals being accessible by thermal or photochemical decomposition of an initiator, or, eventually, also by thermal or photochemical excitation of the monomer itself. Thus the spectral series given in Figs 4.14 and 4.15 may be recorded within quite different reaction periods, varying from a few minutes to several hours. On a Fourier-transform instrument, the time required to record one spectrum of the quality of those in Figs 4.14 and 4.15, and with the cell dimensions given in Fig. 4.3, is about a few seconds.

For the proper selection of optical path lengths for experiments such as those given in Figs 4.14 and 4.15, the same considerations must be made as were put forward in the discussion of Fig. 4.1. Figure 4.16 shows a corresponding plot of the molar absorption coefficient, ε, against $\tilde{\nu}$ for selected ethylene (E) and polyethylene (PE) bands in an extended i.r. and n.i.r. range.[2,3] Roughly speaking, the four absorption regions from about 3000 to about 11 000 cm^{-1} may be assigned to the fundamental C–H stretching mode and to the first, second, and third C–H stretching overtones.

To obtain an absorbance of about $A = 1$ in the ethylene-band maximum in each of the four regions for an ethylene density of 0.5 g cm^{-3}, which is typical of high-pressure polymerization, the optical path lengths should be as

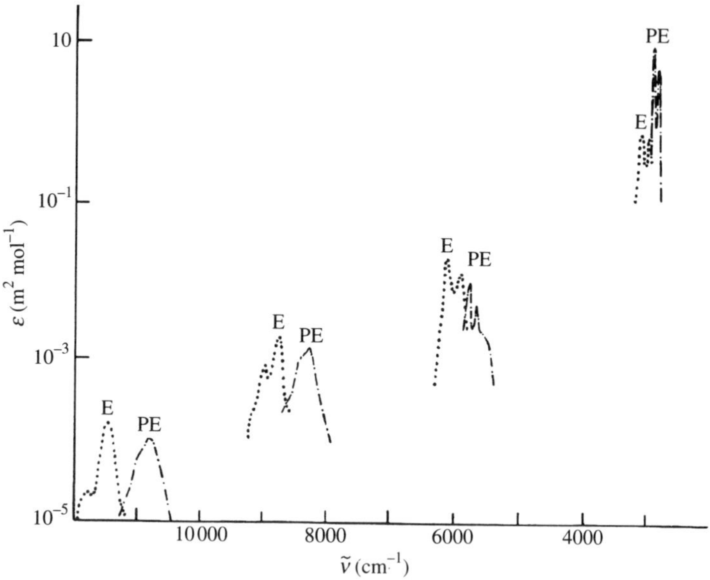

Fig. 4.16 The molar absorption coefficient, ε, for selected bands of compressed fluid ethylene (E) and polyethylene (PE) between 2000 and 12000 cm^{-1}.

follows: for the fundamental mode around 3000 cm^{-1}, l = 0.05 mm; for the first overtone region around 6000 cm^{-1}, l = 1.1 mm; for the second overtone and around 9000 cm^{-1}, l = 12.5 mm; and for the third overtone around 11 500 cm^{-1}, l = 140 mm. Hence the polymerization process can be monitored in autoclaves with widely different dimensions.

Infrared and n.i.r. spectroscopy are also well suited for kinetic (and thermodynamic) investigations of other monomer–polymer systems. Examples of kinetic studies on acrylate[21] and methacrylate[22,23] polymerizations in the fluid state up to high pressures are detailed elsewhere. Moreover, copolymerizations[24] and even terpolymerizations[25] have been monitored via high-pressure vibrational spectroscopy.

The general principle of considerably increasing the dynamical range for quantitative analysis (see Fig. 4.1) by simultaneous measurement of the entire i.r. and n.i.r. range (at the same optical path length) may also be used in the study of polymerization kinetics. This strategy—by which minor concentrations are determined via fundamental modes, whilst the concentration of major constituents is found from overtone vibrations—has favourable applications in chemically induced free-radical polymerizations. The very small concentrations of initiator and of initiator-derived products, and the large monomer and polymer contents, are simultaneously measured in the i.r. and n.i.r. regions, respectively. This procedure has been developed using the azobisiso-butyronitrile-initiated styrene polymerization as a first example.[26]

Quantitative absorption spectroscopy is also perfectly suited for systematic online studies of unimolecular decomposition rates of peroxides and azocompounds in different solvents as a function of pressure and temperature. This kind of experiment is illustrated in Fig. 4.17, where the decay of dibenzoyl peroxide (BPO) is taken as an example.

The spectral series shown in Fig. 4.17(a) traces the decomposition of BPO at 80 °C and 2.25 kbar at a concentration of 0.01 M in dichloromethane. The absorbance due to the dibenzoyl peroxide appears as a doublet at the higher-frequency end of the carbonyl region shown. The doublet is due to coupling between identical characteristic vibrations of the two carbonyl bonds in the symmetric molecule. Of course, it decreases with time. Also shown in this region (on the lower $\tilde{\nu}$ end) is the absorbance due to one of the decomposition products, benzoic acid. The acid appears in both the monomeric (higher $\tilde{\nu}$) and dimeric (lower $\tilde{\nu}$) forms. These bands increase with time and their relative intensities alter with time as the monomer–dimer equilibrium shifts with increasing acid concentration.

In Fig. 4.17(b) absorbance difference spectra for the same BPO decomposition are shown. The first spectrum is subtracted from each subsequent spectrum. A time scale has been added to illustrate the time profile of the 14 h reaction.

Just to show that online vibrational spectroscopy has extended applications in quantitative analysis of fluid systems up to high pressures, spectra recorded during a Diels–Alder reaction will be presented as a final example.

The spectral series depicted in Fig. 4.18(a) were measured during the *hetero*-Diels–Alder reaction of the phthalimidobutenone **1** with propenyl methyl ether **2** to yield the diastereomeric dihydropyrans **3** and **4**.[27] The transformation is monitored via the carbonyl modes of the phthalimido moiety. It should be noted, however, that the two diastereomers, **3** and **4**, cannot be distinguished by the i.r. absorption of these vibrations. The reaction in Scheme 4.1 proceeds in a dichloromethane solution at a pressure of 500 bar and at a temperature of 50°C, when the initial concentrations of **1** and **2** are 0.01 and 1.0 mol per kilogram of solution, respectively. In Fig. 4.18 the spectra are recorded over a reaction time of 9 h. The arrows indicate the direction of the spectral changes with time. In order to eliminate the absorption due to the solvent, and from groups whose absorption does not change during the transformation, absorbance difference, ΔA, spectra (as in Fig. 4.17) are calculated. The sequence of such spectra is shown in Fig. 4.18(b). The traces were obtained from the spectra in Fig. 4.18(a) by subtracting the first spectrum, recorded after reaching the experimental conditions, from each of the spectra measured thereafter. By this procedure, the products give positive absorbances whilst the starting compound **1** gives an increasingly negative absorbance. The concentrations of **1** and of the sum of **3** and **4** are derived from an integration of the corresponding band sections, for example, above and below 1785 cm^{-1}, against the straight line at $\Delta A = 0$. Problems with the validity of the Lambert–Beer law

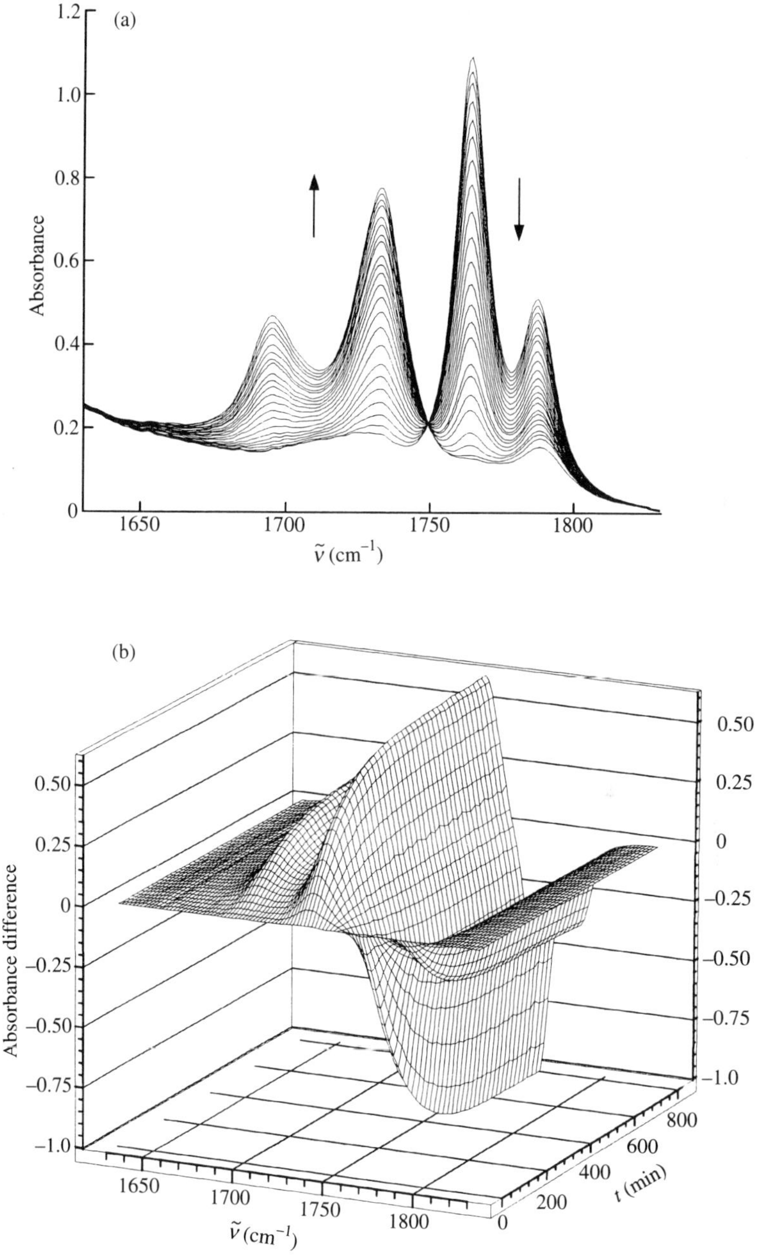

Fig. 4.17 (a) The carbonyl region of the i.r. absorbance spectra measured during the decomposition of dibenzoyl peroxide in dichloromethane at 80 °C and 2.25 kbar. (b) Infrared absorbance difference spectra corresponding to the changes in the spectra in (a) over time.

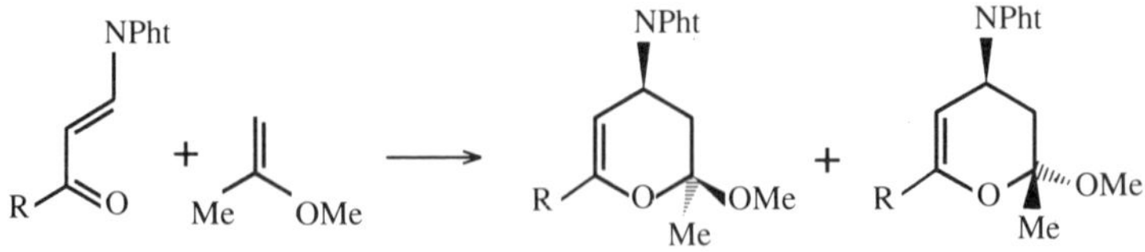

Scheme 4.1 Me = methyl, NPht = phthalimido, and R = CF_3.

at high pressures are less severe in experiments such as that shown in Fig. 4.18 because the molecular surroundings are not significantly changed during the reaction in a dilute solution at constant pressure and temperature. The spectra shown in Fig. 4.18 were recorded using the experimental set-up shown in Fig. 4.6. The entire reaction mixture is thus obtainable after the experiment, and so the product ratio of the two diastereomers **3** and **4** can be found from independent experiments, for example, from high-pressure liquid chromatography (HPLC). The spectroscopic study yields the overall rate coefficient, *k*, which equals the sum of the rate coefficients for the two reaction pathways leading to **3** and **4**. The HPLC analysis, on the other hand, yields the ratio of **3** and **4** in the product mixture, which, in kinetically controlled reactions, equals the ratio of the rate coefficients for the processes to **3** and to **4**. Combining the i.r.-spectroscopy data with the results of the HPLC analysis thus yields the two individual rate coefficients for reaction to the two diastereomers. The kinetics and selectivity of a wide variety of *hetero*-Diels–Alder reactions have been studied along these lines in a common research project with the group of Tietze.[28]

Since pulsed laser radiation may be easily introduced into the optical cells, some of the devices described above are well suited for kinetic studies that are highly resolved as a function of time. In free-radical polymerization, the use of i.r. and n.i.r. spectroscopy for the measurement of concentration, in conjunction with excimer lasers used for initiating the reaction, has proven to be a very effective method for measuring the rate coefficients of propagation and termination up to high pressures.[3] In these experiments, the monomer and polymer concentrations are measured, with a time resolution of a few microseconds, after inducing macromolecular growth by a single laser pulse with a duration of about 20 ns. With the optical cells presently available, and with further enhancement of the time resolution, very detailed kinetic studies may be performed on a wide variety of fluid systems up to high pressures and temperatures.

Acknowledgements

The studies carried out in the laboratories of the University of Göttingen and described in this article were generously supported by the *Deutsche Forschungsgemeinschaft* (Graduiertenkolleg 'Kinetik und Selektivität

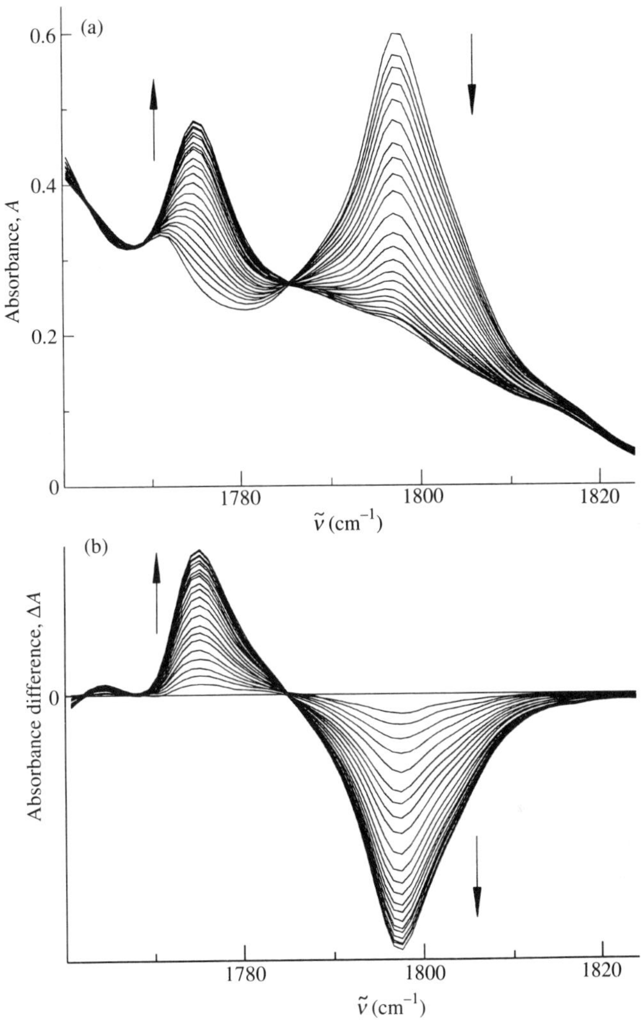

Fig. 4.18 (a) Infrared absorbance spectra measured in the carbonyl-stretching-mode region during the reaction of the phthalimidobutenone **1** with propenylmethylether **2** at a temperature of 50°C and a pressure of 500 bar. (b) Infrared absorbance difference spectra, derived from the absorbance spectra in (a) (see text).

Chemischer Prozesse in verdichteter fluider Phase' and SFB 93) and by the *Fonds der Chemischen Industrie*. C. H. is grateful for a fellowship from the *Deutscher Akademischer Austauschdienst*. The substantial support from U. Vriesen-Lippke, Dr H.-P. Vögele, and by Dr M. Busch in preparing this article is gratefully acknowledged.

References

1. (a) Asano, T.; le Noble, W. J. *Chem. Rev.* **1978**, *78*, 407–489. (b) van Eldik, R.; Asano, T.; le Noble, W. J. *Chem. Rev.* **1989**, *89*, 549–688.
2. Buback, M. Kinetics and selectivity of chemical processes in fluid phases. In: *Supercritical Fluids: Fundamentals for application*, Kiran, E.; Levelt-Sengers, J. M. H., eds. Kluwer: **1994**.
3. Buback, M. *Angew. Chem. Int. Ed. Engl.* **1991**, 641–653.
4. Isaacs, N. S. *Liquid Phase High Pressure Chemistry*; Wiley: Chichester, **1981**.
5. Sherman, W. F.; Stadtmuller, A. A. *Experimental Techniques in High-Pressure Research*; Wiley: Chichester, **1987**.
6. Franck, E. U.; Roth, K. *Discuss, Faraday Soc.* **1967**, *43*, 108–114. Roth, K., Ph.D. thesis, Karlsruhe, **1967**.
7. Buback, M. *Phys. Chem. Earth* **1981**, *13/14*, 345–360.
8. Ferraro, J. R.; Basile, L. J. *Appl. Spectrosc.* **1974**, *28,* 505–517.
9. Poulter, T. C. *Phys. Rev.* **1932**, *40*, 860–869.
10. Wegst, C. W. *Stahlschlüssel*, Verlag Stahlschlüssel: Marbach, **1986**.
11. Le Noble, W. J.; Schlott, R. *Rev. Sci. Instrum.* **1976**, *47*, 770.
12. Buback, M. *Z. Naturforsch*, A, **1977**, *32*, 1295–1300.
13. Tödheide, K. Water at high temperatures and pressures. In: *Water—A comprehensive treatise*; Franks, F., ed., Plenum Press: New York, **1972**, Vol. 1, Chapter 13. Lindner, H. Ph. D. thesis, Karlsruhe, **1970**. Schulz, K. Ph. D. thesis, Karlsruhe, **1974**.
14. Eckel, R.; Buback, M.; Strobl, G. R. *Coll. Polym. Sci.* **1981**, *259*, 326–334.
15. Buback, M.; Busch, M.; Lovis, K.; Mähling, F.-O. *Chem. Ing. Tech.* **1994**, *66*, 510–513. Busch, M. Ph.D. thesis, Göttingen, **1993**.
16. Buback, M.; Busch, M.; Panten, K.; Vögele, H.-P. *Chem. Ing. Tech.* **1992**, *64*, 352–354.
17. Buback, M.; Schweer, J.; Tups, H. *Ber. Bunsenges. Phys. Chem.* **1985**, *89*, 545–550.
18. Buback, M.; Schweer, J.; Tups, H. *Z. Naturforsch.* A, **1986**, *41*, 505–518.
19. Buback, M.; Nees, F. W. *Ber. Bunsenges. Phys. Chem.* **1976**, *80*, 650–656. Nees, F. W.; Buback, M. *Z. Naturforsch.* A, **1976**, *31*, 1690–1695. Bohn, M. A.; Buback, M.; Heym, I. *High Temp.—High Press.* **1977**, *9*, 585–586.
20. Buback, M. *Z. Naturforsch.* A, **1984**, *39*, 399–411.
21. Buback, M.; Degener, B. *Makromol. Chem.* **1993**, *194,* 2875–2883.
22. Beuermann, S. Ph.D. thesis, Göttingen, **1993**.
23. Bergert, U. Ph.D. thesis, Göttingen, **1994**.
24. Brackemann, H.; Buback, M. *Makromol. Chem.* **1989**, *190*, 3073–3084.
25. Buback, M.; Panten, K. *Makromol. Chem.* **1993**, *194*, 2471–2481.
26. Buback, M.; Huckestein, B.; Kuchta, F.-D.; Russell, G. T.; Schmid, E. *Makromol. Chem.* **1994**, *195*, 2117–2140.
27. Kuchta, G. Ph.D. thesis, Göttingen, **1995**.
28. Tietze, L. F.; Hübsch, T.; Voss, E.; Buback, M.; Tost, W. *J. Am. Chem. Soc.* **1988**, *110*, 4065–4066. Buback, M.; Tost, W.; Hübsch, T.; Voss, E.; Tietze, L. F. *Chem. Ber.* **1989**, *122*, 1179–1186. Tietze, L. F.; Hübsch, T.; Oelze, J.; Ott, C.; Tost, W.; Wörner, G.; Buback, M. *Chem. Ber.* **1992**, *125*, 2249–2258. Gerke, K. Ph.D. thesis, Göttingen, **1993**.

2. NMR Measurements

Lothar Helm, André E. Merbach, and D. Hugh Powell

2.1. Introduction

Nuclear magnetic resonance (NMR) spectroscopy has proved to be a versatile and powerful experimental technique for structural and dynamical studies in chemistry and physics. Although the concentration of species, the temperature, and the magnetic field are often varied in NMR studies, the pressure is normally left constant at its ambient value. This is mainly because apparatus must be constructed specially for high-pressure measurements since it is not commercially available. None the less, there is increasing interest in the development of high-pressure NMR instrumentation.

There are many reasons for performing NMR experiments at high pressures. In an experiment where the temperature is varied and the pressure is constant, the density of a liquid also changes. Due to the close packing of molecules in a liquid, even a small change in the density can produce a considerable change in the molecular dynamics of the system. It is only by using both pressure and temperature as variables in an NMR experiment that the effects of density and temperature on molecular motion can be separated. Furthermore, high pressures can be used to extend the temperature range of measurements on liquids below the freezing point (in the case of water) or above the boiling point, as in the study of supercritical dense fluids.

The activation volume, obtained from the pressure dependence of the reaction rate, has become a well-established, and often decisive, criterion for the determination of the reaction mechanism, so that variable-pressure NMR has also become an important tool in the understanding of chemical reactions. High gas pressures can be used to produce increased concentrations of a gas in solution, leading to faster reaction rates, advantageous shifts in chemical equilibria, or even entirely new chemical species. High-pressure NMR is an obvious technique for the study of such reactions. Nuclear magnetic resonance has also proven to be a powerful technique in solid-state physics, where variable-pressure experiments can be used to study phase transitions, transport phenomena, and electronic properties.

The first high-pressure NMR studies were performed by Benedek and Purcell in the mid nineteen-fifties.[1] They discovered that the beryllium–copper alloy Berylco-25 is nonmagnetic and that it is very suitable for high-pressure equipment for magnetic studies. Their pressurizable NMR probe (see Fig. 4.19) was similar to designs used today, and it was used at pressures up to 1 GPa (10 000 atm) to study the variation with viscosity of proton relaxation times and self-diffusion constants in a series of liquids. Since then a variety of systems for variable-pressure experiments have been developed. Three basic approaches have been used:

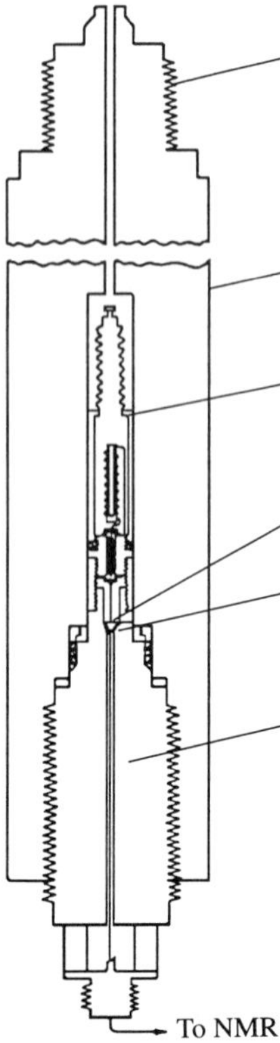

Fig. 4.19 The pressurizable NMR probe designed by Benedek and Purcell.[1] (Reprinted with permission from *J. Chem. Phys.* **1954**, *22*, 2003 © American Institute of Physics.)

(1) pressurizable glass[2,3] and sapphire[4,5] tubes that can be used with commercial NMR probes;

(2) metallic high-pressure NMR probes, in which the sample container and the radio frequency (RF) coil are surrounded by a pressure-transmitting liquid (or gas);[1,6,7]

(3) NMR probes incorporating diamond anvil cells (DACs) for very high pressures in low-resolution solid-state NMR.[8,9]

Figure 4.20 illustrates the differences between the first two techniques. The main advantage of pressurizable glass tubes is that they can be used with commercial, highly optimized NMR probes, so that sample spinning can be used to improve the resolution, and the highest available NMR frequencies can be employed without special problems. The main disadvantage is that pressures of 100 MPa or more require very thick-walled tubes in which the sample is confined to a capillary; this results in a low filling factor and thus poor sensitivity. Furthermore, due to their thick walls, the temperature con-

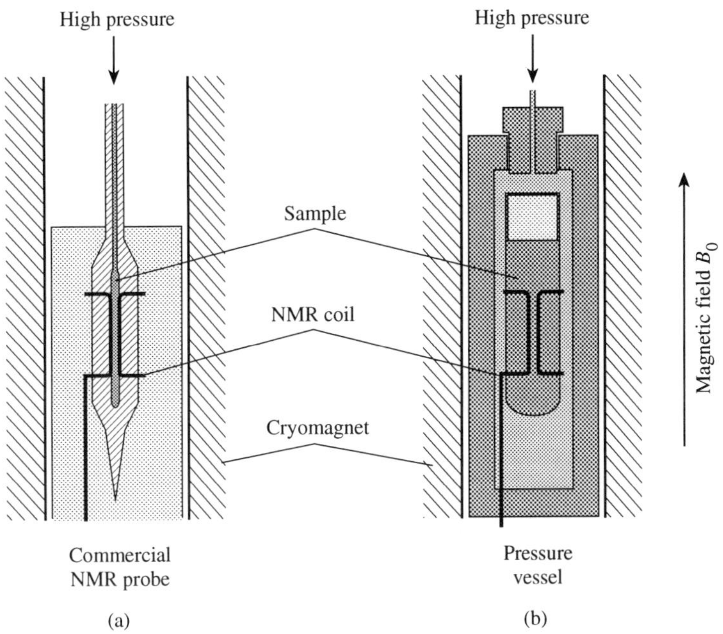

Fig. 4.20 A schematic representation of two different approaches to high-pressure NMR: (a) a pressurizable sample tube, and (b) a dedicated pressurizable NMR probe.

trol of such tubes is not very good. The use of relatively thin-walled tubes manufactured from single crystals of sapphire largely removes these disadvantages. The bursting pressure of these tubes is very variable however, and the maximum pressure that can be used safely is about 13 MPa.[4]

In high-pressure NMR probes, the filling factor of the RF coil can be optimized by mounting the coil on a support placed closely around the sample tube, so that the sensitivity is much higher than for thick-walled glass tubes. Since the sample tube is surrounded by the pressure-transmitting liquid, thermal contact is good, so the sample temperature can be regulated precisely, with a minimal temperature gradient over the sample volume. Another advantage is that probes can be designed to meet specific experimental requirements, such as very high or low temperatures. In designs published up to the present, sample spinning cannot be used. However, with modern superconducting cryomagnets, good field homogeneity over a small volume can be obtained and in a recent high-pressure NMR probe line widths of less than 1 Hz were obtained routinely at 400 MHz using ^1H (that is, the absolute resolution was better than 2.5×10^{-9}).[10] A critical point in the design of such probes is the feedthrough of the RF to the NMR coil; this becomes more problematic as the frequency increases, and the designs published up to now operate with NMR frequencies up to only 400 MHz.

Extremely high pressures can be produced using DAC technology, which has been widely used in solid-state physics. Nuclear-magnetic-resonance probes incorporating DACs have been constructed; a recently published design is capable of producing pressures up to 8.3 GPa.[8] The main disadvantages of the technique are the very small sample volume, leading to a very low sensitivity, and the need for metal gaskets, which shield the sample from the applied RF, and which can disturb the NMR experiment.

In the following we will describe the different technological options available to experimentalists wishing to perform high-pressure NMR experiments. We will consider the different criteria involved in the design of high-pressure equipment for experiments on liquids, gases (liquefied or dissolved), and solids, and we will describe systems designed for various fields of study and give examples of their application.

2.2 High-pressure NMR for liquids

2.2.1 General considerations

In chemistry and biology, NMR is mainly applied to liquid samples. Most experiments consist of the recording of one-dimensional and higher-dimensional spectra and the measurement of longitudinal and transverse relaxation rates. The quality of the NMR results is mainly influenced by the frequency resolution and by the signal-to-noise ratio of the spectra obtained. It is usually necessary, therefore, to pay particular attention to the optimization of these parameters when constructing a high-pressure NMR probe for liquid samples. Whereas a pressure range up to 1000 MPa is sufficient for almost all these applications, temperatures as low as −70 °C may be required. While such general criteria should always be borne in mind, it is evident that the design of an NMR probe for high-pressure experiments on liquids will depend on the intended area of application.

We will restrict ourselves here to high-pressure NMR probes built for modern superconducting cryomagnets. Nonsuperconducting electromagnets perform less well and are now of only marginal utility. Thick-walled, pressurizable glass tubes for use in commercial NMR probes have been described adequately elsewhere.[2,3] A schematic general design for a high pressure, high-resolution NMR probe for a modern NMR cryomagnet is given in Fig. 4.21. The most important parts are the high-pressure bomb, with an NMR insert holding the RF coil and the sample container, and a unit containing the RF tuning and matching electronics. The high-pressure bomb must be constructed using a nonmagnetic material capable of withstanding high pressures: the two metal alloys Berylco-25 and titanium IMI-680 are proven to have highly suitable magnetic and mechanical properties (see Table 4.1). The bomb usually consists of a cylinder closed by two plugs, at the top and at the bottom. The connection for the pressure-transmitting liquid or gas and the electrical feedthroughs for radiofrequency, (RF), and temperature measurement are

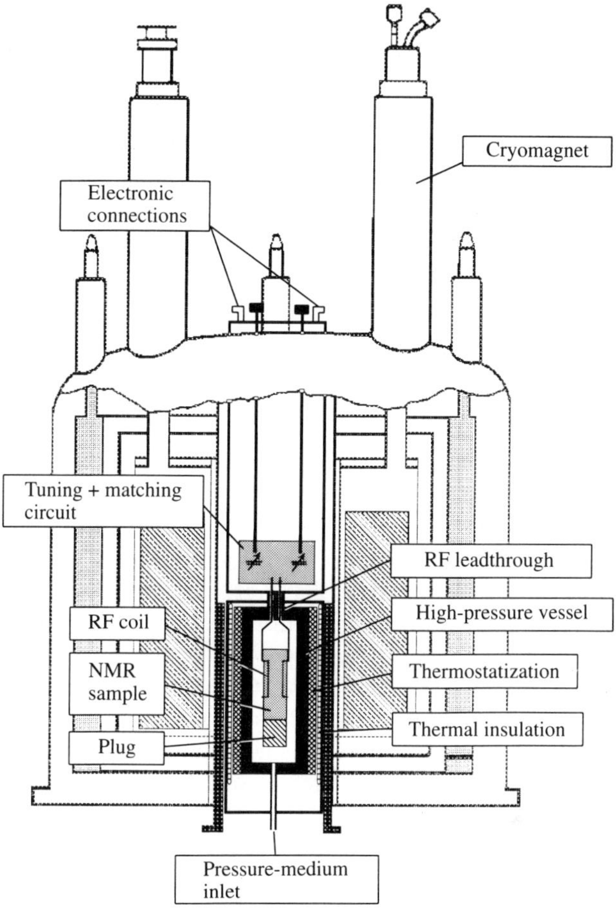

Fig. 4.21 A schematic general design of a high-pressure, high-resolution NMR probe for a modern NMR cryomagnet.

built into these plugs. The design and the quality of the RF feedthrough determines the maximum resonance frequency which can be obtained, and it can also be an important source of signal loss. The choice of pressure-transmitting liquid or gas depends on the experiments to be performed because it must not contain any of the nuclei that are to be observed by the NMR.

The NMR insert serves as a support for the RF coil and the NMR sample

Table 4.1 Nonmagnetic materials for high-pressure NMR

	Berylco-25	Titanium alloy IMI-680
Tensile strength (kg mm^{-2})	120–125	127

container. A temperature-measuring device, such as a Pt-100 resistor or a thermocouple, is often built into the NMR insert (note that these devices change their properties with pressure and so they have to be calibrated[11]). The sample container used inside the high-pressure probe must separate the liquid sample from the pressure-transmitting medium while allowing pressure transmission. In order to ensure a good filling factor for the RF coil, and hence the optimum sensitivity, a thin-walled container such as a standard NMR sample tube should be used. Since the sample to be studied must often be prepared and handled in an oxygen-free or dry atmosphere, it should be possible to close the sample easily and in an airtight fashion in a glove box. The dead volume of the sample should be minimized if the samples are expensive or if they are only available in small amounts, for example, isotopically enriched samples.

Depending on the required specifications, a variety of NMR coils may be used; they are usually of the Helmholtz (saddle) type but may sometimes be of the solenoid type. Test experiments using toroidal coils indicate that this design may be especially suitable for the high-pressure NMR of low-sensitivity nuclei (see Fig. 4.22).[12] Compared to Helmholz coils with similar dimensions, toroidal coils were shown to have greater sensitivity, lower magnetic coupling with the metallic pressure bomb, and good resolutions (1 Hz at 75.6 MHz for ^{2}H, that is, 1.3×10^{-8}). A particular feature of the toroidal-coil system is that it is not possible to insert a sample container into the coil because of the coil geometry: the coil is immersed directly in the sample. This means that the sample volume must be large, and that reaction with the pressure vessel may be a problem, but such a system may have advantages in the study of supercritical fluids.

For the best NMR performance, the RF tuning and matching electronics should be sited as close as possible to the RF coil. The use of nonmagnetic high-quality capacitors is strongly recommended. For ease of use and maximum performance, it should be possible to adjust the resonance frequency (tuning) and the impedance (matching) of the probe while it is inside the NMR magnet.

Finally, it should be mentioned that, if very low or very high temperatures are to be used, the (expensive) cryomagnet must be protected, either by simple thermal isolation of the pressure bomb or in extreme cases by cooling or heating the outside of the probe, for example, by circulating water.

2.2.2 Pressure-generating system

The type of pressure-generating system normally used for pressures up to 5 kbar is well described in literature.[13,14] A typical design, used for NMR experiments and with a liquid pressure-transmitting medium is shown in Fig. 4.23.

Two pressure-transmitting fluids are normally used. A mixture of motor oil and kerosene is used from the pump up to a separator. The fluid used after

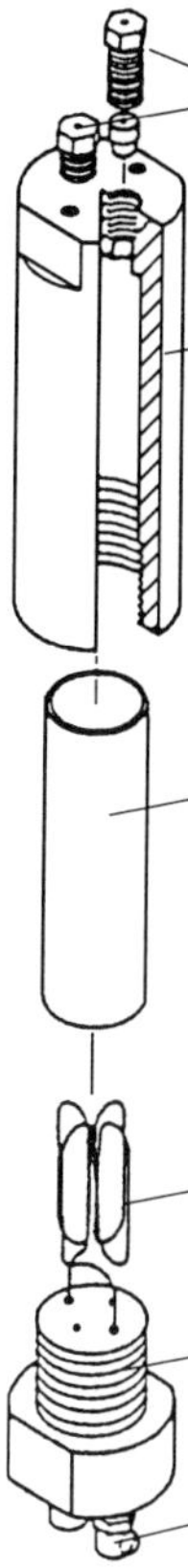

Fig. 4.22 A schematic of a toroidal detection probe, as used by Rathke.[12] (Reprinted with permission from *J. Magnet. Res.*, **85**, 150, (1989).)

the separator and that entering the NMR probe should not contain any of the nuclei which are to be observed in the NMR experiment. For ^1H NMR, various fully halogenated hydrocarbons (for example, C_2Cl_4) or CS_2 are widely used. A liquid, rather than a gas, is preferred as the pressurizing medium because of the lower leakage and safety problems. At very high temperatures, however, corrosion may be a problem with liquids so that inert gases may have to be used.

The pressure is measured by means of a Heise–Bourdon pressure gauge. The components of the pressure-generating system (hand pumps, high-pressure tubing, valves, gauges, high-pressure separators) are available from several commercial sources.

2.2.3 Probes for kinetic measurements

Several-high pressure NMR probes have been developed in our laboratory since 1978 for the study of the pressure dependence of the kinetics of chemical

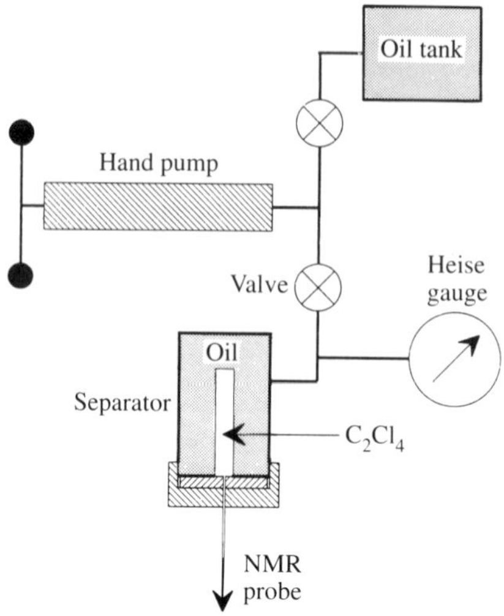

Fig. 4.23 A schematic representation of a pressure-generating system for hydrostatic pressures up to 200 MPa.

reactions. The first probes were designed for use with a 1.4 T electromagnet.[15] The more recent probes fit into wide-bore or normal-bore cryomagnets, and they have the following general features:

(1) usable in the medium-pressure range (up to ~ 200 MPa);
(2) high-frequency resolution of the NMR spectra (1 Hz or less);
(3) good temperature stability and small temperature gradient inside the probe (± 0.2 °C);
(4) multinuclear capability.

Two designs of probes developed in our laboratory will be presented as typical examples. Both probes are designed for working pressures up to 200 MPa and the temperature range −40 to 150 °C.

The first probe (Fig. 4.24) fits into a wide-bore (diameter 89 mm) cryomagnet and works up to frequencies of 400 MHz, corresponding to a ^1H resonance in a 9.4 T magnetic field.[10] The probe is constructed in a modular way; this allows the RF coil and the tuning/matching electronics to be easily changed, for optimum performance with different observed nuclei, such as ^1H, ^{13}C, ^{31}P, or ^{17}O (see Fig. 4.24(a)). A ^2H field lock can be used with ^1H observation for improved long-term stability and easier shimming. The Vespel (Du Pont de Nemours) NMR insert (Fig. 4.24(c)) fits inside the pressure bomb (Fig. 4.24(b)). The sample tube is a 5 mm outer diameter (o.d.) glass NMR tube cut to the correct length and closed with a movable seal, consisting of a piston and a cap made of machinable glass, which allows pressure transmission to the sample. The sample volume is about 1 cm^3. The sample tube fits into the 7 mm o.d.

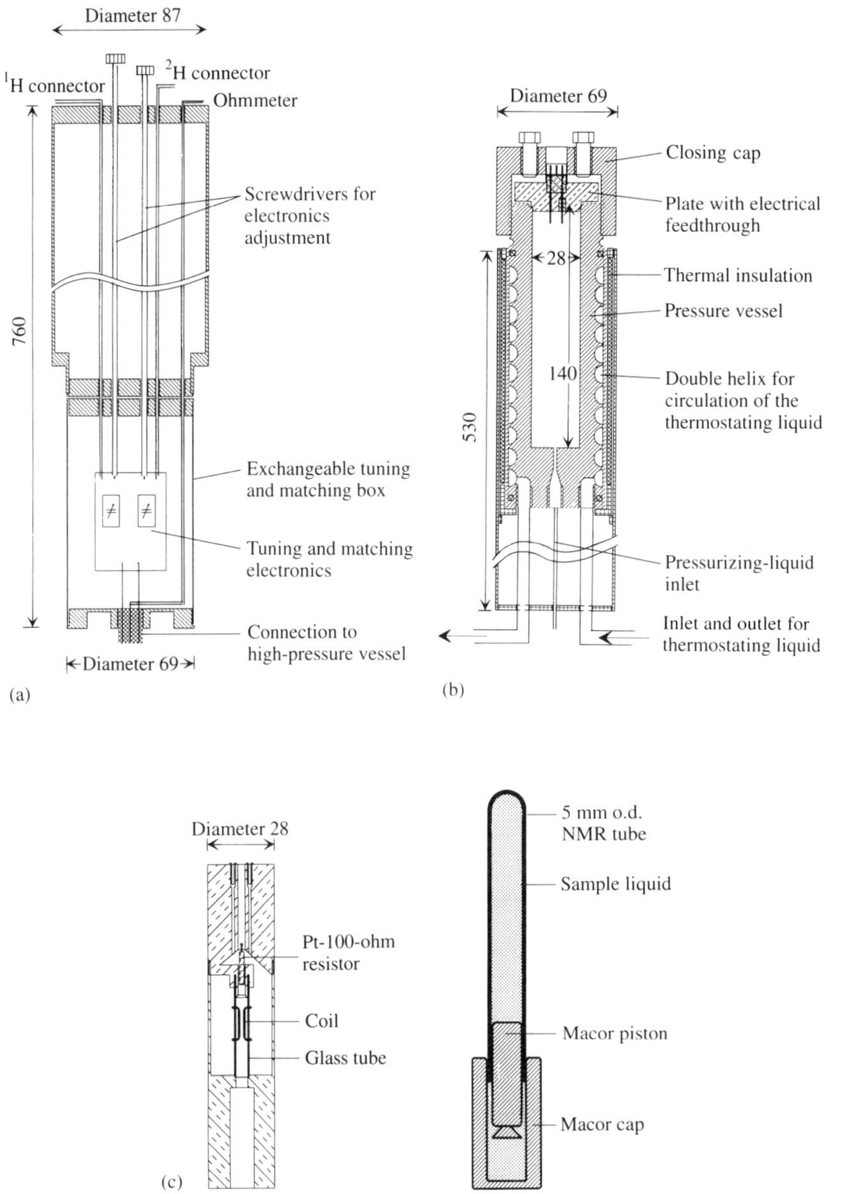

Fig. 4.24 A schematic drawing of a wide-bore high-pressure, high-resolution multinuclear probe used for frequencies up to 400 MHz: (a) the upper part, including the tuning and matching electronics; (b) the high-pressure vessel; and (c) the NMR insert and sample tube. (The dimensions are all in millimetres.)[10] (Reprinted with permission from *High Press. Res.*, **2**, 237, (1990).)

glass tubing on to which the RF coil is wound in a saddle-shaped form. A platinum resistor is placed just above the sample tube for temperature measurement. The temperature is regulated by pumping liquid from a thermostated bath through the double helix cut into the outside of the bomb. The temperature gradient inside the bomb between −40 and 90 °C is found to be less than ±0.2 °C over the sample region. Thermal equilibration after a pressure change is achieved within 15 min, and the temperature stability is better than ± 0.2 °C over the whole pressure range.

The second probe (Fig. 4.25) is the first high-pressure, high-resolution

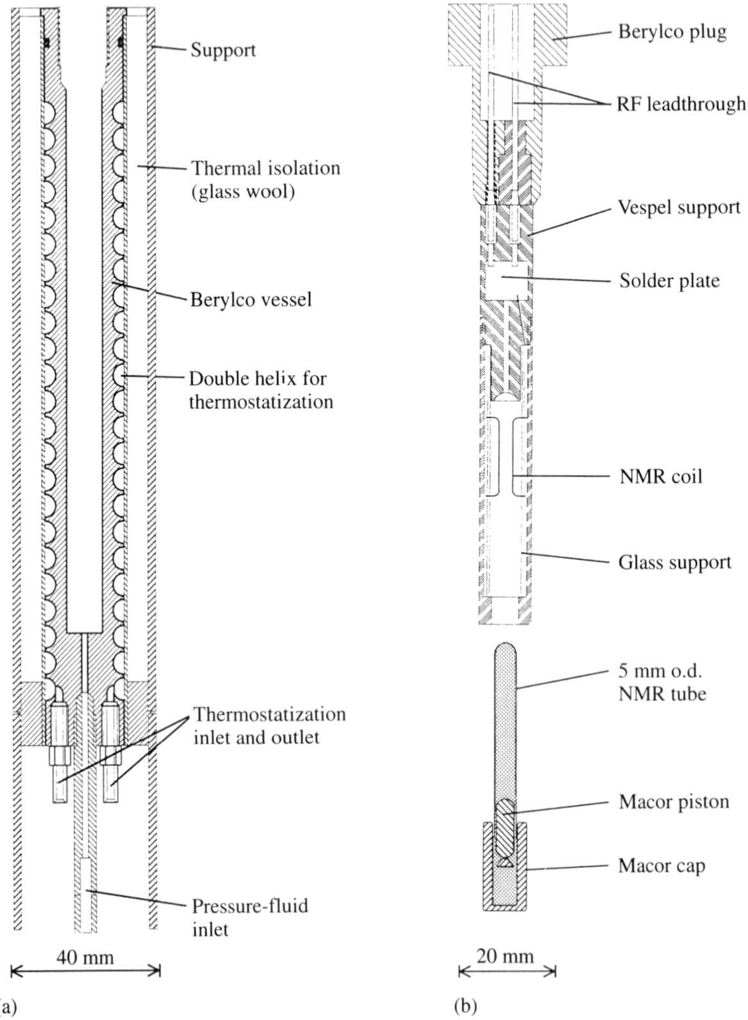

Fig. 4.25 (a) The high-pressure vessel, and (b) the NMR insert with a sample tube of a high-pressure, high-resolution NMR probe, fitting into a narrow-bore magnet[29]. (Reprinted with permission from *High Pres. Res.*, **12**, 285, (1994).)

NMR probe fitting into a normal-bore (40 mm diameter) cryomagnet. It works at frequencies up to 200 MHz (for a ^{1}H resonance in a 4.7 T magnetic field) and allows the use of a ^{2}H lock.[29] This probe is lighter than the wide-bore model, and it is therefore much more convenient to handle. Furthermore, it opens up the possibility that high-pressure probes could be developed for use with very-high-field magnets, which are currently only available with normal bores. The pressure bomb is machined from Berylco and has a minimum o.d. of 16 mm, an inner cavity of diameter 11 mm, and a length of 140 mm. A double helix is cut at the outside of the vessel for circulation of the thermostating liquid. Two NMR inserts (Fig. 4.25(b)), with NMR coils with two and four turns have been built to obtain the optimum sensitivity at different observation frequencies. The sample container has the same design as that described above for Fig. 4.24.

During the last two decades the activation volume, $\Delta V^{\ddagger}$, determined from the pressure dependence of the reaction rate, has become the primary criterion for the assignment of mechanisms of substitution reactions of the general type of eqn (4.2.1), where X is the leaving and Y the entering ligand:

$$ML_nX + Y \rightarrow ML_nY + X. \tag{4.2.1}$$

Langford and Gray[16] have proposed a system of classification of substitution-reaction mechanisms based on operational criteria. If kinetic tests show the presence of an intermediate with an increased coordination number, an associative mechanism (A) is assigned. Conversely, if an intermediate with a reduced coordination number is detected, a dissociative (D) mechanism is assigned. When no intermediate can be found, a concerted interchange (I) mechanism is assigned. This last category may be subdivided further, depending whether evidence is found for important incoming group influences (I_a) or not (I_d). If one neglects electrostrictive effects, then $\Delta V^{\ddagger}$ (the difference between the partial molar volumes of the transition state and the reactants), will be a direct measure of the degree of associativity or dissociativity at the transition state.[17] Dominant-bond breaking in a dissociatively activated reaction results in a positive value for $\Delta V^{\ddagger}$, while dominant-bond making in an associatively activated reaction gives a negative value for $\Delta V^{\ddagger}$.

A particularly illustrative example is solvent exchange on Be(II).[17] The exchange rate can be determined from the ^{1}H spectra of Be(II) solutions: at low temperatures the signals of any free solvent and any solvent bound to Be(II) are resolved, but at high temperatures the signals coalesce due to solvent exchange. The effect of pressure on the spectra for dimethyl sulphoxide (DMSO) and tetramethylurea (TMU) solvents is shown in Fig. 4.26. It is evident that the rate increases with pressure for DMSO; this corresponds to a negative activation volume ($\Delta V^{\ddagger} = -2.5$ cm^3 mol^{-1}) and an associatively activated exchange process. For TMU, the rate decreases with pressure, so that the activation volume is positive ($\Delta V^{\ddagger} = +10.5$ cm^3 mol^{-1}) and the exchange process must be dissociatively activated. This is a direct result of

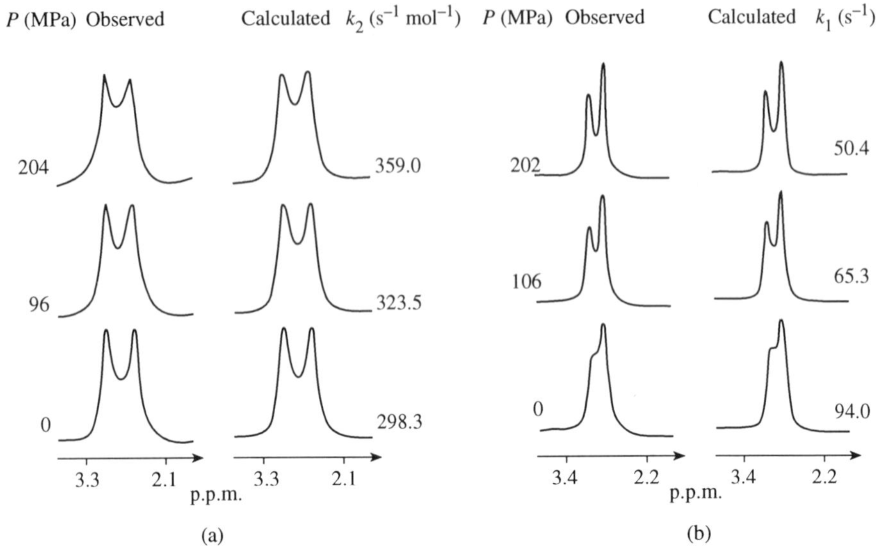

Fig. 4.26 Experimental and calculated pressure-dependent 200 MHZ ^{1}H NMR spectra[17] for BeS_4 in CD_3NO_2, where: (a) S is DMSO at a temperature of $T = 300$ K, and (b) S is TMU at a temperature of $T = 371$ K.

the greater bulkiness of the TMU ligand, which prevents the formation of a five-coordinate transition state.

2.2.4 High-resolution high-pressure probes for physicochemical studies

Pressure variation is used in dynamical studies in physical chemistry to separate the effects of density and temperature. To achieve this, pressures up to 500 MPa are desirable. Jonas *et al.* have developed a high-pressure, high-resolution NMR probe working at a field of 7.05 T (300 MHz, ^{1}H) and pressures up to 500 MPa in the temperature range −50 to 150°C (see Fig. 4.27).[6] The pressure bomb is machined from Berylco-25, and the sample container is an 8 mm diameter glass NMR tube connected to a Teflon piston arrangement for pressure transmission. A particular feature of this probe is the design of the RF feedthroughs, which are capable of withstanding many pressure cycles up to 500 MPa. A resolution of 2 Hz for ^{1}H NMR (that is, 4×10^{-9}) can be obtained with this probe.

An example of pressure effects on the dynamic properties of liquids is given in the study by Hasha *et al.* of conformational isomerization in liquid cyclohexane.[18] In contrast to classical transition-state theory, stochastic models predict that for such reactions the transmission coefficient, κ, should depend on the *collision frequency* between the solvent and the solute molecules, which is a measure of the coupling of the reaction coordinate with the

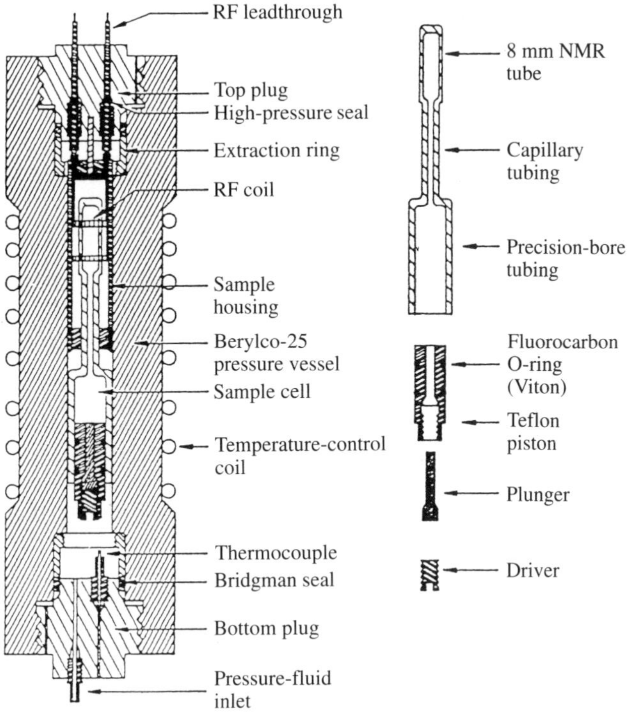

Fig. 4.27 A diagram of a high-resolution, high-pressure NMR probe operating in the pressure range 0.1–500 MPa.[6]

surrounding medium. Hasha *et al.* measured the rate of the isomerization reaction as a function of temperature and pressure for several solvents. The activation volume was found to have a large collisional contribution that was strongly pressure dependent. The correlation of κ with pressure and viscosity (see Fig. 4.28) is in good agreement with the stochastic models, and it was the first experimental proof of the predictions of these models for isomerization reactions in condensed phases.

2.2.5 High-pressure, high-temperature probes

In order to perform NMR studies on compressed supercritical steam, which exhibits a number of unusual properties, de Fries and Jonas developed an NMR probe working in the temperature range 25–700 °C and at pressures up to 200 MPa.[19] The design of this probe, consisting of two high-pressure bombs, is given in Fig. 4.29. Both bombs have a 76 mm outer diameter and a 38 mm inner diameter and are made of IMI-680 titanium alloy (Imperial Metal Industries). The primary bomb at the top contains a furnace, the RF coil and two thermocouples. The sample extends from the ceramic tube in

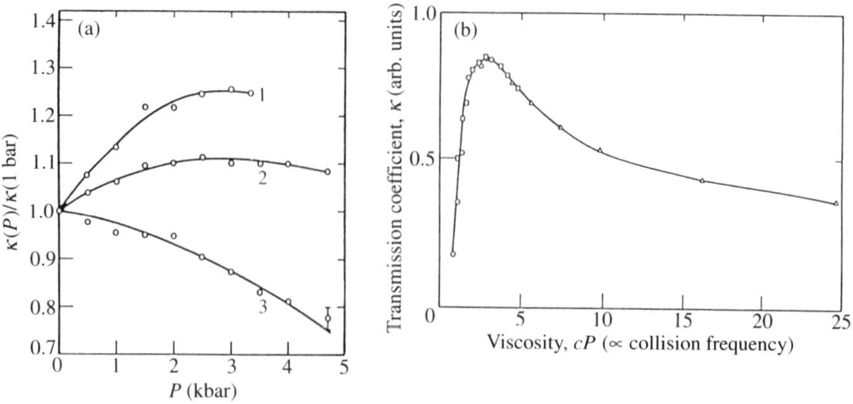

Fig. 4.28 (a) The reduced transmission coefficient, κ, as a function of pressure and (b) the relationship between the transmission coefficient, κ, and the viscosity, η, for conformational isomerization of cyclohexane[18] at 225 K in: (1) acetone-d_6, (2) carbon disulphide, and (3) methylcyclohexane-d_{14}. (Reprinted with permission from *The American Chemical Society*.)

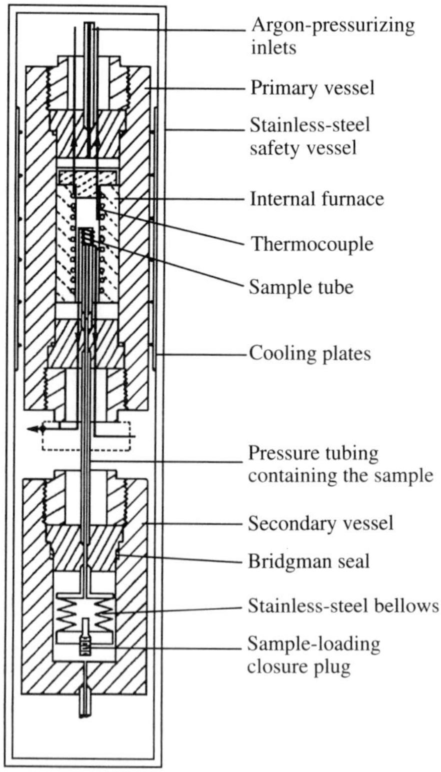

Fig. 4.29 A high-pressure NMR probe,[19] designed for temperatures up to 700 °C. (Reprinted with permission from *J. Magnet. Res.*, **35**, 11, (1979).)

the primary bomb via pressure tubing to the secondary bomb, which is at room temperature. A stainless-steel-bellows arrangement in this secondary bomb compensates for volume changes on superheating of the sample. The bombs are pressurized using argon gas, which enters at the far ends of each vessel. A major concern when working with high-temperature steam at high-pressures is corrosion, and the materials used must be chosen carefully. The two-bomb arrangement has the advantage of minimizing the components that are in contact with the steam. It also allows the high-temperature bomb to be made as small as possible—an important consideration, since the tensile strength of the bomb material decreases at high temperatures. The use of a furnace internal, rather than external, to the bomb makes it easier to make gastight feedthroughs for the RF, thermocouple, and furnace power connections, since these are located at the far end of the bomb, which is cooler than the interior of the furnace. The design of the furnace was of particular importance, because of the high power involved and because temperature and magnetic-field gradients had to be avoided. A resolution of the order of 8 Hz at 60 MHz ^{1}H (that is, 1.3×10^{-7}) was achieved with this system, independently of the power level delivered to the furnace. The pressure-leak rates of the entire set-up are as low as 70 kPa h^{-1} at 51.7 MPa and 500 °C.

The proton spin-lattice relaxation time, T_1, of water was measured using this probe as a function of pressure and temperature (see Fig. 4.30).[20] The study focused on the supercritical region (above T_c = 374 °C) where the spin-

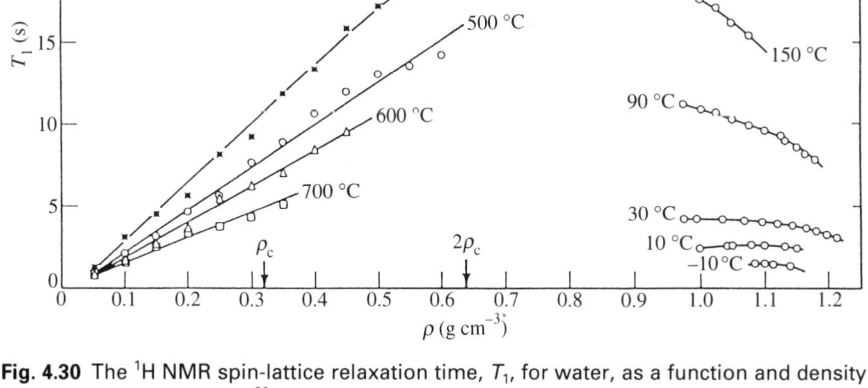

Fig. 4.30 The ^{1}H NMR spin-lattice relaxation time, T_1, for water, as a function and density for various temperatures.[20] (Reprinted with permission from *J. Chem. Phys.* **1981**, *74*, 913. © American Institute of Physics.)

rotation relaxation rate, $1/T_1^{SR}$ dominates the observed proton relaxation rates. The relaxation rates were analysed in terms of the angular-momentum correlation time, which was interpreted using a rough hard-sphere model, T_1^{SR} was found empirically to have a linear dependence on the logarithm of the shear viscosity in supercritical compressed water.

2.3 High-pressure NMR for gases

2.3.1 General considerations

Nuclear-magnetic-resonance apparatus used to study systems involving gases under pressure can be divided roughly into two classes, those working at pressures below 10 MPa, and those working at high pressures. Many gases liquify, or can be made to dissolve in liquids, at room temperature for pressures below 10 MPa. The most convenient way to study such systems is by using a pressurizable sample cell that can be used with a commercial NMR probe. Such cells are easy to handle, and, since they are generally transparent, they allow the liquefaction of the gas to be verified. A major drawback of such cells is that they can and do explode, and users must be protected adequately. For working pressures above 10 MPa, dedicated high-pressure probes are used. These allow safer operation than pressurizable tubes, but are less convenient, especially for occasional use.

If gases dissolved in liquids are to be studied, the sample must be agitated under pressure, because the diffusion across the gas–liquid interface is very slow if this is not the case. This is straightforward with a pressurizable sample tube, but it presents technical difficulties for the design of a pressurizable NMR probe for gases.

2.3.2 Pressurizable sapphire tubes

The first design for pressurizable NMR tubes made from sapphire was published by Roe.[4] The tubes are grown as single crystals of white sapphire (aluminium oxide) and can be obtained commercially (Tyco Saphikon Inc.). The high tensile strength of the single crystal means that the tubes can be made with relatively thin walls, so that the sample-volume problems associated with pressurizable glass sample tubes are avoided. In Roe's design, the sapphire tube is glued to a titanium-alloy flange with an epoxy adhesive (Fig. 4.31). A titanium-alloy valve assembly allowing pressurization is then screwed into the flange. The sample volume is about 0.3–0.4 cm^3 with 5 mm outer diameter tubes. The entire assembly weighs about 110 g and can be made to spin in a commercial NMR spectrometer. This design was found to work well in the temperature range -100 to 150°C, and a maximum safe working pressure of ~ 13 MPa was estimated. The resolution obtained on a 360 MHz instrument is about 1 Hz (that is, 2.8×10^{-9}) with some line-shape distortion.

Horvath and Ponce[21] published an improved, lighter design for the valve assembly that can be used with both 5 mm and 10 mm outer-diameter sapphire

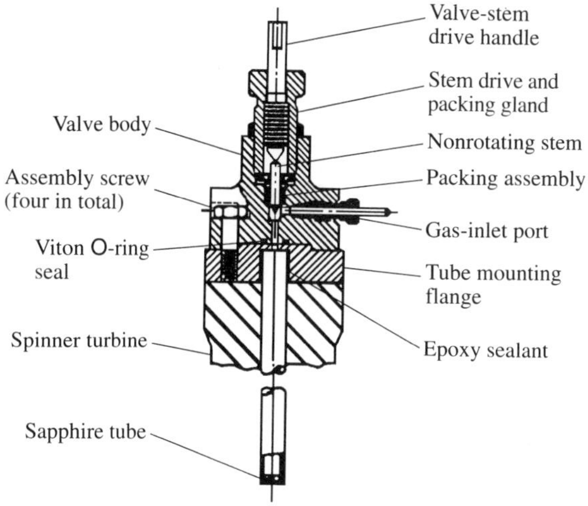

Fig. 4.31 A diagram of Roe's titanium-alloy-and-sapphire-tube assembly.[4] (Reprinted with permission from *J. Magnet. Res.*, **63**, 388, (1985).)

sample tubes. The 10 mm outer-diameter tube provides greater sensitivity, and it is light enough with the valve assembly to spin in most commercial NMR spectrometers. Further modifications have been made in our laboratory (Fig. 4.32). In particular, our valve assembly can be completely unscrewed from the tube for easier filling of solid material and for cleaning of the sapphire tube. The line widths obtained with our device and a 10 mm tube are less than 0.4 Hz at 200 MHz ^{1}H (that is, 2×10^{-9}) and spinning side bands are less than one per cent.

Safety considerations are of special importance in the use of these tubes, since their breaking pressure varies and may depend on microscopic faults in construction. We subjected our tubes to numerous pressure tests. In two cases the tube assemblies broke: in both cases the epoxy adhesive (Araldite AV 138M and a HV 998 hardener, from Ciba-Geigy) detached at about 50 MPa. These tubes have been in use for more than one year in several laboratories, with a maximum working pressure of 10 MPa. No accident has occurred to date. Nevertheless, the tubes are always used with a protection device (Fig. 4.33) developed in our laboratory that ensures that the user is never exposed directly to the tube while it is pressurized.

This system was used in the study of the catalytic dimerization of ethylene in aqueous Ru^{2+} solutions.[22] Three identical ^{17}O-enriched aqueous solutions of $[Ru(H_2O)_6]^{2+}$ in 10 mm outer diameter sapphire tubes were pressurized with ethylene at 6 MPa and mixed by shaking for different times. The chemical reactions were followed by ^{17}O, ^{13}C, and ^{1}H NMR, which showed the successive formation of $[Ru(CH_2{=}CH_2)(H_2O)_5]^{2+}$ and $[Ru(CH_2{=}CH_2)_2(H_2O)_4]^{2+}$. After shaking under ethylene pressure for three days, an organic phase

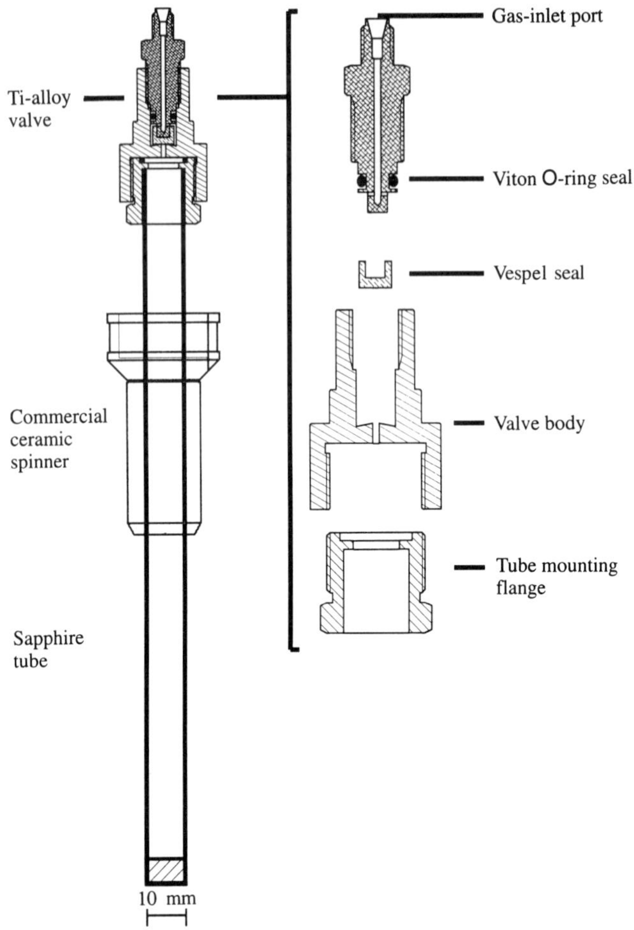

Fig. 4.32 The titanium-alloy-and-sapphire-tube assembly developed in Lausanne.

formed above the solution. The ^{13}C spectra of this new organic phase separated under pressure, showing the characteristic signals of butenes (Scheme 4.2). This aqueous catalytic dimerization reaction is not stereoselective, and all the possible butenes are formed except 2-methylpropene. These results show the utility of NMR studies under gas pressure for the study of catalytic reactions in aqueous solutions, which are increasingly attractive because of ecological considerations.

2.3.3 High-resolution pressurizable NMR probes for homogeneous-catalysis studies

A design for a high-pressure, high-resolution probe for NMR studies of

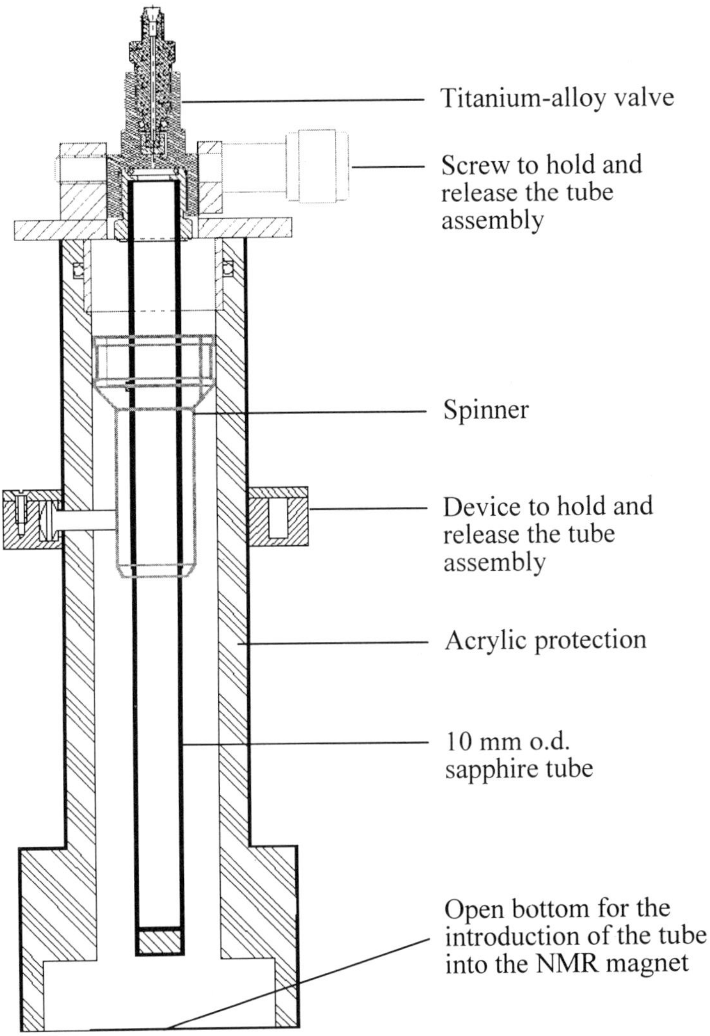

Fig. 4.33 The safety system designed in Lausanne for use with sapphire tubes.

homogeneous catalysis was published by van der Velde and Jonas (Fig. 4.34).[23] Temperatures between −50 and 150 °C are obtained by circulating a liquid through copper tubing coiled on a metal cylinder which is tightly fitted to the pressure vessel. An outer set of coils carry water to protect the magnet bore from extreme temperatures. The Berylco-25 pressure vessel (with an outer diameter of 6.35 cm, and an inner diameter of 3.18 cm), chosen for its resistance to hydrogen embrittlement and for its nonmagnetic properties, is closed at each end with a Berylco plug and driving nut. The vessel has a

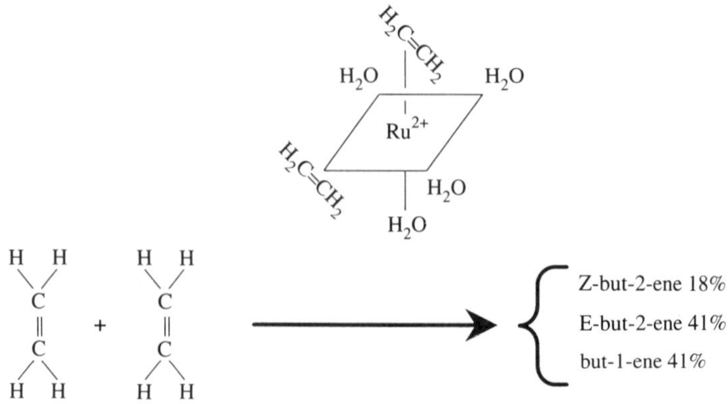

Scheme 4.2 The solution-phase dimerization of ethylene by Ru^{2+}.[22]

calculated bursting pressure of ~500 MPa and has been pressure tested to 200 MPa. The NMR insert is mounted on the lower closure plug. It contains an 8 mm sample tube and crossed Helmholtz coils: one for ^{13}C or ^{31}P, and one for ^{1}H observation.

The most innovative feature of the probe is a built-in stirring mechanism to

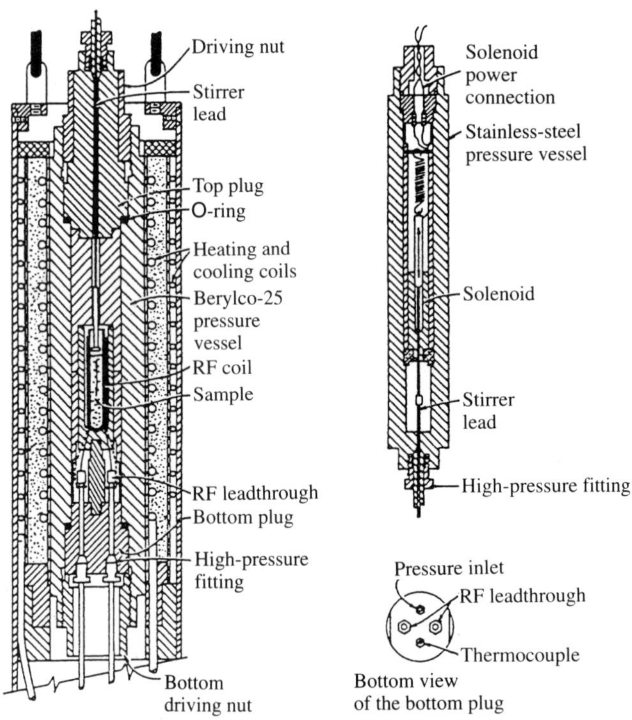

Fig. 4.34 A high-resolution probe for NMR studies of homogenous catalysis.[23]

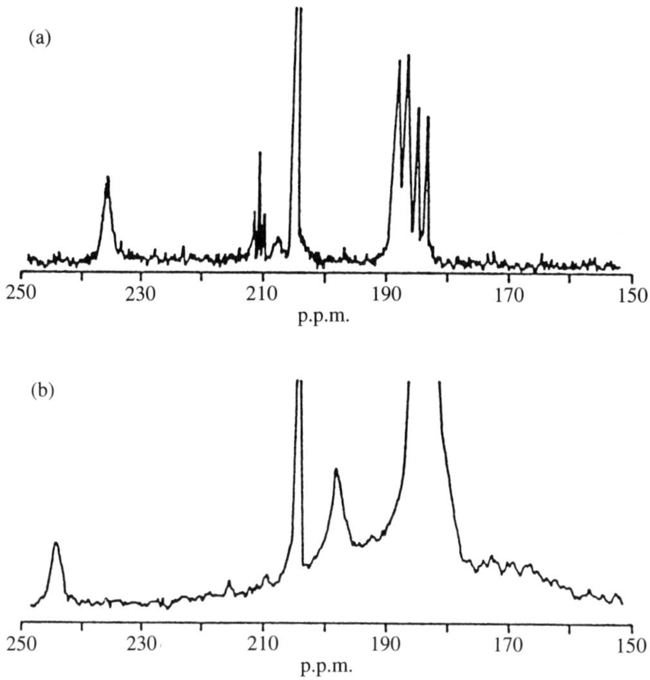

Fig. 4.35 (a) The spectrum of $[NMe_3Bz]_2[Rh_{12}(CO)_{30}]$ in acetone and, (b) the same sample after ^{13}CO pressurization and stirring. The peak at 184 p.p.m. is ^{13}CO gas; peaks at 245 p.p.m. and 198 p.p.m. are $[Rh_5(CO)_{15}]$; peaks at 215 p.p.m. and 209 p.p.m. are side products.[23]

mix catalyst solutions with reactant gases. The stirring assembly is contained in a separate pressure vessel located outside the magnet. It consists of a solenoid and a spring which retracts the stirring rod when the solenoid is off. The stirrer is connected to the probe by a 3 ft length of stainless-steel pressure tubing. A separate power supply activates the solenoid, moving the stirring rod up and down by 2.5 cm. An optimum resolution of 4 Hz in 180 MHz ^1H spectra (that is, 2×10^{-8}) was obtained.

This pressurizable probe was used to study the carbonylation of $[NMe_3Bz]_2[Rh_{12}(CO)_{30}]$ under CO pressure to form $[Rh_5(CO)_{15}]^-$. The ^{13}C spectrum of a 28 mM acetone solution of $[NMe_3Bz]_2[Rh_{12}(CO)_{30}]$ is shown in Fig. 4.35(a). After 4 h of ^{13}CO pressurization without stirring no reaction could be detected. Fifteen minutes of stirring produced the spectrum in Fig. 4.35(b) which shows quantitative conversion to $[Rh_5(CO)_{15}]^-$. These experiments clearly show the necessity for a mixing mechanism if reactions involving dissolved gases are to be studied in a dedicated high-pressure NMR probe. Recently, a probe with stirring capability fitting into a normal-bore cryomagnet was designed in our laboratories.[30]

2.4. High-pressure NMR for solids

2.4.1 General considerations

The design criteria for high-pressure probes for solid-state NMR are, in general, different to those for liquid- or gas-state NMR. On the one hand, more extreme conditions of temperature and pressure are often needed to study, for example, structural phase transitions. On the other hand, the NMR resonances in solids are normally extremely broad compared to those in liquids, so that high resolution is no longer a primary consideration. We present here two designs of probe that demonstrate these different criteria, and show state-of-the-art high-pressure solid-state NMR apparatus.

2.4.2 Low-temperature, high-pressure probes

A design for an NMR probe operating at pressures up to 700 MPa and at temperatures from 77 to 300 K was published by Huber *et al.* (Fig. 4.36).[24] The hydrostatic pressure is produced in a two-stage pressure generator (from Nova Swiss AG). Because of the very low temperatures used, the pressure is transmitted to the NMR probe using helium gas. The pressure vessel is machined from unannealed 1/1 hard Berylco-25 and contains a sample space which is 8 mm in diameter and 10 mm long. Before machining, the Berylco-25 rod was ultrasonically tested then annealed at 315 °C for 150 min. The RF solenoid coil is mounted on the lower closing plug. The tuning capacitor is mounted as close as possible to the RF coil, in order to optimize the quality

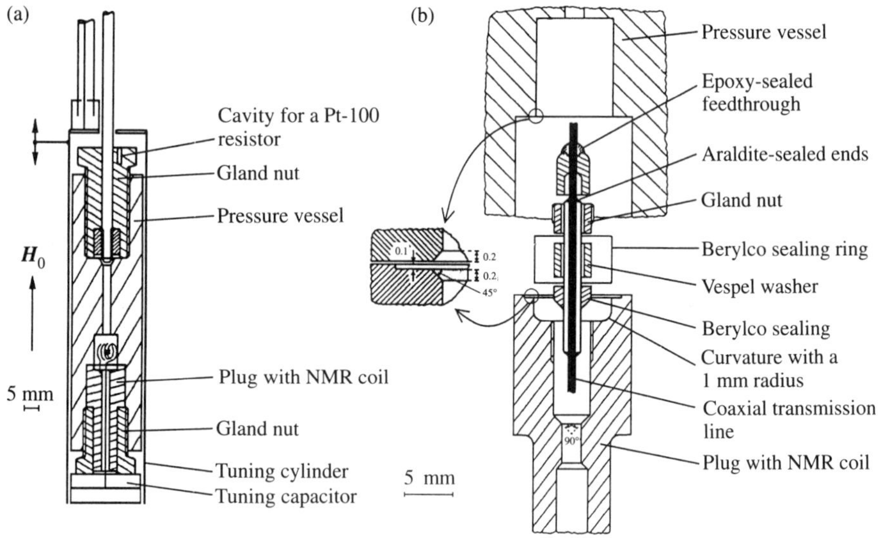

Fig. 4.36 A low-temperature, high-pressure probe for solid-state NMR: (a) an overall cross-sectional view and (b) details of the high-pressure seal and feedthrough.[24]

factor of the resonant circuit. Low temperatures are achieved by flowing cold nitrogen or helium gas around the probe, and they are measured with a platinum resistor. The difference in temperature at the platinum resistor and at the sample position is less than the temperature stability of the cryostat, that is, ± 0.2 K. The rate of pressure leakage is less than 10 MPa h^{-1} at 0.7 GPa, and the leakage is mainly from the pressure generator. A crucial point in the probe design is the high-pressure sealing plug (Fig. 4.36(b)). The seal is made by a 5 mm high, 0.25 mm wide, unannealed 1/1 hard Berylco-25 ring. Under internal pressure, the ring forms a ridge along the bevelled edges between the vessel and the plug. Once the ridge has formed, the ring can be removed and reused many times without loss of sealing performance. For high-pressure measurements at the lowest temperatures, it is essential first to generate the pressure and then to cool the vessel to the desired temperature, in order to avoid rupture of the ring. The 1 mm radius curved surface in the plug shown in Fig. 4.36(b) is essential: sharp edges lead to rupture of the plug.

This probe was used to study the isostructural superionic conductors $RbAg_4I_5$ and KAg_4I_5, where Ag^+ is the mobile ion.[24,25] These compounds exhibit two superionic high temperature phases, α and β, separated by a first-order phase transition at a temperature T_c and a low-temperature, low-conductivity phase, γ. The quadrupolar splitting of the ^{87}Rb, or the ^{39}K, NMR line can be used to determine the pressure-versus-temperature phase diagram of these compounds. The appearance of a new line in the NMR spectrum of $RbAg_4I_5$ at high pressures is indicative of a transition into a new phase with a lower symmetry, the δ-phase (Fig. 4.37(a)). Since the spin-lattice relaxation time increases drastically at the β–δ phase transition, the new phase is probably a low-conductivity phase. The β–δ and γ–δ phase boundaries were determined by measuring the spectra as a function of temperature and pressure (Fig. 4.37(b)).

2.4.3 Very-high-pressure probes

Extremely high static pressures (up 100 GPa) can be produced in, for example, X-ray diffraction and optical spectroscopy, using DACs.[26] The basic principle of a DAC is that pressure is generated in a hole drilled in a metal gasket, which is compressed between two opposed diamond anvils. The main difficulties of adapting a DAC for high-pressure NMR are the very small sample size and the need for metallic gaskets, which shield the sample from the applied RF field. Only studies of nuclei with very high NMR sensitivities can be envisaged. None the less, designs of high-pressure NMR probes incorporating DACs have recently been published by several research groups:[8,9,27] we present here the design of Bertani *et al.*[8]

Figure 4.38 gives a schematic overview of an NMR DAC apparatus designed for operation up to 10 GPa. The direction of the force on the diamonds was chosen to be perpendicular to the static magnetic field, in order to retain the option of generating an RF field perpendicular to the static mag-

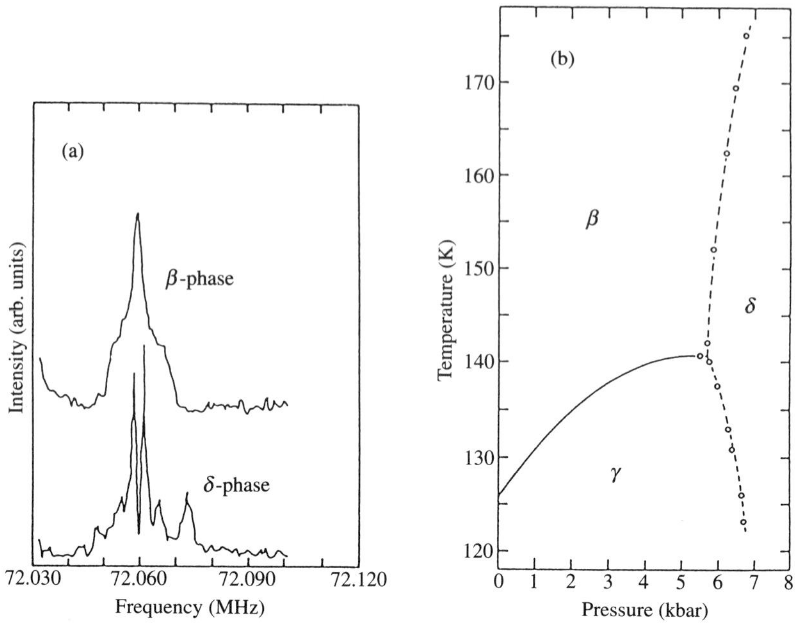

Fig. 4.37 (a) The NMR central-line spectra of ^{87}Rb in the β-phase and δ-phase of $RbAg_4I_5$ at 169.5 K and pressures of 640 and 650 MPa, respectively; and (b) the temperature-versus-pressure phase diagram of $RbAg_4I_5$.[24]

netic field, using either a coil or the gaskets themselves as the RF resonator. Such a coil could be wound simply around the diamond cones; it would have a good filling factor, but it would not work with metal gaskets. Tests with insulating gasket materials were unsuccessful. A preassembly view of a DAC with the split-pair coil actually used is shown in Fig. 4.39. The coil was wound on a frame, glued with epoxy resin, and removed after curing. It produces an RF field parallel to the gasket surface, which dips into the sample hole. The one-third of a carat diamonds, which have a culet area of about 1 mm^2 with 16 sides, are glued with epoxy resin to adjustable silicon-nitride semicylindrical rockers connected to the lever-based thrust mechanism. Berylco-25 gaskets with 0.5 mm diameter sample holes were used for pressures up to 5 GPa. For higher pressures, and for ^{23}Na NMR—where the ^{63}Cu resonance could interfere with the spectra—rhenium gaskets with 0.7 mm holes made by electroerosion were used. The pressure is measured *in situ* by the ruby-fluorescence technique, using the optical system shown in Fig. 4.38, and with a precision of ± 0.1 GPa.

Bertani *et al.*[28] measured the Knight shift and diffusion coefficients of ^{7}Li (up to 8 GPa) and of ^{23}Na (up to 6 GPa) in metallic lithium and sodium, respectively, at room temperature. The skin depths of the RF fields for the

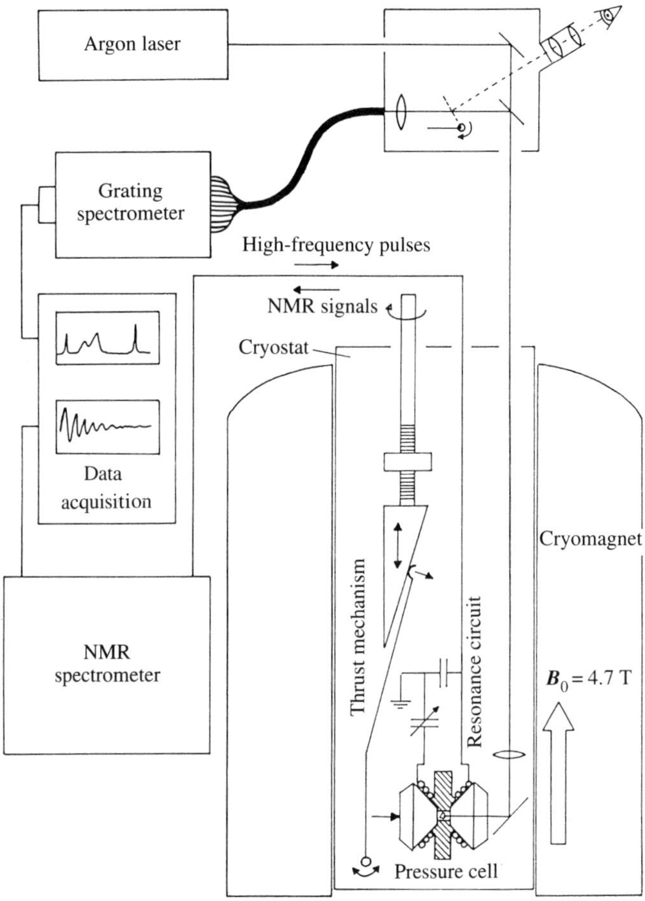

Fig. 4.38 A DAC apparatus used for very-high-pressure NMR.[8]

lithium and sodium at their respective resonance frequencies (77.7 and 52.9 MHz) are ~15 μm. The samples were therefore used in the form of spheres, with diameters of less than 30 μm, dispersed in mineral oil that serves as the pressure-transmitting medium and prevents corrosion. Up to 6 GPa no significant broadening of the ruby-fluorescence lines was observed, indicating truly hydrostatic (that is, homogeneous) pressures. The pressure dependence of the diffusion coefficient of ^{7}Li (Fig. 4.40) shows that pressure hinders the diffusion process. The transition at high pressures from a body-centred-cubic (b.c.c.) to a face-centred-cubic (f.c.c.) phase seems to be diffusion controlled: diffusion in the f.c.c. phase is faster than in the b.c.c. phase. Finally, we draw attention to the fact that, although DACs are primarily tools for solid-state research, they have been successfully applied to the measurement of ^{1}H NMR chemical shifts of organic liquids at pressures up to ~1.3 GPa.

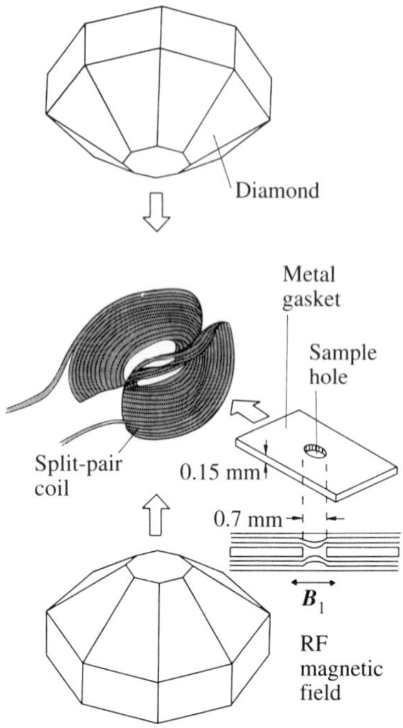

Fig. 4.39 A preassembly view of an NMR DAC with a split-pair coil and a metal gasket.[8]

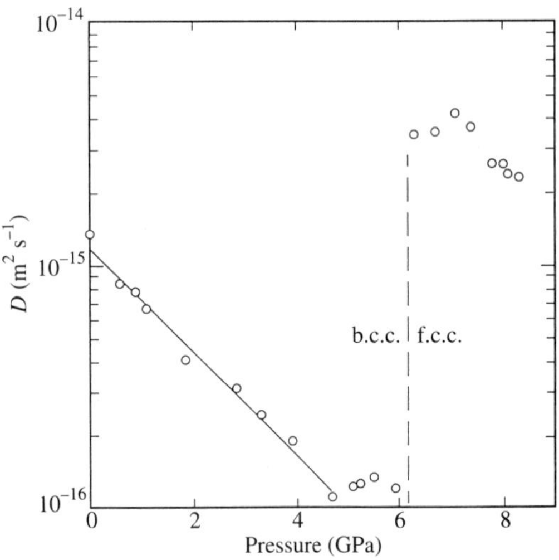

Fig. 4.40 The pressure dependence of the ^{7}Li self-diffusion coefficient, *D*, at room temperature.[28]

2.5 Conclusions

We have attempted to show, by giving representative examples, the range of technologies available for high-pressure NMR. Since NMR has become such a wide and varied field, and the requirements of different high-pressure NMR experiments are very different, we cannot give simple recipes for the design of a high-pressure NMR system. We hope, however, that experimentalists wishing to construct such a system will find here the information necessary to decide what type of system is best suited to their particular needs, as well as useful tricks of the trade and warnings of possible pitfalls.

2.6 Protocols

Protocol 5.
Preparation of the high-pressure, high-resolution NMR probe

Caution! For safety procedures see Appendix 2.

1. Start the thermostat (not yet connected to the pressure bomb).
2. Prepare 1 mL of solution for analysis (where necessary, in a glove box to prevent exposure to moisture and/or oxygen).
3. Check the solution by performing an NMR experiment in an ordinary NMR probe under atmospheric pressure.
4. Fill the high-pressure NMR cell with the solution; close it with the Macor piston; fill the Macor cap with solution and put it on the cell (these operations can be made in a glove box to prevent exposure to moisture and/or oxygen).
5. Check the high-pressure-generation system (oil separator, hand pump, and valves; see Fig. 4.23).
6. Fill the high-pressure NMR probe with the pressurizing liquid.
7. Connect the pressure-transmitting capillary to the high-pressure-generation system, taking care not to introduce air bubbles into the capillary.
8. Put the high-pressure NMR cell into the NMR insert (Fig. 4.24 (c)), connected to the plate with the electrical feedthrough (Fig. 4.24(b)).
9. Place the insert in the pressure bomb.
10. Close the bomb with the closing cap; seal the assembly by tightening the screws with a torque wrench to 30 N/m.
11. Connect the thermostat to the pressure bomb and switch on the pump for the thermostating liquid.
12. Connect the upper part of the NMR assembly (Fig. 4.24(a)) to the lower part, and check the tuning and machining of the RF circuits (outside the NMR magnet).

Protocol 5. *Continued*

13. Take out the upper part of the commercial internal NMR magnet assembly, including the spinner turbine.
14. Bring both parts of the assembly into the superconducting NMR magnet (upper part from the top, lower part from below).
15. Connect the RF circuit to the preamplifier of the NMR spectrometer.
16. Connect the Pt-100 resistor to a high-quality ohmmeter and wait for temperature stabilization.
17. Tune and match the probe using ordinary procedures.
18. Shim the NMR magnet using a deuterium lock or free induction decay, (FID), signal.

Protocol 6
Performance of a high-pressure, high-resolution NMR experiment

Caution! For safety procedures see Appendix 2.

1. Prepare the high-pressure, high-resolution probe (as described in Protocol 5).
2. Measure the temperature with a Pt-100 resistor; correct for the cable resistance; always disconnect the ohmmeter during NMR measurements.
3. Perform the first NMR experiment at the ambient pressure; compare with the result from a commercial NMR probe.
4. Measure the temperature with a Pt-100 resistor; correct for the cable resistance and the dependence of the Pt-100 resistance on pressure.
5. Increase the pressure.
6. Wait for temperature stabilization.
7. Measure the temperature with a Pt-100 resistor; correct for the cable resistance and the dependence of the Pt-100 resistor on pressure.
8. Where necessary, reshim, and retune the spectrometer.
9. Perform an NMR experiment.
10. Measure the temperature with a Pt-100 resistor; correct for cable resistance and the dependence of the Pt-100 resistance on pressure.
11. Continue to increase the pressure up to the maximum pressure desired (for example, 200 MPa).
12. Start to decrease pressure and perform NMR experiments in order to check the results for reproducibility.
13. Perform the last NMR spectrum at ambient pressure; compare the result with the first spectrum.

Protocol 7
Performance of NMR experiments with gases under pressure using a sapphire tube system

Caution! For safety procedures see Appendix 2.

1. Prepare 2–4 mL of solution.
2. Check the sapphire tube visually before filling the solution (after using twenty times, a full check by a suitable workshop is recommended).
3. Unscrew the valve body (Fig. 4.32) and load the sample solution into the tube; take care not to lose the O-ring.
4. Close the sapphire tube and tighten gently using two spanners.
5. Place the sapphire tube into an NMR spinner and adjust the height.
6. Place the tube and spinner into the safety system (Fig. 4.33).
7. Connect the sapphire tube to a gas-distribution system (equipped with a pressure-limiting safety valve), and open the tap one turn.
8. Fill the tube, taking care not to exceed the maximum pressure.
9. Close the tap and disconnect the assembly from the gas-distribution system.
10. Transport the sapphire tube inside the safety system to the NMR magnet; take care to hold it vertically, because the protection is open at the bottom.
11. Place the safety system with the tube inside on the top of the NMR magnet.
12. Put on the sample lift, release the sapphire tube and lower it into the magnet like an ordinary commercial tube (increase the lift-gas pressure to about 5 bar).
13. Lock, shim, and tune the NMR probe as usual; perform the NMR experiments.
14. Put on the sample lift to extract the sapphire tube from the NMR magnet.
15. Once the tube is inside the safety system, fix it in place and stop the sample lift.
16. Open the gas valve slowly to release the pressure (if the pressure is released too quickly, the liquid may spurt out).
17. Unscrew the valve body and remove the solution.
18. Clean the sapphire tube and valve in the normal way (acids and bases can be applied for short times: less than one hour).

Acknowledgements

We thank all those who have been involved in the development of high-pressure NMR in Lausanne, especially Dr Urban Frey and Mr Bertrand

Moullet who kindly allowed us to show one of their designs prior to publication. We thank the Swiss National Science Foundation for their continued support of our work.

References

1. Benedek, G. B.; Purcell, E. M. *J. Chem. Phys.* **1954**, *22*, 2003.
2. Yamada, H. *NMR Basic Principles and Progress*, **1990**, *24*, 234.
3. Lang, E. W.; Lüdemann, H.-D. *NMR Basic Princip. Prog.* **1990**, *24*, 130.
4. Roe, D. C. *J. Magn. Reson.* **1985**, *63*, 388.
5. Horváth, I. T.; Millar, J. M. *Chem. Rev.* **1991**, *91,* 1339.
6. Jonas, J. *NMR Basic Princip Prog.* **1990**, *24*, 86.
7. Merbach, A. E.; Akitt, J. W. *NMR Basic Princip Prog*, **1990**, *24*, 189.
8. Bertani, R.; Mali, M.; Roos, J.; Brinkmann, D. *Rev. Sci. Instrum.* **1992**, *63*, 3303.
9. Lee, S.-H.; Luszcynski, K.; Norberg, R. E.; Conradi, M. S. *Rev. Sci. Instrum.* **1987**, *58*, 415.
10. Frey, U.; Helm, L.; Merbach, A. E. *High Press. Res.* **1990**, *2,* 237.
11. Meyer, F. K.; Merbach, A. E. *J. Phys.* E, **1979**, *12*, 185.
12. Rathke, J. W. *J. Magn. Reson.* **1989**, *85,* 150.
13. Bridgeman, P. W. *The Physics of High Pressures*. Bell: London, **1949**.
14. Holloway, J. R. In *Research techniques for High Pressure and High temperature*; Ulmer, G. C., ed., Springer Verlag: Berlin, **1971**.
15. Vanni, H.; Earl, W. L.; Merbach, A. E. *J. Magn. Reson.* **1978**, *29*, 11.
16. Langford, C. H.; Gray, H. B. *Ligand Substitution Processes;* Benjamin: New York, **1965**.
17. Merbach, A. E. *Pure Appl. Chem.* **1987**, *54*, 161.
18. Hasha, D. L.; Eguchi, T.; Jonas, J. *J. Am. Chem. Soc.* **1982**, *104*, 2290.
19. De Fries, T. H.; Jonas, J. *J. Magn. Reson.* **1979**, *35*, 111.
20. Lamb, W. J.; Jonas, J. *J. Chem. Phys*, **1981**, *74*, 913.
21. Horvath, I. T.; Ponce, E. E. *Rev. Sci. Instrum.* **1991,** *62*, 1104.
22. Laurenczy, G.; Merbach, A. E. *J. Chem. Soc. Chem. Comm.* **1993,** 187.
23. van der Velde, D. G., Jonas, J. *J. Magn. Res.* **1987,** *71*, 480.
24. Huber, H.; Mali, M.; Roos, J.; Brinkmann, D. *Rev. Sci. Instrum.* **1984,** *55*, 1325.
25. Huber, H.; Mali, M.; Roos, J.; Brinkmann, D. *J. Physique* **1984,** *45*, 75.
26. Jayaraman, A. *Rev. Mod. Phys.* **1983,** *55*, 65.
27. Halvorson, K. E.; Raffaelle, D. P.; Wolf, G. H.; Matkze, R. F. In *Frontiers of High-Pressure Research*, Hochheimer, H. D. ed.; Plenum: New York, **1991**.
28. Bertani, R.; Mali, M.; Roos, J.; Brinkmann, D. *J. Phys.:Condens. Matter.* **1990**, *2*, 7911.
29. Moullet, M., Frey, U.; Helm, L.; Tschans, T.; Merbach, A.; Ith, M. *High Press. Res.* **1994**, *12,* 285.
30. Frey, U.; Helm, L.; Roulet, R. In *Advanced Application of NMR to Organometallic Chemistry*, Vol 1. Wiley: Chichester, **1996**.

3. Polymer studies at high pressures

D. C. Bassett

3.1 Introduction

High-pressure studies have a special relevance to synthetic polymers. In addition to the normal interest in the way familiar materials alter as the pressure is raised, it has become evident that some important polymers form novel structures at high pressures—usually high temperatures are also required. This is a consequence of their long-chain constitution, which promotes orientational order so that stable phases exist which are orientationally, but not fully spatially, ordered, in the manner of liquid crystals. The name columnar crystals has been suggested.[1]

Such ordering is not restricted to macromolecules with rigid chains but is present, for example, in polyethylene, the archetypal flexible polymer. The nature and properties of such novel phases and their possible role as metastable intermediates in growth at ambient pressure are all matters of considerable interest and current research activity.[1,2]

In consequence, the predominant emphasis of high-pressure research on polymers continues to be on phase transitions. Typically, pressures up to only ~5–10 kbar (0.5–1.0 GPa) are required, because the melting points have then reached values of ~300 °C, a temperature where spontaneous cross-linking occurs, and the integrity of the constituent macromolecules is no longer retained.

Although pressures of this magnitude are achievable using readily available commercial equipment, apparatus which is suitable for defining phase diagrams in this range (0–10 kbar, 0–300 °C, conditions in which the melting points may increase initially by ~50 K $kbar^{-1}$) must be able to specify a pressure to ± 20 bar to be equivalent to a precision of ± 1 K for a phase transition. This is a more exacting requirement than is commonly available in very-high-pressure studies. There are corresponding differences in the design of equipment when polymers are studied in large sample sizes where possible, to allow subsequent study of pressure-treated materials.

3.2 The gasketed DAC

As an illustration of the specific problems which arise in the study of polymeric phase transitions at high pressures we consider gasketed DACs.[3] Twenty years ago, the author and B. Turner proposed that the anomalous crystallization behaviour of linear polyethylene at ~5 kbar was due to the intervention, during crystallization, of a then unknown high-pressure phase.[4] A test of this prediction using *in situ* X-ray diffraction required apparatus capable of operating at 5 kbar and 250 °C. At that time these conditions were difficult to meet, not least because they were too severe for cameras with gaskets made from plastics, with the possible exception of PTFE. In the event, the postulated phase was discovered using a gasketed DAC operating for the first time at high temperatures.[5] Even so, the normal means of monitoring the pressure, ruby fluorescence,[3] was inadequate for the sensitivity required. Instead, the melting point of polyethylene itself (known from previous *in situ* measurements using differential thermal analysis (DTA)) was used together with optical examination with a polarizing microscope to determine the state of the sample in the cell.[4] Other workers have included hexamethylene tetramine with the polymer and used its X-ray spacings as an internal calibrant for the pressure on the sample.[6]

In spite of the difficulties of measuring the internal pressure of gasketed DACs, precisely, especially at high temperatures, they remain a useful and versatile tool. High pressures and temperatures are attainable easily and quickly, while changes, especially of pressure, can be accomplished promptly. The specimen generally fills the cell completely so that no encapsulation is necessary. The facility of interchanging the electromagnetic wavelengths used for the examination is especially valuable. Suitably chosen diamonds are transparent not only to optical wavelengths but also to X-rays (MoK_α) and to infrared (i.r.) radiation, allowing spectroscopic examination,[7] including Raman techniques.[8] One may, therefore, study one sample in the same state, by two or more of these methods, as may be appropriate.

Diamond-anvil cells have, nevertheless, some inherent disadvantages. Their small sample size severely limits any subsequent examination of samples prepared for them, although ingenious experimentation can still accomplish much. For example, the observation of falling spheres in a fluid enables viscosity data to be measured,[9] while individual crystals grown under pressure may be extracted and studied with an electron microscope.[10] But much larger samples, centimetres across or more, are required for, say, mechanical measurements of polyethylene crystallized in the high-pressure phase. The need for precise knowledge of the internal pressure remains a problem, and this problem is aggravated by the way DACs operate. Because the cell is subjected to a constant load, the sample is effectively constrained to have a constant volume. In a first-order transition, therefore, the sample's specific volume and pressure must adjust so that the total volume is constant. Thus, the pressure has to be monitored as it changes during the transition and not just measured once as in an isobaric experiment.

Isobaric cells are preferable for experimentation, primarily because their constraints are the intrinsic thermodynamic variables of pressure and temperature. One such modern facility has been described by Hikosaka *et al.* (Fig. 4.41).[3] But if there is no internal pressure sensor it is still necessary, for a precise knowledge of the internal pressure, to know the friction in the system, or at least to ensure that it retains the same sense so that it is not a factor when the pressure is changed. The Hikosaka cell relies upon the low friction of PTFE to keep the internal pressure close to the controlled applied pressure. There is no reason why larger isobaric cells could not be made with internal sensors for

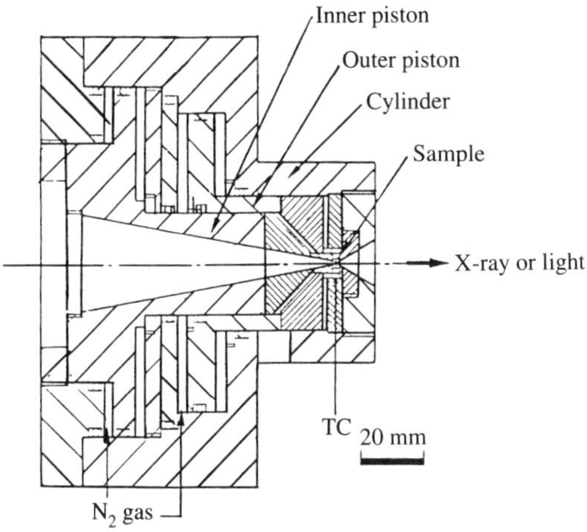

Fig. 4.41 A schematic drawing of a piston–cylinder high-pressure cell using a modified diamond anvil.[3] (Reprinted with permission from *J. Macromol. Sci.* **1992**, *1*, 91, by courtesy of Marcel Dekker Inc.)

use at pressures up to 10 kbar and employing, for example, sapphire rather than diamond windows. However, this desirable development has yet to be adopted.

3.3 Measurement of phase diagrams

Both (polarizing) optical microscopy and X-ray diffraction may be used to indicate when a phase transition occurs. The latter may also provide information about the nature of the crystalline phases involved, although, as for polyethylene, when two-dimensionally ordered columnar phases are involved the information may be meagre and incapable of defining the full nature of the molecular ordering.

In these circumstances it is useful to return to basic thermodynamics and the Clausius–Clapeyron equation for a first-order transition.

$$\frac{\mathrm{d}T}{\mathrm{d}p} = \frac{\Delta V}{\Delta S},$$

which relates the slope of the equilibrium line on a phase diagram in (T, p)-space to the differences in specific volume and specific entropy across it. The location of the phase lines may be made directly by measurements either of the specific volume or of the specific entropy (or preferably both) and this equation can be used to determine the nature of the phases and the reliability of the data. The anabaric or high-pressure phase of polyethylene was predicted in this way.[4]

Measurement of volume changes by dilatometry tends to be awkward and of low precision. Traditionally it has required either a piston–cylinder

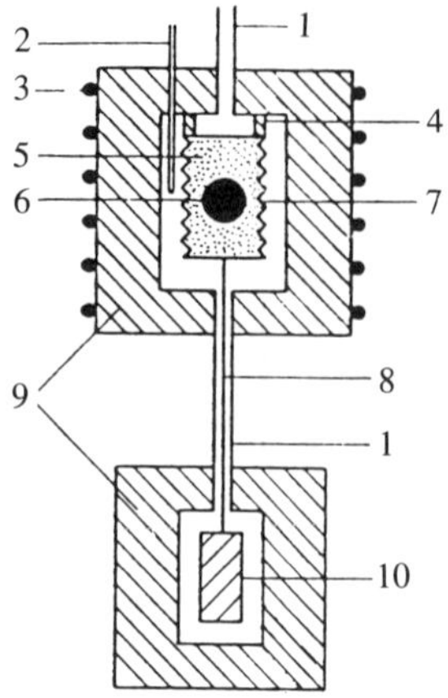

Fig. 4.42 A schematic drawing of a high-pressure dilatometry apparatus: (1) high-pressure pipe, (2) a sheathed Alumel–Chromel thermocouple, (3) a sheathed heater, (4) a support, (5) a liquid medium (silicon oil), (6) the sample, (7) a bellows cell, (8) steel wire, (9) high-pressure vessel, and (10) a differential transformer.[11] (Reprinted with permission from *Polymer J.* **1992**, *24*, 778.)

apparatus to record the piston displacement or a sample contained within collapsible bellows, as in Fig. 4.42.[11] Modern transducers have considerably enhanced the sensitivity with which a displacement can be measured, but real gains in accuracy also require quasicontinuous movement with minimal intermittent stick/slip jumps caused by frictional forces.

Thermal measurements, on the other hand, are comparatively simple to make and to use to identify when transitions occur, but it is much harder to convert thermal measurements into thermodynamic quantities, such as ethalpy and entropy. This distinction is the essential difference between the two techniques of measurement known as differential thermal analysis (DTA) and differential scanning colorimetry (DSC). Both methods are highly sensitive, allowing small (milligram) quantities of sample to be used. Both methods place the sample and a reference in the same thermal environment—usually with a uniform rate of temperature change—and the temperature difference between the sample and the reference is measured. At a transition the absorption or evolution of heat causes a corresponding fall or rise in temperature of the sample and this is the output for DTA, whereas DSC converts this change in temperature into an energy change by careful control over, and knowledge of, the thermal environment.

Measurement of the temperature difference is simple, and this is why DTA is more than a century old. With modern electronics one can achieve a greatly increased sensitivity, but the extension to accurate calorimetry is far from easy because of the uncertainty of the thermal environment. Although this problem has been overcome in commercial equipment which is designed

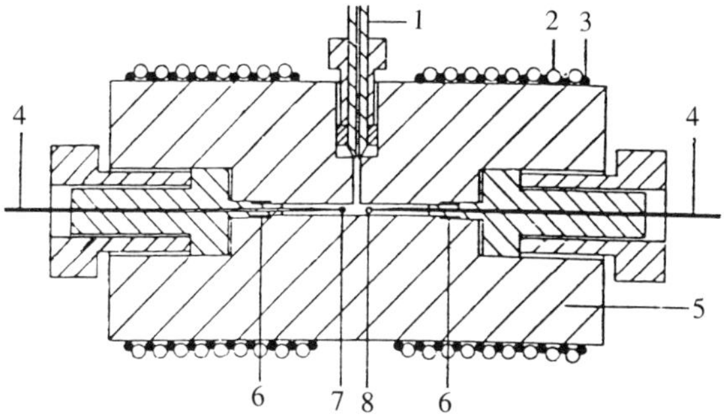

Fig. 4.43 A schematic drawing of a high-pressure DTA apparatus: (1) high-pressure pipe, (2) copper tube, (3) sheathed heater, (4) sheathed Alumel–Chromel thermocouple, (5) high-pressure vessel, (6) copper seal, (7) sample, and (8) reference.[15] (Reprinted with permission from *Polymer J.* **1992**, *24*, 422.)

to operate at the ambient pressure, it is not straightforward simply to transfer the sample holders to a high-pressure environment. On only a few occasions (see, for examples, reference 12) has this been accomplished. At present, most workers still use DTA techniques;[13–16] a modern DTA apparatus for identifying transitions from temperature changes is shown in Fig. 4.43. However, when entropy or enthalpy changes are measured, a much fuller description of the nature of the changes involved is possible. The implementation of DSC techniques thus remains highly desirable.

Of the many other ways of detecting phase transitions, including cloud points for systems with mixed phases,[17] particular mention should be made of the use of dynamic–mechanical methods, and specifically measurement of the shear compliance to ~1 kbar.[18] When applied to columnar crystals of poly-diethylsiloxane) such a method was able to detect the melting point in conditions where the volume change was close to zero.

3.4 General apparatus

The modest pressures of ~5 kbar (0.5 GPa) necessary to cause transitions to columnar phases in polymers make the design of apparatus using large specimens relatively straightforward and it is easy to accommodate associated characterization techniques such as thermal conductivity[19] or nuclear magnetic resonance (NMR) probes.[20] Much of the author's early work was with a piston–cylinder apparatus with a working length of 5 cm and a diameter of 1.5 cm, controlled to within ±30 bar and ±1 K for pressures and temperatures up to 10 kbar and 300 °C.[21] Mechanical test specimens are readily contained, and dilatometry or DTA measurements can be made. Such general-purpose equipment retains its utility and illustrates how properties which are of interest for polymers can be investigated under pressure in comparatively simple ways.

3.5 Protocols

Protocol 8
Crystal-growth kinetics

Caution! For safety procedures see Appendix 2.

1. Use equipment which allows the sample to be inspected with a polarizing optical microscopy such as a DAC or, if possible, similar isobaric equipment.
2. Fill the cell with the sample, preferably in powder form.
3. Apply pressure and heat the sample until it just melts (the melting temperature should be consistent with the pressure applied: if the melting temperature is much lower than expected, the sample will probably not fill the cell completely, for example, because of an air bubble, so that its internal pressure will be below the applied pressure).
4. Adjust the pressure to the selected value (for example, a pressure which gives a desired melting temperature). Melt the sample, then lower the temperature rapidly to the value chosen for crystallization, and record how the crystals grow with time.

Protocol 9
The Preparation of pressure-treated samples for subsequent examination

Caution! For safety procedures see Appendix 2.

1. Choose apparatus with a suitably large working volume with adequate controls for temperature and pressure.
2. Consider the application of pressure to the sample. (For many purposes it is satisfactory to immerse the sample in silicone oil, which is used as a pressure-transmitting fluid. This oil is relatively inert, and it can be removed subsequently by washing with acetone (propanone). In circumstances where penetration by silicone oil is deleterious then alternative pressure-transmitting liquids such as water may be useful. If, however, it is inadvisable to have water in direct contact with the pressure vessel then secondary encapsulation, for example, in thin-walled gold tubes, may be an effective alternative. The use of high-pressure gases is often unsatisfactory because they tend to cause various polymers to foam.)
3. Choose a pressure/temperature programme which avoids solidification of the silicone oil with increasing pressure,

4. Carry out the desired pressure treatment, for example, annealing or crystallization, then reduce the temperature to the ambient value or to a suitably low value while maintaining pressure.
5. Release the pressure, remove and clean the sample prior to testing.

References

1. Ungar, G. *Polymer* **1993**, *34*, 2050–2059.
2. Hikosaka, M.; Rastogi, S.; Keller, A.; Kawabata, H. *J. Macromol. Sci.* **1992**, *1*, 87–131.
3. Barnett, J. D.; Block, S.; Piermarini, G. J. *Rev. Sci. Instrum.* **1973**, *44*, 1–9.
4. Bassett, D. C.; Turner, B. *Phil. Mag.* **1974**, *29*, 925–955.
5. Bassett, D. C.; Block, S., Piermarini, G. J. *J. Appl. Phys.* **1974**, *45*, 4146–4150.
6. Newman, B. A.; Sham, T. P.; Pae, K. D. *J. Mat. Sci.* **1977**, *12*, 1064–1068.
7. Variano, B. F.; Sandroff, C. J.; Baker, G. I. *Macromolecules* **1991**, *24*, 4376–4380.
8. Zhao, Y.; Wang, J.; Cui, Q.; Liu, Z.; Yang, M.; Shen, J. *Polymer* **1990**, *31*, 1425–1448.
9. Cook, R. L.; King, H. E.; Peiffer, D. G. *Macromolecules* **1992**, *25*, 629-635, see also pp. 122–150.
10. DiCorleto, J. A.; Bassett, D. C. *Polymer* **1990**, *31*, 1971–1977.
11. Tsubakihara, S.; Higashi, K.; Taki, S.; Matsushige, K.; Yasuniwa, M. *Polymer J.* **1992**, *24*, 777–781.
12. Wortmann, F. J.; Deutz, H. *J. Appl. Polym. Sci.* **1993**, *48*, 137–150.
13. Maeda, Y.; Toriumi, H. *Makromol. Chem.* **1993**, *194*, 3123–3134.
14. Nakafuku, C. *Polymer* **1993**, *34*, 4166–4167.
15. Tsubakihara, S.; Yasuniwa, M. *Polymer J.* **1992**, *24*, 421–426.
16. Takamizawa, K.; Toratani, H. *Polymer* **1992**, *33*, 1333–1334.
17. Hasch, B. M.; Meilchem, M. A.; Lee, S. H.; McHugh, M. A. *J. Polym. Sci. Polym. Phys. Ed.* **1992**, *30*, 1365–1373.
18. Pechhold, W.; Schwarzenberger, P. In *Frontiers of High-Pressure Research*; Hocheimer, H. D.; Etters, R. D. eds; Plenum: New York, **1991**, pp. 57–71.
19. Yen, C. L.; Tseng, H. C.; Yang, Y. Z.; Hsieh, K. H. *J. Appl. Polym. Sci.* **1991**, *42*, 1179–1184.
20. Kulik, A. S., Prins, K. O. *Polymer* **1993**, *34*, 4629–4648.
21. Bassett, D. C., Carder, D. R. *Phil. Mag.* **1973**, *28*, 513–533.

5

NMR and diaphragm-cell techniques for the study of molecular dynamics in fluids

W. E. PRICE and H.-D. LÜDEMANN

1. High-pressure NMR techniques

1.1 Introduction

Nuclear-magnetic-resonance (NMR) spectroscopy has, arguably, become the most powerful and important analytical tool for structural elucidation of chemical compounds. It has completely revolutionized whole areas of chemistry and biochemistry. So it is perhaps not surprising that modern Fourier-transform (FT-NMR) spectroscopy also plays a central role in studies which focus upon intermolecular interactions, dynamics, and their effects upon the fluid state. The importance of FT-NMR spectroscopy in high-pressure molecular-dynamics experiments is due to its noninvasive nature and its versatility; a number of different parameters may be determined to describe various aspects of the motions and intermolecular forces within a system. With a number of modifications to facilitate the application of elevated pressures the whole gamut of modern FT-NMR spectroscopy is potentially available. A detailed description of the theory and the methods will not be given here, since an outline is sufficient for the present purpose; many scholarly expositions are available on the subject, to which readers are directed.[1–3] In addition, Section 3 gives a brief description of the range of measurements which can be made using high-pressure NMR techniques based on recent work.

The NMR technique is based upon the quantum-mechanical property of nuclear spin. A nucleus with a spin I has $2I + 1$ energy levels, which are normally degenerate. For example, ^{1}H has a spin of $I = \frac{1}{2}$ and thus has two energy levels. These degenerate states, however, separate on the application of a strong magnetic field (commonly now up to 10–15 T). The difference between these levels is in the radio-frequency range, and perturbation of the equilibrium Boltzmann distribution of nuclear energies in the sample over the different spin states may be induced by radiation of the appropriate frequency (that is, by photons with energies equal to the difference in energy between two particular spin states). After emitting a pulse of radio-frequency radiation (which may be detected) the resulting distribution will tend to relax back to its equilibrium state. The overall relaxation is usually a first-order process, called longitudinal- or spin-lattice relaxation, which has a time constant T_1. In addition, a second relaxation process takes place in which the spins are exchanged within a given distribution. This is also a first-order process, the transverse or spin–spin relaxation characterized by a time constant T_2. Both of these relaxation times are affected by dynamic processes in the fluid phase which cause localized time-dependent fluctuations of the electric and/or magnetic fields. These are the two fundamental parameters of the NMR phenomenon. Determination of T_1 is straightforward, and is now commonly carried out by the inversion recovery method.[1]

Diffusion measurements are very important probes of fluid-phase molecular dynamics. Nuclear magnetic resonance is currently the most convenient and versatile method for measuring self-diffusion (intradiffusion) coefficients in fluids and mixtures.[4] In addition, this technique can be readily adapted to high-pressure studies. Diffusion measurements have advantages over other NMR techniques because they yield a basic molecular quantity without recourse to a model and they can be directly compared with values from other techniques. The interpretation of diffusion coefficients is also often easier. The method by which diffusion is measured is dependent upon T_2, the transverse relaxation time; T_2 is usually obtained by measuring the time decay of the transverse magnetization of the sample, which is orthogonal to the applied field B_0. The variation in B_0 across the sample, and the diffusion of nuclei through regions of different field strength, provides a mechanism by which the orthogonal magnetization can decay that is additional to spin relaxation. The effect of diffusion on the decay may be magnified by setting up a small, but known, field gradient ΔB across the sample. This must be significantly larger than the variation in B_0 over the sample volume. The intradiffusion coefficient is then obtained by measurement of the change in the relaxation rate (as characterized by T_2) and its dependence on the magnitude of the gradient. This is the so-called pulsed-gradient spin–echo technique.[4,5] The amplitude, A, in a Hahn spin-echo signal in the presence of a pulsed field gradient is given by

$$A(2\tau) = A(0)\exp(-2\tau/T_2)\exp[-\gamma^2 D\delta^2(\Delta-\tfrac{1}{3}\delta)g^2]. \qquad (5.1.1)$$

In eqn (5.1.1) τ is the time interval between the 90° and the 180° radio-frequency pulse. The 90° pulse has the effect of aligning the spin-magnetization vector along the direction transverse to the applied field. This signal decays as the spins dephase. The 180° pulse then rotates the vector so that the signal is again coherent at a time 2τ; this is the spin echo. γ is the gyromagnetic ratio of the nucleus studied; D is the intradiffusion coefficient of the nucleus concerned; δ is the duration of the magnetic-field gradient pulses; Δ is the delay between the two gradient pulses; and g is the strength of the field gradient. The gradient is given by $g = kI$, where I is the gradient coil current, and k is a constant which is usually determined by calibration using substances with known diffusion coefficients[4,6] (see Section 2.2 in this chapter). The determination of D is most conveniently carried out by performing 10–15 spin-echo experiments, increasing the current, I, in each experiment but keeping all the other variables constant.

The interaction of the magnetic moment of a nucleus with an applied field which gives rise to the splitting of the spin states also induces changes in the orbital angular momentum of the electrons. This results in a small additional contribution to the magnetic field. This means that the effective magnetic field and, consequently, the frequency of the radiation emitted by the relaxing nucleus following a perturbation are both slightly dependent on the local chemical environment of a particular nucleus. This 'blurring' of the resonance phenomena, together with couplings between neighbouring nuclei, results in a fine structure within a spectrum that constitutes the main tool for chemical-structure elucidation in modern high-resolution FT-NMR.[7,8] The peak-intensity position, or chemical shift, may also be used directly as a temperature- and pressure-dependent experimental parameter to probe molecular associations and conformational equilibria in fluids. However, this has rarely been achieved[9–11] because quantitative interpretation of the (p, T)-dependence is difficult since it generally requires a model for the system. The pressure- and temperature-dependence of the chemical shift is, however, still of importance in a discussion of NMR studies of the dynamics of fluids. In mixtures, the resolution of different component peaks may be pressure and temperature dependent. Consequently, this may hinder (or prevent) the measurement of diffusion or relaxation processes since expensive remedies—such as partial isotopic substitution (for example, of deuterium for hydrogen) of one of the components—may be required.

The modifications that are necessary to carry out high-pressure studies of diffusion and relaxation processes in fluids are not major, and they can be incorporated into standard NMR spectrometers. In essence, one requires a special probe containing a gradient coil (to allow additional magnetic fields to be produced across the sample of diffusion measurements), which allows temperature control of the sample, together with modified sample holders, which facilitate the safe application of high pressures within the spectrometer. The spectrometer has to be designed in such a way that perturbations of the magnetic field are minimized, say by using nonmagnetic materials, and so that there is no undue interference with the radio-frequency transmitter or receiver, to maintain the signal quality and sensitivity. We give practical advice below on the changes necessary for the study of molecular motions in a variety of fluids and mixtures as a function of pressure and temperature. In addition, typical results are given for the different system types, to demonstrate the technique, and to amplify some of the theoretical interpretations which can be drawn from such measurements.

1.2 Types of sample cell

In order to apply high pressures to a fluid sample the container must be strengthened. The most common type of container is a toughened glass cell, which was introduced by Yamada.[12] This is the most general-purpose cell used for high-pressure NMR studies in the laboratories in Regensburg, Germany. Special requirements, such as supercooling liquids, necessitate modifications to this simple design.

1.2.1 Yamada glass cells for liquid probes

Figure 5.1(a) depicts the basic design of a Yamada glass cell (together with the autoclave assembly). The glass cells are drawn from borosilicate-glass capillaries; the size of the end sample cell is somewhat dependent on the construction of the magnet and spectrometer. Prior to use, the capillaries are etched with 3 per cent HF (aq.) on both the inner and outer surfaces. The outer surface is subsequently protected from scratches with a thin layer of nitrocellulose. Surface marks are likely to weaken the glass, and they can, therefore, result in failure when high pressures are applied. The capillary is then placed within a hollow copper–beryllium nipple with a cone at one end. It is secured by drawing an aluminium-oxide-filled two-component epoxy

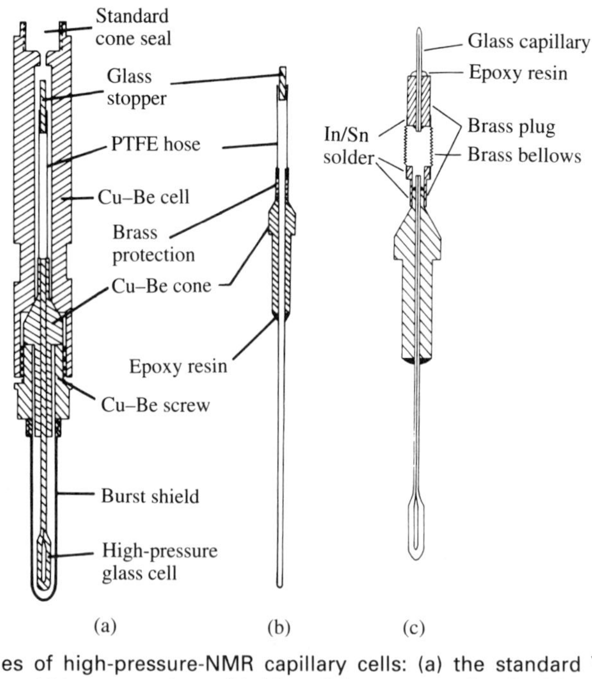

Fig. 5.1 Samples of high-pressure-NMR capillary cells: (a) the standard Yamada-type capillary, shown within an autoclave; (b) thin cell for supercooling liquids; (c) a modified cell for use with liquids which would be gaseous at ambient conditions.

resin (Eccobond 104, Grace Electronic Material, Heidelberg, Germany) into the cavity between the glass and the bore of the metal tube, by the use of a vacuum line. The resin is cured at 150 °C for 8 h. A piece of poly(tetrafluoroethene) (PTFE) shrink tubing is attached to the open end of the capillary. A small piece of glass capillary (1 cm) is secured to the other end of the shrink tubing. After filling the cell (see Section 1.5) the system can be closed by flame sealing the end of the glass capillary. Alternatively, a simple glass stopper (made from a piece of glass rod) can be used to seal the end. The flexible PTFE tubing allows the pressure to be transferred from the pressure fluid to the sample fluid. Consequently, it is important to ensure that this tubing is not damaged during the flame sealing, and so care should be taken during this procedure.

Capillaries of this design have been used with the following cell dimensions: an inner diameter of 1.2 mm and an outer diameter of 5 mm, or an inner diameter of 1.8 mm and an outer diameter of 7 mm. Pressures of up to 200 MPa have been routinely achieved.[13–15] It should be noted that it is an excellent practice to pressure test all glass capillaries after manufacture in order to weed out those which would fail under the application of high pressures in the spectrometer.

The main disadvantage of the strengthened-glass design for many applications is the low effective sample volume, and thus the lower signal strength and sensitivity. It is thus desirable to employ the maximum possible cell diameter given the constraints of the bore of the particular spectrometer used. However, the use of glass as a pressure container does have definite advantages. Glass is completely transparent to radio-frequency radiation and it is also obviously nonmagnetic. This means that all spectrometers with either electromagnets or cryomagnets can be used without any major modifications, and so the complete range of modern FT-NMR is potentially at the user's disposal. The only major consideration in using glass cells is to ensure that the sample cell is aligned with the magnetic field. This is readily achieved by appropriate modifications to the design of the probe head.

1.2.2 Modification for gaseous samples

In order to fill a sample cell with a fluid which is gaseous at room temperature and pressure, the cell detailed above needs to be altered slightly. A suitable modification is shown in Fig. 5.1(c). The shrink-fit tubing is replaced by a small brass bellows. A piece of glass capillary fixed into a brass plug sealed with epoxy resin completes the design; this allows flame sealing of the system after filling. The brass bellows facilitates the complete filling of the glass cell by condensation under reduced pressure.

1.2.3 Thin glass cells for supercooling liquids

There has been much interest in the unique properties of water and aqueous solutions at supercooled temperatures,[16,17] including its transport properties.

In order to carry out NMR studies on aqueous solutions at supercooled temperatures and high pressures a modified cell has to be designed. This is shown in Fig. 5.1(b). In the metastable state of a supercooled liquid the probability of unwanted heterogeneous nucleation is proportional to the cell volume.[16] Consequently, the only design change necessary is a reduction in the dimensions of the glass cell. Capillaries of this design have been made with an internal diameter of 150–220 μm or 40–80 μm, and with an outer diameter of 1.5 mm. These capillaries have routinely been used up to pressures of 250 MPa,[18,19] and careful selection of the capillaries by pressure testing allows high pressures (up to approximately 450 MPa) to be achieved.[13] It should be noted that for many studies with supercooled liquids the liquid is in emulsion form to help further prevent nucleation.[16,20]

1.3 Modification to the probe-head design for high-pressure spin-echo measurements

Although commercial probe heads are available with gradient coils, suitable for spin-echo measurements, generally they are not specifically designed for work at elevated pressures. A number of research groups have published designs of probe heads using either home-built or modified commercial units.[3,14] Such commercial probe heads have been altered and used in the Regensburg laboratory in Germany.[21]

However, there are often practical problems with the modification of existing probe heads, particularly if the applications and experimental conditions are extreme. For this reason, home-built probe heads have superseded hybrid designs. The main features of probe heads designed for high-pressure diffusion studies are adequate temperature control and a method of protecting the probe head and spectrometer if the glass sample capillaries fail at elevated pressures. Figure 5.2 depicts a design for a probe head[21] used on Bruker MSL 300 and 100 spectrometers. Nitrogen flows through the central inlet over the sample space, which also contains a heating element and a temperature sensor connected to a controller. For studies below room temperature, the nitrogen is first cooled by passing the flowing gas (contained within a metal coil) through an insulated vessel of liquid nitrogen. The construction of the probe-head interior directs the flow of nitrogen up and down the entire length of the sample compartment several times, thereby ensuring good temperature stability. The actual temperature of the sample is measured by a thermocouple placed alongside the capillary before and after a spin-echo experiment. The wire is withdrawn from the vicinity of the sample during the experiment to avoid interference with the RF signal.

The magnetic-field gradients are provided by a Helmholtz coil wound on to the outside of an insulating glass Dewar flask. This Dewar flask is kept at a constant temperature by the flow of room-temperature air which is forced into the annular space between the outer wall of the Dewar flask and the glass support. As mentioned before, keeping the gradient coils at a constant (room) tempera-

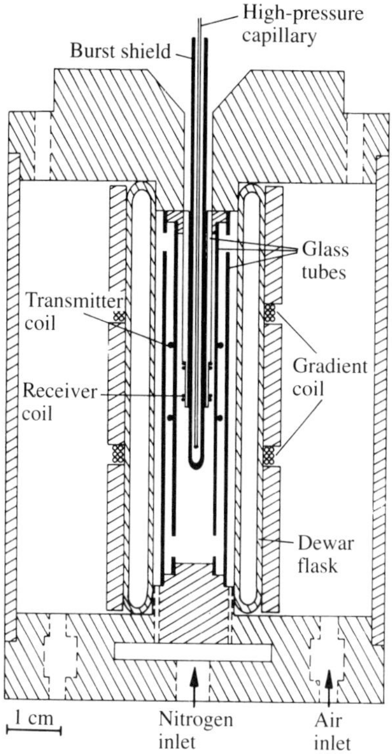

Fig. 5.2 An NMR probe-head design.

ture avoids problems of temperature effects on the field gradient strength and hence avoids difficulties due to temperature-dependent calibration constants.

In order to prevent damage to the probe head in the event of capillary failure under pressure, the glass capillary is sheathed with a glass burst shield. This stops any fluid or shards of glass getting into the probe head or magnet. The burst shield is generally made of standard borosilicate tubing with a diameter which ensures that the capillary fits snugly inside. It should be noted that if the capillary has too much room to move then problems of signal reproducibility can result. This can be rectified by using Teflon inserts inside the end of the burst shield resting on the sample bulb to keep it in a fixed position. The burst shield is held in place by a brass ring at the open end, and a brass compression nut, which screws into the top of the autoclave, holds the sample capillary in the spectrometer.

1.4 Autoclave design and modifications for very high pressures

In order to carry out high-pressure NMR measurements it is necessary to generate and maintain a pressure and to transfer it through the pressure-transmitting fluid to the sample cell. The design of the autoclave which is used will depend upon the application and the experimental conditions, but a number of features should be noted. The material from which the autoclave is made should be nonmagnetic, so that it can be used inside the spectrometer without affecting the signal. The autoclave must, obviously, be robust enough to be able to maintain the pressure safely, without danger of failure. In addition, the temperature range of operation must be considered with the expansion coefficients of the various parts of the pressure apparatus, so that inadequate maintenance of the pressure at low or high temperatures does not result.

1.4.1 Standard copper–beryllium autoclaves

The standard autoclave[10,13,14] used in the Regensburg, Germany, and other laboratories is based on a copper–beryllium alloy, Berylco-25 (Deutsche Beryllium, D-61440 Oberursel); this is in common use in pressure vessels, because of its strength and low coefficient of expansion (see Fig. 5.1(a), where a sample cell is also in place). The autoclave itself has an internal diameter of about 3 mm, which is marginally larger than the PTFE hose used to transfer the pressure from the generating fluid to the sample. The external diameter should be sufficient to make it robust, but it should also fit comfortably inside the internal magnet space.

At the lower end of the autoclave is a wider opening into which the copper–beryllium (Cu–Be) cone of the sample cell is seated. This has to be machined precisely because it forms the pressure seal. The sample cell is secured in place by a Cu–Be screw which acts as a compression nut. This screw has two threads (one at each end of the screw); the thread is needed to hold the burst-shield device in place (see Section 1.3). The upper end of the autoclave is fitted with a standard cone seal which connects the autoclave to standard pressure-generating equipment.

Autoclaves with this design have routinely been used up to pressures of 200–250 MPa (with pressure-tested sample cells) and in the temperature range from −140 to 200 °C. This is sufficient for a large number of applications. The standard Cu–Be autoclave is also suitable for use with other sample cells (e.g. Fig. 5.1(b)). A different design is more appropriate at higher pressures.

It should be noted that a second general approach to the production of high-resolution NMR spectra at elevated pressures has been developed by the groups of Jonas[22] and Merbach.[11] In these groups, the sample is completely enclosed within the autoclave. With these designs, pressures up to 1 GPa can be generated and large sample volumes can be used, which ensures a relatively high sensitivity. The major difficulties with this approach are in the construction of the autoclave in order to achieve minimal perturbation of the magnetic field and in the large modification necessary to the standard electronics of a commercial spectrometer. The design detailed here circumvents these problems at the cost of a smaller sample volume and a lower maximum applied pressure, but it is excellent for a wide range of uses. An alternative approach when higher pressures are required is discussed below.

1.4.2 Titanium-alloy-based autoclaves

In order to study fluids under more extreme and exacting experimental conditions of temperature and pressure another approach to pressure generation and maintenance is required. The dynamics of fluids under high pressures is of importance not only from a fundamental fluid-structure viewpoint but also because of its relevance in oil exploration and in polymer production (for example, in polyethylene melts). The approach used in the Regensburg laboratory in Germany in contrast to that used by the groups of Jonas[22] and Merbach,[11] has been to try to use most of the ambient-pressure features of commercial spectrometers (in this case the Bruker MSL 300 and MSL 100 instruments); this minimizes the changeover times and spectrometer modification. However, in order to achieve a wider working range of temperature (150–450 K) and pressure (up to 600 MPa) the autoclave is machined from a high-strength titanium alloy (TiAl6V4, Weingärtner, Hamburg, Germany). Fig. 5.3 shows the basic design.[23] A slightly modified sample tube is used. The sample tube is made from borosilicate glass, and a sealed brass bellows serves as a separator between the probe and the pressurizing fluid and the pressure transmitter. This has two advantages: it allows the bellows to be kept at the ambient temperature and only the glass–metal seal of the top end is exposed to high pressure. The lower autoclave is connected to the upper by TiAl6V4 pressure tubing, and it is heated and cooled with gaseous nitrogen. It is insulated against the inner wall of the superconducting magnet by several layers of ceramic paper, and finally it is wrapped in aluminium foil. Room-temperature air may be pressed into the annular space between the heated compartment and the magnet to help remove any dissipated heat and to protect the magnet wall from large temperature fluctuations. Figure 5.4 shows details of the probe-head design. The advantage of this design is that the low-impedance electrical feedthrough permits all the tuning and matching circuits to be used at ambient temperature and pressure. This enables the probe-head electronics to be changed and retuned to another resonance frequency without releasing the pressure. This is useful, not only to combat drift problems but also to allow different nuclei to be followed during the same experimental run. This probe has been successfully operated to frequencies between 14 and 300 MHz.

The most convenient gradient coil is an opposite-wound Helmholtz coil fixed on to the outside of the autoclave body. Extensive tests have been carried out[24] which have shown such a design is superior—in terms of its reproducibility of pulsed field gradients—to both quadrupole coils and Helmholtz coils wound on the inside of the autoclave. In order to generate these higher pressures, commercial 1 GPa equipment may still be used. However, a different pressurizing medium from that described previously is employed. Either

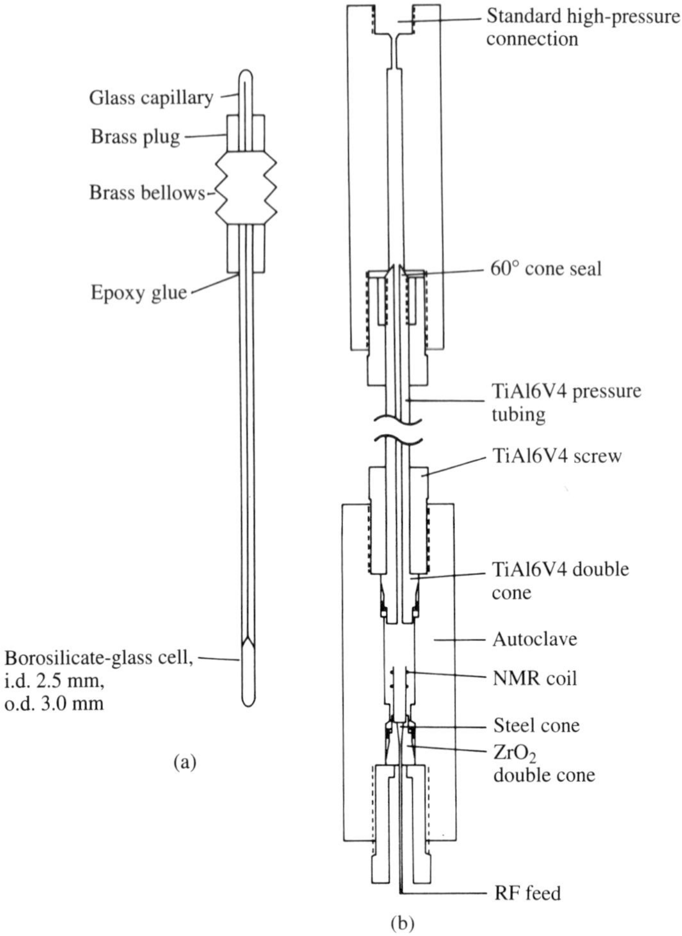

Fig. 5.3 A titanium-alloy autoclave used at very high pressures (not to scale): (a) the probe container, and (b) the Ti Al6V4 autoclave.

the Freon tetrafluorodibromoethane or compressed argon (or neon) are suitable, although this Freon will corrode the walls of the titanium autoclave at temperatures above 350 K.

An alternative improved design[24] for the separation system of the sample from the pressurizing fluid for use with this autoclave is shown in Fig. 5.5. A glass capillary with a precision bore of 2.0 ± 0.01 mm is filled with the substance under study. A series of three polished glass pistons with diameters of 1.993 ± 0.005 mm is pressed into the capillary. These pistons are chosen, prior to filling the capillary, to ensure a tight fit. Short columns of the liquid, interspersed by the pistons, prevent contamination by the pressurizing fluid of the sample in the lowest compartment. This design has been found to work extremely well, even after seven days of continuous high-pressure operation.

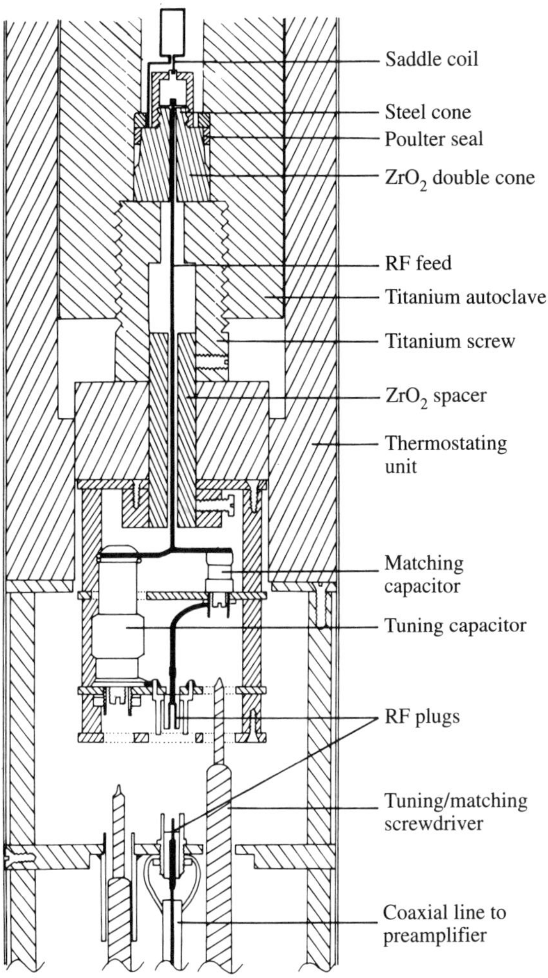

Fig. 5.4 Details of the probe-head design used with the titanium autoclave.

Probe temperatures that are accurate and stable to better than ± 1 K are readily obtainable by this method using a suitable temperature-control unit.

1.5 Filling of samples: Tricks and procedures

In order to deal with a wide range of different types of sample (such as liquids at room temperature, gases, or mixtures), a number of different filling techniques have been developed. This is an important aspect of the use of NMR for high-pressure diffusion measurements, since the quality of the data is dependent on the integrity of the sample. This is particularly true for mixtures, where inappropriate filling can lead to substantial changes in the composition, producing inaccurate and irreproducible results.

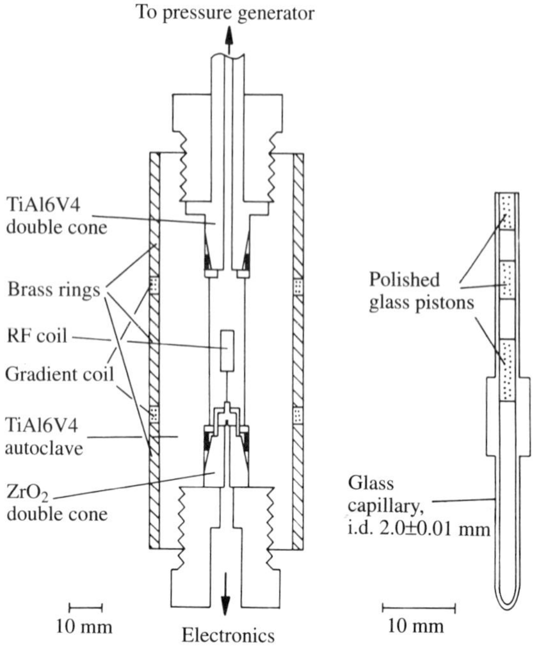

Fig. 5.5 A system to separate the sample from the pressurizing fluid in a titanium autoclave.

The most important points to consider when filling a glass cell sample are to ensure complete filling without compromising the composition and to deoxygenate the sample. The second factor is crucial when accurate relaxation times are being measured. Both T_1 and T_2 are seriously affected by the presence of oxygen; dissolved oxygen provides an additional mechanism for the relaxation of the nuclei, leading to shortened, inaccurate values.

1.5.1 Neat liquids

The simplest type of sample to deal with is a pure substance which is liquid at room temperature. The sample is transferred to a standard strengthened glass cell using high-vacuum techniques, as depicted in Fig. 5.6. A flask containing the liquid is attached to the vacuum line. Any major sample purification can be done prior to this, but molecular sieves can be added to the flask to aid the removal of impurities, such as water. The sample is then subjected to freeze–thaw cycles with prolonged vacuum pumping during the freezing cycle in order to remove dissolved gases, such as oxygen and nitrogen. A final vapour pressure of less than 10^{-3} Pa is normally adequate. The sample is then sublimed into the evacuated compartment containing the sample cell and allowed to liquify. The glass cell is connected to a spindle by a fishing line, and the glass-capillary end is protected from damage by a glass sleeve. The

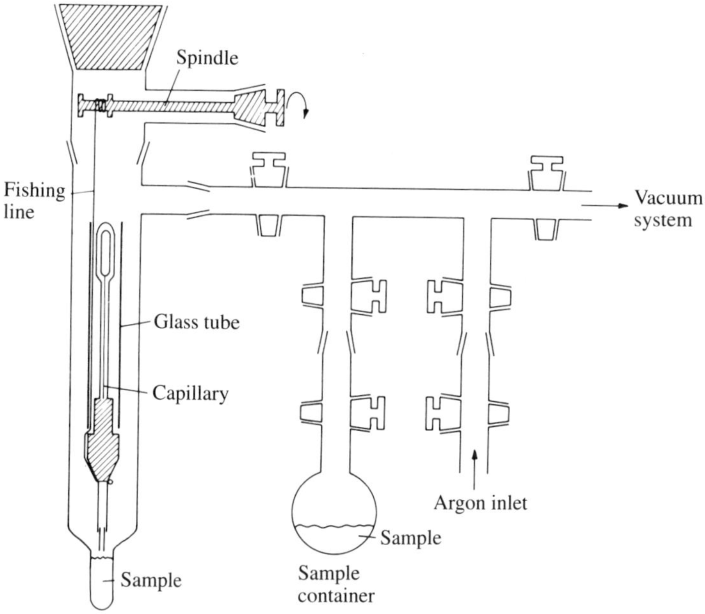

Fig. 5.6 A capillary-filling technique for liquid samples.

open glass tubing at the end of the PTFE tubing is lowered into the liquid, and the cell is filled by flushing the vessel with argon. The apparatus is dismantled and the glass end of the sample cell is then flame sealed.

1.5.2 Volatile liquids and neat gases

For volatile samples, such as ammonia[25] and completely oxygen-free samples, a modified procedure is necessary. The PTFE tubing which acts as a separator between the sample and the pressurizing medium is not sufficiently vacuum-tight to allow filling of oxygen-free liquid. In addition, the low boiling point of a volatile liquid such as ammonia means that all filling and condensing operations have to be performed at reduced temperatures. This procedure is shown in Fig. 5.7. A quantity of dried sample is condensed from a reservoir into a degassing container by cooling the container with an acetone and dry ice slush bath. This sample is then thoroughly degassed by at least five freeze–pump–thaw cycles to a final vapour pressure of approximately 10^{-3} Pa (the value depends upon the nature of the sample). The sample is frozen during this process using liquid nitrogen and melted with a suitable coolant such as acetone/dry ice. The choice of coolant is again dependent upon the boiling and freezing point of the sample. After degassing, the high-pressure capillary is immersed in a suitable coolant to the level shown in Fig. 5.7. The temperature of the bath is chosen to ensure that transfer and condensation of the sample into the capillary occurs when the degassing chamber is warmed.

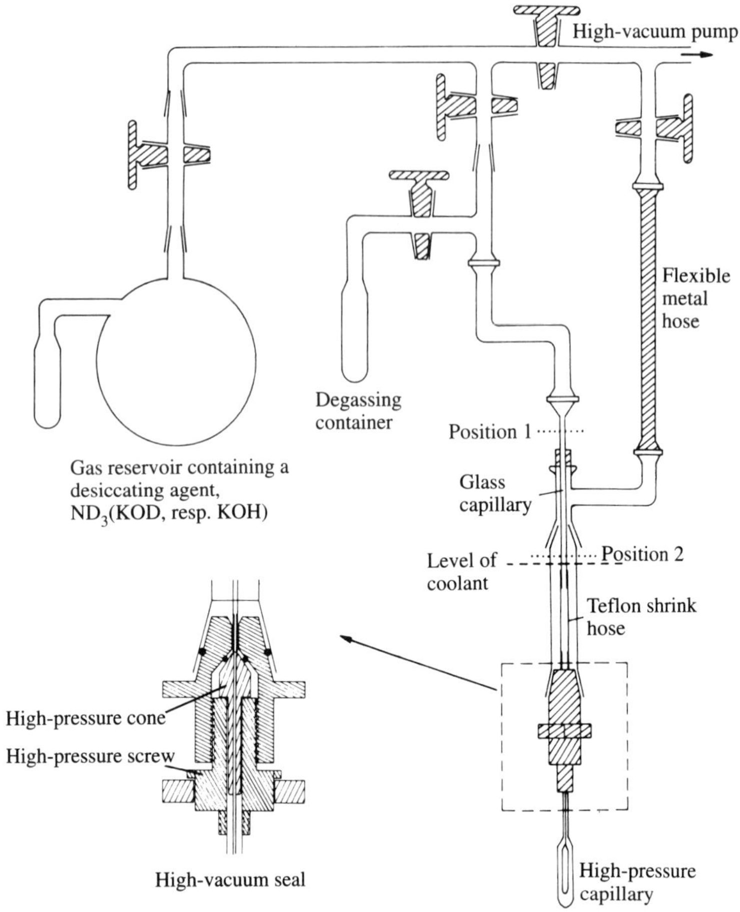

Fig. 5.7 A filling apparatus for volatile-liquid samples.

After this has been achieved the sample is flame sealed at position 1. All the glass parts and the high vacuum seal are removed under continuous cooling from the high-pressure capillary, and the filling capillary is resealed at position 2. The capillary glass sample cell is then screwed into the copper–beryllium autoclave, which is precooled, while the assembly remains in the cold bath until the assembly is complete.

An alternative method which is particularly suitable for gases such as methane[26] has been developed. This is shown in Fig. 5.8. After evacuating the high-pressure apparatus, the gases are condensed directly from the steel gas cylinder into a liquid-nitrogen-cooled autoclave made of a titanium alloy. Freeze–pump–thaw cycles are then performed until a final constant pressure at liquid-nitrogen temperatures is achieved. This might take 10–12 h for each cycle, because of the small cross-sectional area of the high-pressure tubing,

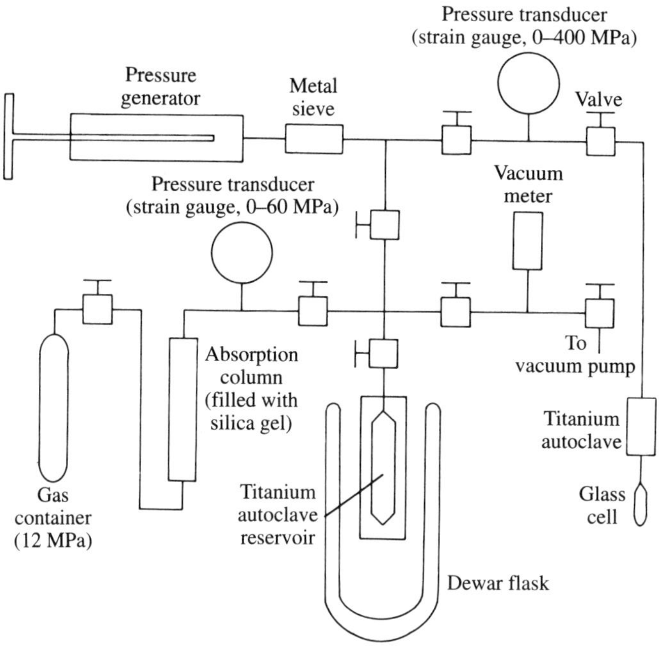

Fig. 5.8 A high-pressure apparatus used for filling capillaries with a gaseous sample.

and it also usually results in a considerable loss of the volatile substance under study.

1.5.3 Gas–liquid mixtures

The most suitable type of glass cell for use with gas–liquid mixtures is illustrated in Fig. 5.9.[27] The gas component is dried and degassed in the high-pressure apparatus shown in Fig. 5.8. The gas–liquid mixture is prepared in the autoclave in Fig. 5.9(a). The autoclave is flushed with the gas at atmospheric pressure. Then a defined amount of the degassed liquid is introduced into the autoclave with a syringe fitted with a hypodermic needle. The autoclave is closed immediately, and the gas is forced into the autoclave at the lowest pressure used later in the experiments. During this process the bronze bellows has to be kept in a contracted position which is controlled by linear-position transducers and regulated by the flow of standard-pressure liquid into the bellows.

After being filled, the autoclave is connected by a three-way valve and an $\frac{1}{8}$ in outer diameter high-pressure steel capillary with the strengthened glass cell shown in Fig. 5.9(b). This part of the system is evacuated and flushed with the gas several times before the final filling, in order to remove paramagnetic oxygen and other gaseous impurities. After pressure cycling between 25 and 200 MPa and after waiting for approximately 12 h, the

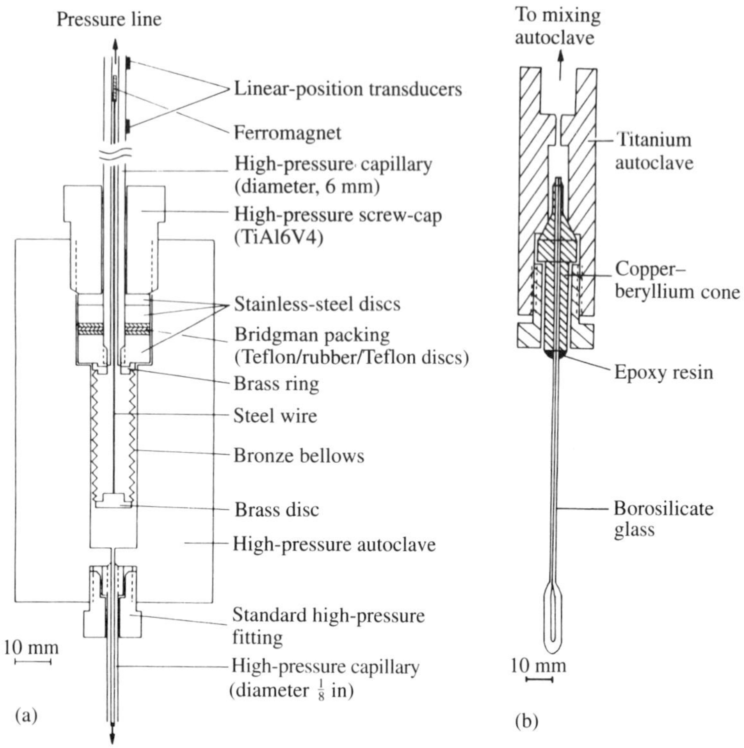

Fig. 5.9 An apparatus used for the preparation of samples of gas–liquid mixtures.

concentrations in the glass cell were found to be stable. They are determined from integration of the Fourier-transformed proton NMR spectrum. One of the inherent problems with this method is that it is very difficult to add a precise amount of gas. Therefore, the most reliable estimate for the composition of the mixture is that obtained from the NMR-signal ratios. Good spectra and signal-to-noise ratios are desirable in order to obtain values for the composition.

1.5.4 Emulsions

Much work has been carried out on pressure-dependent studies of properties (including diffusion measurements and spin-lattice relaxation times) in water and aqueous solutions.[13,14] There has been particular interest in supercooled metastable states. Heterogeneous nucleation is dependent on the sample volume. Consequently, a small volume capillary cell is required for high-pressure NMR studies on aqueous solutions under these conditions. This was discussed in Section 5.1.2.3. In order to stabilize aqueous solutions under

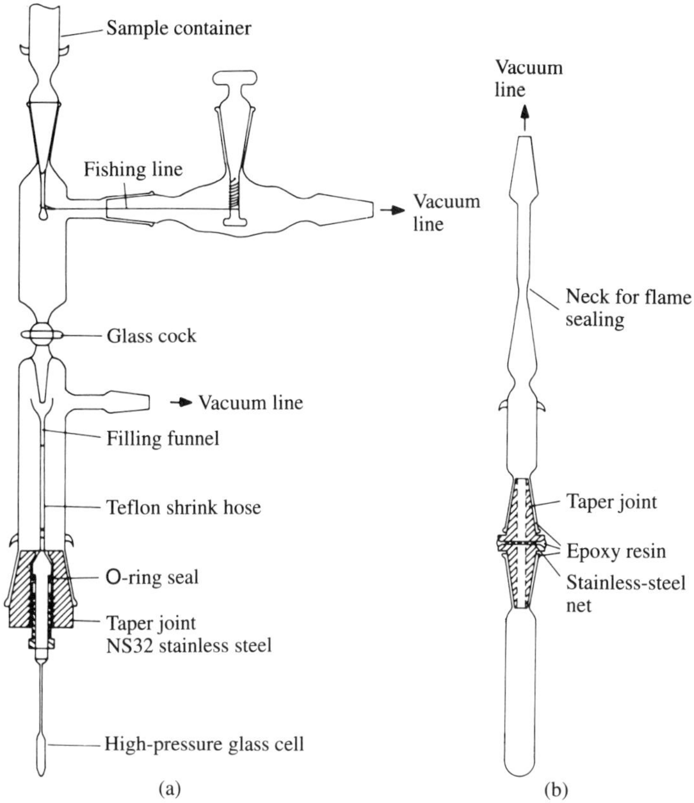

Fig. 5.10 The filling technique used for emulsified samples.

supercooled conditions further and to measure the spin-lattice relaxation times, an emulsion is often made of the sample. For example, a mixture of the aqueous sample with 1:1 methylcyclohexane–methylcyclopentane has been used.[16] The emulsion may itself be stabilized by the addition of 3 w/w% sorbitane tristearate.

In order to use these emulsions in NMR experiments a different filling procedure is required. This is detailed in Fig. 5.10. The components of the emulsion are weighed into the sample container in turn and then degassed by several freeze–pump–thaw cycles. The container is then flame sealed at the neck. The emulsion is produced by forcing the contents repeatedly through the stainless-steel net in the system. This might take a hundred or more passes. The emulsion is subsequently transferred to the high-pressure capillary in the filling apparatus after removing the neck of the sample container with the piece of fishing line.

1.5.5 Solutions

Liquid solutions and mixtures may be used in a similar manner to that described for neat liquids. The major difference is that the nature of the mixture must be considered, and steps have to be taken to retain the compositional integrity of the sample. For example, in experiments studying alkane–alcohol mixtures,[28] the capillaries may be filled in the way shown in Fig. 5.6. In this case it is necessary to ensure that all the sample volume has sublimed into the compartment containing the glass capillary. However, because of the fact that there is still the possibility of compositional change, it is wise to conduct a trial filling of the samples and to check the final composition by a suitable analytical method. In addition, it is desirable to place a small magnetic stirrer in the chamber containing the sublimed mixture so that the sample can be thoroughly mixed just prior to filling, because it is likely to be heterogeneous after the sublimation and melting processes.

In cases in which there is a nonvolatile solute in a solvent (for example, salt in water)[14,15] it is not possible to sublime the solution into the left-hand chamber (shown in Fig. 5.6), and oxygen removal is best accomplished by carrying out the freeze–pump–thaw cycle *in situ* before filling. Again, it is necessary to add a magnetic-stirrer bar to the bottom of the compartment to ensure that a representative sample of the mixture is drawn up into the capillary.

2. Diaphragm-cell techniques for high-pressure diffusion measurements

2.1 Introduction

The diaphragm cell is the simplest method of determining intradiffusion coefficients under elevated-pressure conditions with an accuracy of ± 2 per cent or better. The technique used for measuring diffusion at ambient-pressure conditions was developed from Graham's original work[29] in the 1860s.

The method which is currently most often used for ambient-pressure measurements is based upon work by Stokes[29] and by Mills and coworkers.[30] The set-up in this method has two compartments of a cell separated by a diaphragm (generally a glass frit) through which diffusion occurs. The two compartments contain the same solution at the beginning of the experiment. After reaching thermal equilibrium a small amount of a radio-labelled tracer is added to one of the volumes. For example, carbon-14-labelled dichloroethane may be added to a mixture of dichloroethane and hexane. The small amount of tracer added (approximately 10^{-11} mol mL^{-1}) should be sufficient to allow accurate analysis of the concentrations of the tracer species in both the cell volumes by standard liquid-scintillation techniques,[30] but it only produces a single concentration gradient. This means that the experiment measures the intradiffusion coefficient of dichloroethane in dichloroethane–

hexane at the particular mole fraction of mixture made. In the case of a pure liquid, the amount of tracer added is sufficiently small that the measured quantity is the self-diffusion coefficient, although care has to be taken because of isotopic-mass effects. The measurement simply involves measuring the concentrations of the tracer in the two volumes accurately ($\pm$ 0.1 per cent is possible) at a certain time after the tracer was added (generally measured in days). The unknown diffusion coefficient is calculated using a calibration constant obtained from diffusion measurements on a system with a known diffusion coefficient.

By stirring the solutions in the two compartments with magnetic stirrers, the concentration gradient is confined to the diaphragm pores. This makes the mathematical analysis considerably easier,[30,31] and it aids in the attainment of accurate diffusion coefficients. It should be noted that this technique may also be used for the measurement of interdiffusion coefficients—by having different concentrations of the same system on either side of the diaphragm and not using a tracer but measuring the change in concentration accurately after an appropriate time interval. This is of relevance to the calibration of the cell, and consequently to the calibration of a NMR spectrometer for spin-echo diffusion measurements.

The ambient-pressure diaphragm-cell method was adapted and developed for high-pressure measurements by Woolf and co-workers in the 1970s.[32,33] It retains the simplicity of the ambient-pressure glass diaphragm cell and has been used to obtain self-diffusion and intradiffusion coefficients over a wide temperature and pressure range to an accuracy of 1–2 per cent, which is in excellent agreement with self-diffusion measurements made with NMR techniques.

2.2 Cell design and procedures for high-pressure work

A high-pressure diaphragm cell is shown in Fig. 5.11. The design consists of three sections: two compartments (referred to as the *top* and *bottom compartments*) and a membrane (the stainless-steel-sinter section in Fig. 5.11). The three parts screw together on to sealing rings made of Viton with a square cross-section.

The membrane piece is a large cylinder made from a nonferrous alloy that contains a stainless-steel sinter welded into a cylinder of the same material. This stainless-steel cylinder is then shrink fitted into the larger cylinder. The advantage of this separable feature of the design is that a variety of different membrane sections may be interchanged (with a range of permeabilities), making the cell suitable for accurate diffusion-coefficient measurements over a wide range of values.

The stirrers have two sets of blades and are made of magnetic stainless steel. They rest on supports located on the ends of membrane-containing cylinder; thus they do not come into contact with the membrane, avoiding abrasion of the sinter surface.

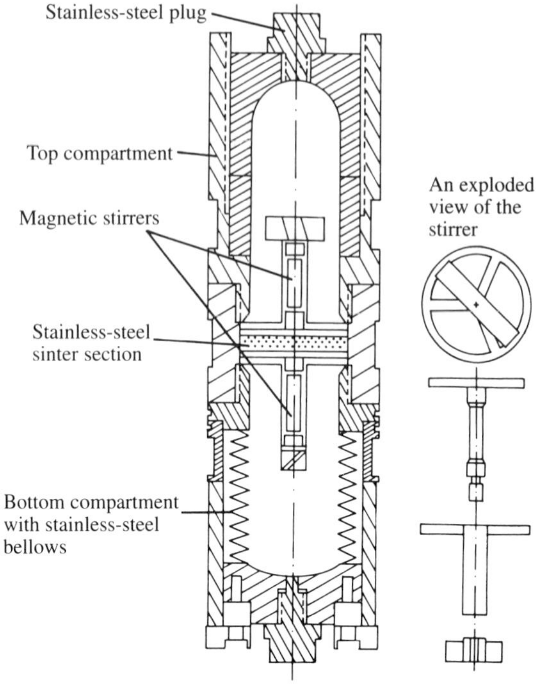

Fig. 5.11 A high-pressure diaphragm cell.

The bottom compartment incorporates a stainless-steel bellows, which transmits the pressure to the cell solutions. The top-compartment volume is adjusted before the experiments using inserts which allow the volumes of the cell compartments to be matched at the operating temperature. It is important to note that, as a consequence, for accurate work, compressibility data for the liquid system is necessary in order to match the volumes successfully.

The metal cell is evacuated, the solution being drawn up through a metal filter attached to the opening of the bottom compartment. During filling, retaining screws may be used to hold the bellows in place to avoid damage. A glass separating bulb is screwed into the top compartment, allowing the cell to be overfilled. The other end of the glass bulb is connected to the vacuum. Once the cell has been filled (overfilled) it is essential to ensure that no gas bubbles are trapped within the sinter or any other parts of the cell. This is achieved by removing the bellows' retaining screws and repeatedly flushing liquid through the sinter and up into the glass reservoir. Then the bellows' retaining screws are replaced, and the two compartments are sealed with stainless-steel screws fitted with Viton O-rings. The cell is then placed into a thermostated bath to equilibrate at ambient pressure. The top compartment is not sealed tightly during this process. The labelled material is then added (always) to the top compartment before the pressure is applied.

The retaining screws in the bottom compartment are then removed, the top compartment is sealed, and the cell is placed inside a standard, capped, copper–beryllium pressure vessel. The flow of liquid from the bottom compartment into the top, because of the increase in pressure, helps to minimize diffusion across the sinter before thermal equilibrium is restored by the compression-induced temperature rise. Strong magnets, mounted on a rotating shaft outside the vessel, turn the stirrers.

The pressure is released after a suitable time interval, and the contents of the chambers are sampled using syringes and PTFE tubing. It is important to keep the cell thermally stable during the sampling to avoid temperature-induced flows. The drop in pressure has two effects. It causes a cooling of the cell contents and a small but unavoidable amount of bulk flow of material from top to bottom. It is therefore essential to stir the cell for several minutes after reaching ambient pressure, to attain thermal stability and to ensure a uniform composition of the bottom compartment. The concentrations of labelled material in the compartments are determined by liquid-scintillation techniques.[30] The effects of the bulk flow must be taken into account, and equations have been developed for this purpose.[32,33]

This design has successfully measured intradiffusion coefficients up to 400 MPa over a temperature range of 278–338 K.

2.3 Equations and experimental criteria

It is not the purpose of this section to develop fully the theoretical background for the diaphragm-cell method: interested readers are referred elsewhere.[4,30,34] For an intradiffusion experiment in which labelled material undergoes a net transfer from the top chamber to the bottom, the intradiffusion coefficient may be obtained from the equation[4,30]

$$D_{\mathrm{i}} = \frac{1}{\beta t}\ln\left(\frac{({}^{0}C_{\mathrm{T}} - {}^{0}C_{\mathrm{B}})\,(1 - \lambda/6)}{C_{\mathrm{T}} - C_{\mathrm{B}}}\right), \tag{5.2.1}$$

where

$$\beta = \frac{A}{L}\left(\frac{1}{V_{\mathrm{T}}} + \frac{1}{V_{\mathrm{B}}}\right)\left(1 - \frac{\lambda}{6}\right) \tag{5.2.2}$$

and

$$\lambda = \frac{2\,V_{\mathrm{D}}}{V_{\mathrm{T}} + V_{\mathrm{B}}}. \tag{5.2.3}$$

Here, ${}^{0}C_{\mathrm{T}}$ and ${}^{0}C_{\mathrm{B}}$ are the initial concentrations of component i in the top and bottom cell compartments, while C_{T} and C_{B} are the concentrations after the experiment, when time t has elapsed; A is the effective area of the sinter; L is the sinter's effective thickness; V_{B}, V_{T}, and V_{D} are the volumes of the bottom compartment, the top compartment, and the sinter.

The volumes at ambient pressure are determined experimentally by weighing. The particular insert used in the top compartment to match the top and bottom volumes at the operating pressure is taken into account. The bottom-compartment volume at the operating pressure is calculated from knowledge of the compressibility of the liquid.[4,33] The final concentrations C_T and C_B, determined by accurate radiochemical analysis, are adjusted because of small amounts of bulk flow into the bottom compartment.[32,33] The concentrations at the end of the experiment enable ${}^0C_T$ to be calculated by the mass-balance equation

$${}^0C_TV_T + {}^0C_BV_B = C_TV_T + C_BV_B + \tfrac{1}{2}(C_T + C_B)V_D. \tag{5.2.4}$$

This equation assumes that the concentration gradient in the sinter is linear at the conclusion of the experiment. In the case of high-pressure measurements ${}^0C_B$ is generally zero (or negligibly small). Careful radiochemical analysis ($\pm$ 0.1 per cent) allows the diffusion coefficients to be determined with an accuracy of 1–2 per cent.

Although the value of β, the cell constant, can be calculated in principle from geometrical considerations and measurements, a more accurate approach[4,34] is to obtain values for A/L from a calibration experiment using a system at ambient pressure with a known diffusion coefficient. The system most commonly used[35] is the intradiffusion of tritiated water in water at 25 °C. The accepted value is 2.236×10^{-9} m^2 s^{-1}. This value is accurate to $\pm$ 0.5 per cent, and is itself obtained via calibration of a glass (ambient pressure) diaphragm cell using the interdiffusion of 0.5 M potassium chloride into water at 25 °C (known from interferometric measurements).[30]

The diaphragm-cell technique is very simple in principle, but a great deal of care needs to be exercised in order to obtain maximum accuracy. Important points which need to be considered include:

1. The cell volumes need to be matched as closely as possible, and the volume of the sinter should be small ($V_D < 0.02V_T$).
2. Temperature variations in the cell need to be minimized ($\pm$ 0.01 K or better).
3. Stirring must be adequate to ensure mixing but must not cause bulk flow across the membrane. A value of 60 r.p.m. (revolutions per minute) is commonly used; however, it is best to test the system to determine an acceptable range.
4. Regular calibration is required to guard against drifting cell constants.
5. The precision of an experiment can be optimized[30,36] by making its duration satisfy the approximate condition: $t_{opt}/s = 1.1 \times 10^{-4}/\beta D$.

2.4 Use of diaphragm-cell measurements to calibrate spin-echo NMR diffusion measurements

It is worth making a brief mention of the use of accepted diffusion-coefficient values from ambient pressure diaphragm-cell experiments which are suitable for calibration of a gradient-field spectrometer. The commonest value used for the self-diffusion coefficient is the value for water (2.3×10^{-9} m^2 s^{-1}) at 25 °C obtained by Mills[35] using diaphragm-cell measurements and isotopic-mass-effect considerations. This is accurate to ± 0.5 per cent. When a calibration value is needed at temperatures other than 25 °C water is again the best medium to use because accurate (± 0.5–1 per cent) self-diffusion coefficient values are available for the entire ambient-pressure liquid range.[34,37] For much smaller diffusion-coefficient measurements this value is not appropriate, and consequently the value for self-diffusion in cyclo-octane at 25 °C is commonly used.[6]

2.4.1 Comparison of NMR diffusion measurements with diaphragm cells

With the advent of Fourier-transform NMR and powerful superconducting magnets, spin-echo NMR is now the method of choice for high-pressure intradiffusion measurements. Individual measurements take of the order of few (2–30) minutes to collect rather than the 24–50 h needed for diaphragm-cell measurements. Because of the higher sensitivity and the faster collection of data it is now possible to study by NMR systems with diffusion coefficients which are as low as 10^{-14} m^2 s^{-1}, whereas the time scale for diaphragm-cell measurements really prevents accurate measurements much below 10^{-11} m^2 s^{-1}.

The temperature and pressure range that can be covered by NMR measurement is also greater than that by diaphragm cells due to the noninvasive nature of the measurement. However, except in studying very extreme pressure systems or deeply supercooled ones, this is not a serious limitation for the diaphragm-cell method.

The major advantages of the diaphragm-cell method are its low cost and its slightly superior accuracy (under ideal conditions). It is also possible to use the technique to look at diffusion in complex mixtures which are problematical by NMR. In order to obtain a valid spin-echo measurement for a particular species in a mixture, it is necessary to resolve a signal in the spectrum, with no underlying peaks. However, this may be overcome by isotopic substitution (that is, by deuteration) of a sample. Where this is not a problem, NMR is again the preferred technique, because the diffusion coefficient of all species in a mixture can potentially be obtained in one experiment.[38]

3. Typical results from high-pressure NMR studies

3.1 The activation volume of intramolecular dynamics

Analysis of exchange-broadened NMR spectra (that is, the simulation of temperature-dependent spectra as a function of the exchange rate) has become the method of choice for the determination of the free activation enthalpy for intramolecular conformational equilibria and for intermolecular exchange processes.[39] Studying, in addition, the pressure dependence of the exchange-broadened spectra permits the determination of the activation volume, defined by

$$-\frac{d(\ln k)}{dp} = \frac{\Delta V^{\ddagger}}{RT} \qquad (5.3.1)$$

Most scientists involved in high-pressure kinetics believe that activation volumes can be determined more accurately, and that they are easier to interpret, than activation entropies.[40,41] Our group has studied, with the strengthened glass cell, a fairly large series of amides and *N,N*-disubstituted amides by this technique. The results are described in more detail in a recent review.[14] Only one example will be given here, an example in which two independent conformational equilibria are observed in the spectra.

Figure 5.12 shows a series of high-resolution spectra for the *n*-alkyl moieties of *N,N*-Di-*n*-propyl-1-naphthoamide. In this and similar naphthoamides two types of conformational equilibria can be observed in the *N*-alkyl part of the proton spectra:

I. The rotation around the amide C—N partial double bond; this process broadens all six groups of protons.
II. At lower temperatures, where the amide rotation has become too slow to influence the *J*-resolved spectra, only the —N—CH_2-proton signal from the group in the *cis*-position to the naphthalene rings is still exchange broadened, and it separates on further cooling into a complex AB—C_2 pattern. This is ascribed to the sterically hindered rotation of the complete amide group around the carbonyl/aryl bond over the plane of the naphthalene rings.[42] The activation parameters for these two equilibria are given in Table 5.1 for three different naphthoamides.

The aryl/carbonyl rotation rate is practically independent of the pressure, while a significant positive $\Delta V^{\ddagger}$ is seen for the 180° flips of the amide group. Considering that (especially in the case of *N,N*-dibenzyl-1-naphthoamide) really bulky and rigid substituents have to be moved simultaneously neither of the activation volumes can be correlated with the 'free volume' of the solvent necessary for this displacement to occur. Steric requirements can, however, explain $\Delta V^{\ddagger}$ quantitatively for the *N,N*-dimethylamides and the unsubstituted amides. The results have been tentatively explained by comparing the time scales of the solvent-diffusion processes with the time required by the bulky substituents for rotation.[43,44] Estimates show that for the higher amides internal mobility is slower than the diffusivity of the solvent molecules (because of the large moments of inertia), and thus both conformational transitions occur under a continuous rearrangement of the solvent molecules. The complicated solvent mixtures were chosen for this study because large temperature changes, and solubility effects, made it impossible to study the substances in one pure solvent over the whole temperature range necessary without partial crystallization of the solvent or solute.

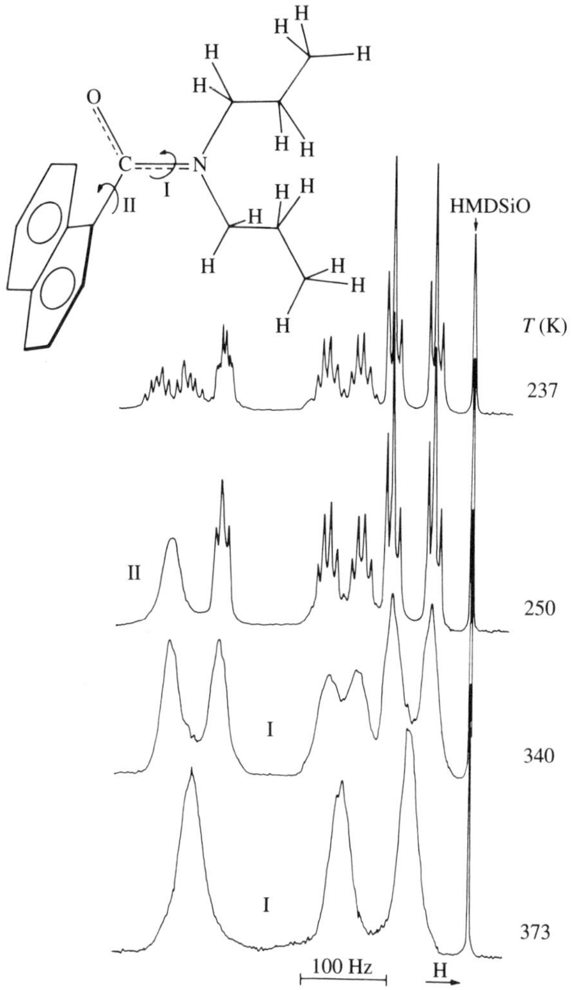

Fig. 5.12 High-resolution NMR spectra for the *n*-alkyl moieties of *N,N*-di-*n*-propyl-1-naphthoamide as a function of temperature and pressure.

3.2 Self-diffusion and spin-lattice-relaxation-time measurements

The self-diffusion coefficients described below were measured by the pulsed-field technique proposed by Stejskal and Tanner.[5] The pulse sequence applied is a modification of the classical Hahn spin-echo experiment for the determination of the spin–spin relaxation time, T_2. Fourier transformation of the second half of the spin-echo permits the simultaneous study of complex mixtures. The limit for the number of compounds that can be measured in one experiment is set solely by the requirement that there is at least one

Table 5.1 The free enthalpies of activation and the activation volumes for the two conformational transitions described in Fig. 5.12 in a series of *N,N*-dialkyl-1-naphthoamides.

Substance	Concentration, w/w (%)	Solvent, w/w (%)	ΔG‡ (kJ mol^{-1}) I	ΔG‡ (kJ mol^{-1}) II	ΔV‡ (cm^3 mol^{-1}) I	ΔV‡ (cm^3 mol^{-1}) II
N,N-dimethyl-1-naphthoamide	10	Acetone 20 CD_2Cl_2 70	74	—	7.5±1	—
N,N-diethyl-1-naphthoamide	20	Acetone 30 Benzene 30 Toluene 20	72	61	6.5±0.5	0±1
N,N-di-*n*-propyl-1-naphthoamide	20	CD_2Cl_2 60 Acetone 20	74	63	6.5±1	0±1
N,N-dibenzyl-1-naphthoamide	10	Acetone 10 Toluene 53 Pyridine 27	73	63	8.5±1	0±1

chemical-shift separated NMR signal for each substance in the mixture. The echo amplitude, A, in a Hahn spin-echo experiment, characterized by the spin–spin relaxation time, T_2, is given by the first exponential term in eqn (5.1.1). It is reduced in the presence of a pulsed field gradient by the amount given in the second exponential term.

The determination of D is most conveniently accomplished by increasing the current, I, and thus raising g while keeping all other variables in eqn 5.1.1 constant. The coil constant was determined using the known self-diffusion coefficient of water at ambient pressure at 298 K obtained by Mills and Wolf;[30] this is generally acknowledged to be the most reliable value. The coil constant was controlled with a redetermination of D for benzene. In our design (given in Fig. 5.3), the gradient coil is kept at atmospheric pressure and room temperature; thus the coil constant is completely independent of the pressure and temperature of the sample. The small diameter of the glass tube, though responsible for the low sensitivity, effectively eliminates thermal convection of the liquid under study. Such convection is a real danger in diffusion studies since it will contribute to the molecular mobility, and thus it will produce erroneously large self-diffusion coefficients.

3.3 The quadrupole relaxation of deuterons (D) and oxygen-17 (^{17}O)

The spin relaxation of most quadrupolar nuclei is completely given by the interaction of the nuclear quadrupole moment, eQ, with the electric-field gradient (e.f.g.) at the nucleus. To a good approximation the e.f.g. is generated in most molecular liquids exclusively by intramolecular charges. The interaction fluctuates in time because of the reorientation of the X—D or the X—^{17}O bond of the diffusing molecules. The spin-lattice ($R_1 = 1/T_1$) relaxation rates of the deuterium nucleus (whose spin $I = 1$) can be expressed as[45]

$$R_1 = \left(\tfrac{3}{2}\pi^2 \frac{2I+3}{I^2(2I-1)}\right) \chi_{\text{eff}}^2 \,[2g(\omega_D\tau) + 8g(2\omega_D\tau)], \qquad (5.3.2)$$

where the effective quadrupole coupling constant $\chi_{\text{eff}} = e^2 \langle q_D \rangle Q/h$; here Q is the nuclear quadrupole moment and $\langle q_D \rangle$ is the component q_{OD} of the e.f.g. along the D-bond averaged over the fast librations of the water molecules. The slower rotational diffusion of the molecules provides the main contribution to the spectral intensity, $g(\omega)$, at the Larmor frequency, ω_D. If the tumbling of the molecules is isotropic and characterized by a single correlation time τ, the spectral intensity is given by

$$g(\omega\tau) = \tau/[1 + (\omega\tau)^2]. \qquad (5.3.3)$$

For low-molecular-weight liquids of normal viscosity the correlation time, τ, is much shorter than ω^{-1}. In this region T_1 is independent of the frequency. It is only in this region, commonly called the extreme-narrowing limit, where eqn 5.3.2 may also be applied to nuclei with $I > 1$ (for ^{17}O, $I = \frac{5}{2}$). Outside this range the magnetic relaxation is no longer characterized by a single exponential.[13,14]

3.4 The magnetic-dipole relaxation of the protons

The magnetic dipole interaction between two protons (I, S) depends on their relative distance $r_{I,S}$ and on the orientation of the molecule in a laboratory-fixed frame. Both protons vary in time due to rotational and translational diffusion depending on whether the two spins reside on the same molecule (intra) or on different molecules (inter). The total dipole relaxation rate for any neat liquid thus consists of two terms,

$$R_1 = (R_1)_{\text{intra}} + (R_1)_{\text{inter}}. \qquad (5.3.4)$$

The intramolecular spin-lattice relaxation rate between two protons ($I = \frac{1}{2}$) may be written as

$$(R_1)_{\text{intra}} = \tfrac{3}{20}\, \gamma_H^4\, \hbar^2 < R_{HH}^{-3} >^2 F(\omega_H\tau), \qquad (5.3.5)$$

where $\langle R_{HH} \rangle$ is the H–H distance within a molecule averaged over the fast librations of the molecules, γ_H is the gyromagnetic ratio of the proton, and $\hbar$ is the Planck constant divided by 2π. Hence $(R_1)_{\text{intra}}$ is a quantitative measure of the reorientation of the H–H vector within a molecule.

The intermolecular spin-lattice relaxation rate of the protons will be written as

$$(R_1)_{\text{inter}} = \tfrac{6}{5}\pi\gamma_H^4\, \hbar^2\, (N_H/2D_H d_{HH})\, F(\omega_H\tau_d), \qquad (5.3.6)$$

where N_H is the number density of the protons, D_H is the self-diffusion coefficient of the protons, and d_{HH} the average distance of closest approach of two protons. The spectral density $F(\omega_H\tau_d)$ can be derived from motional models.[45]

Figure 5.13 shows the dependence on pressure, for various temperatures, of the rotational and translational molecular mobility for trifluoromethane. Rotational mobility is characterized by the deuteron spin-lattice relaxation time, T_1, and translational mobility is characterized by the self-diffusion coefficient. In all nonpolar liquids, and also in most polar liquids, changes in temperature and pressure have a stronger influence upon the translational mobility than upon the rotational mobility.

Figure 5.13 also shows the self-diffusion coefficients measured for octamethylcyclotetrasiloxane, a large, heavy, approximately spherical nonpolar molecule.

The dependence of self-diffusion on density, ρ, and temperature can be described in all nonpolar liquids with good accuracy by modifications of the hard-sphere model.[46] This model also provides quantitative descriptions for self-diffusion of the normal alkanes;[47] and these studies could be extended to polyethylene waxes with an average chain length of 154 carbon atoms.

Figure 5.14 shows high-pressure isobars for the self-diffusion data of the *n*-alkanes in a plot of log D against log M_w where M_w is the molecular weight. These are the first data obtained with the titanium autoclave described in Section 1.4.2.[48] Such results are commonly described by the Rouse model or by the reptation model, which both predict a linear correlation in this type of plot at constant pressure and temperature; this linear correlation is clearly established in Fig. 5.14. Judging from the chain length of the polymethylenes the Rouse model should apply. This model predicts, that D should be proportional to M_u^{-1}, while the experiments give a $D \approx M_w^{2\pm0.2}$ correlation, which is still retained if the data for the lower alkanes (pentane to tetradecane) are included in the plot. This correlation has been discussed in detail previously,[48] and it should provide a valuable method of interpolation for *n*-alkanes with intermediate chain lengths.

Figure 5.15 shows high-pressure self-diffusion coefficients for CH_3OH and H_2O, two hydrogen-bond-containing liquids.[49,18,19] These data extend far into the supercooled range. They were obtained in the thin glass capillaries (with an inner diameter of 100–200 μm) shown in Figs 5.1 and 5.2. The small diameter is essential for reliable undercooling of the neat liquids. The temperature dependence of D for both substances is much more pronounced than for the nonassociated liquids described above. This is typical for all liquids that can form two- or three-dimensional hydrogen-bond networks. In this class of liquids, removal of thermal energy leads to the formation of larger hydrogen-bond networks, and simultaneously it enhances the lifetime of the single hydrogen bonds.

Cold water has, in addition, a unique pressure dependence for D. The initial application of pressure enhances self-diffusion. The effect becomes more pronounced in the metastable supercooled region. This pressure

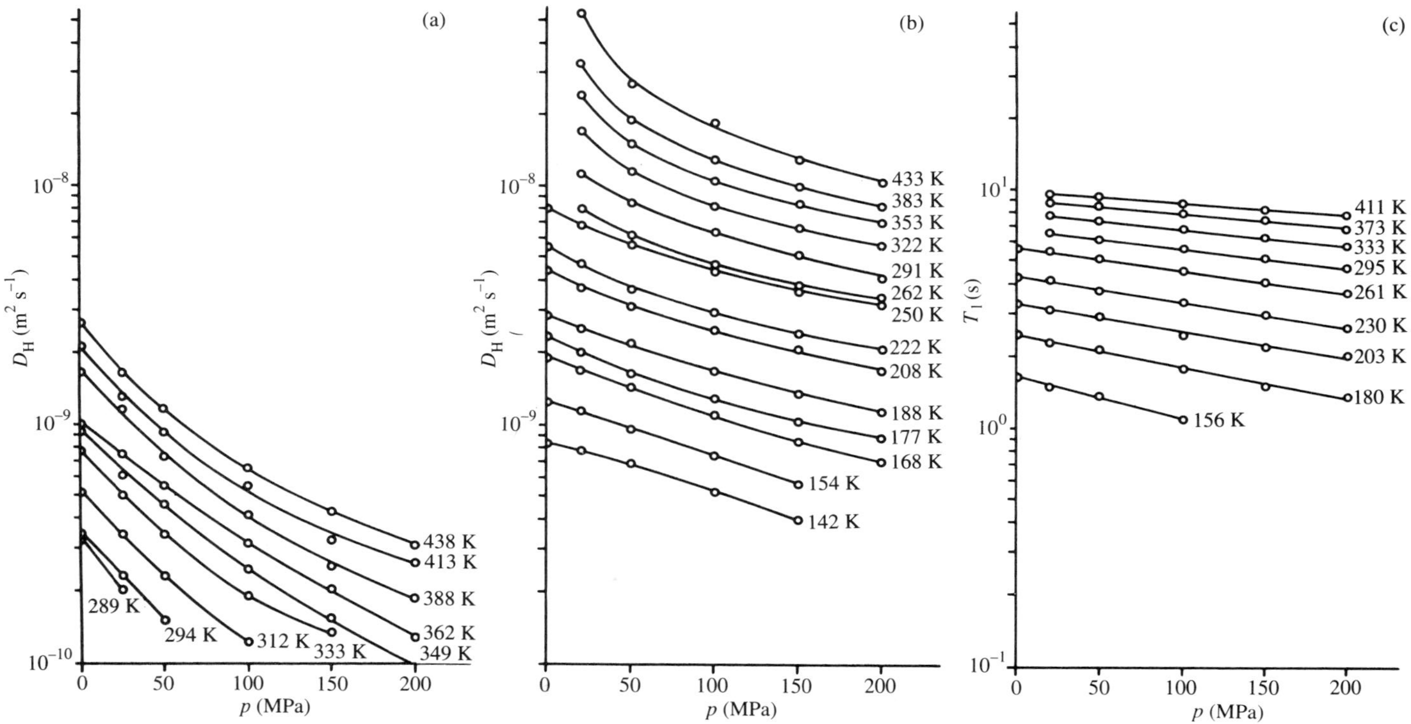

Fig. 5.13 The pressure dependence at various temperatures of: (a) the translational mobility of octamethylcyclotetrasiloxane (OMCTS); and (b) the translational mobility of CF_3H (trifluoromethane), and (c) the rotational mobility of CF_3D.

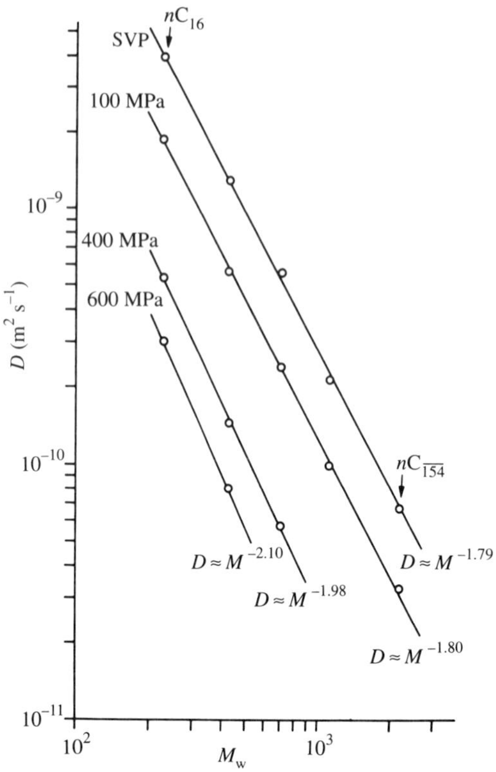

Fig. 5.14 High-pressure (saturated vapour pressures—SVP) isobars for self-diffusion at T = 473 K of a number of n-alkanes plotted as a function of the molecular weight, M_w.

anomaly also occurs for the rotational mobility in D_2O which is enriched with oxygen-17. The experimental spin-lattice relaxation times, T_1, of the deuterons and of the oxygen-17 nuclei are given in Fig. 5.16. Both sets of data reveal an identical (ρ,T)-dependence for T_1. Since the main electric-field gradients of the two nuclei are perpendicular to each other, this result is a strong indication that the reorientation of the water molecules is isotropic on the NMR time scale and can be determined by eqns (5.3.3) and (5.3.4). The quadrupole coupling constant (QCC) for the oxygen-17 nuclei in water is approximately larger by a factor of thirty than the deuteron QCC.[13,14] Thus the oxygen-17 value for T_1 is (at identical density and temperature) almost three orders of magnitude shorter than the deuteron value for T_1. The low-temperature limit of the oxygen-17 value for T_1 is thus set by the technical limitations of current NMR technology, which do not permit a reliable determination of T_1 for less than around 5×10^{-4} s. An unexpected anomaly is apparent from a comparison of the water data given in Figs 5.15 and 5.16. The anomalous increase in the mobility observed in cold water at the initial

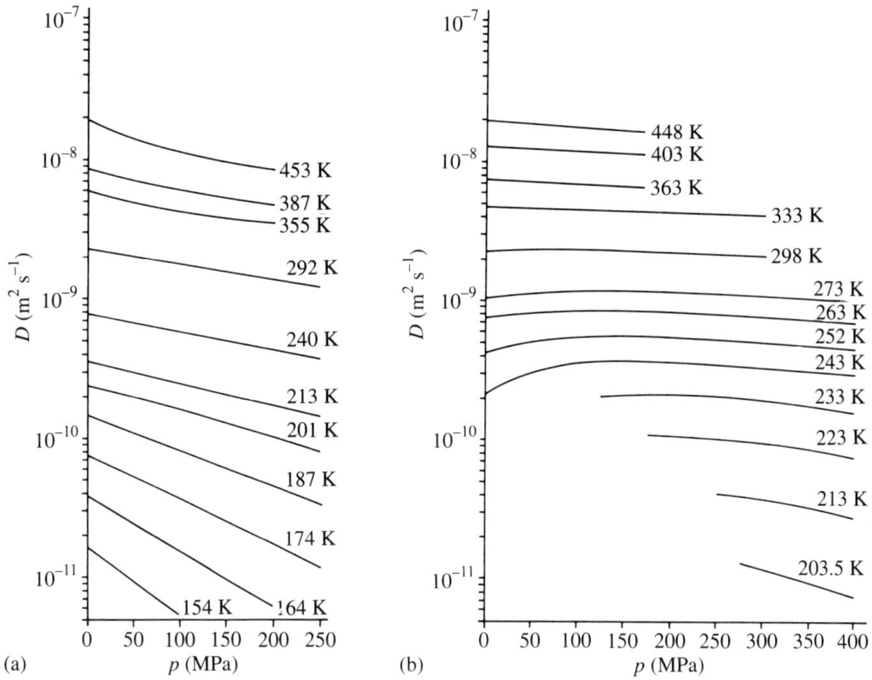

Fig. 5.15 High-pressure self-diffusion data for: (a) methanol (CH_3OH) and (b) water (H_2O).

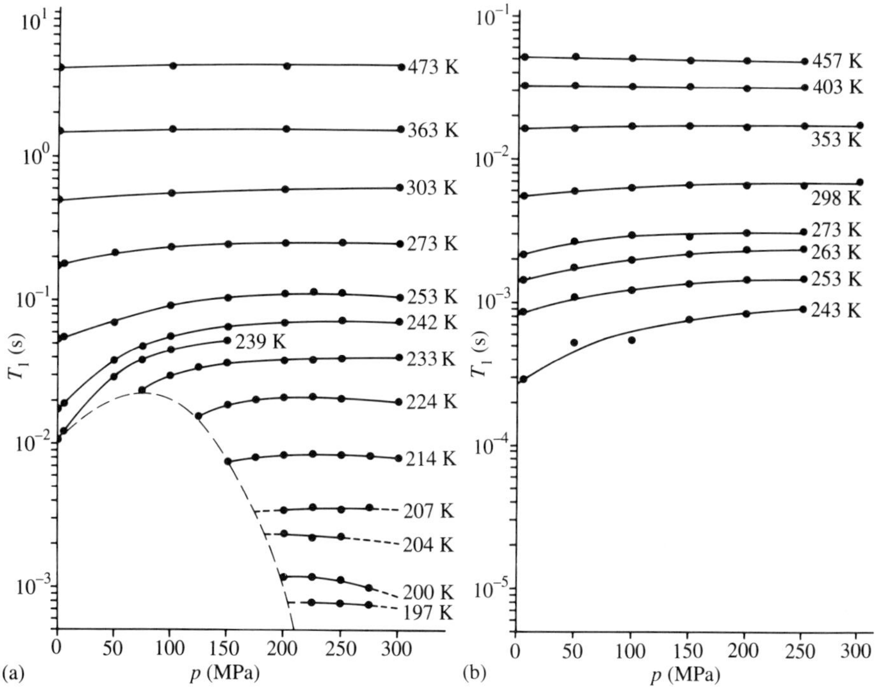

Fig. 5.16 Spin-lattice relaxation times, T_1, of (a) deuterons in D_2O, and (b) ^{17}O in $D_2{}^{17}O$.

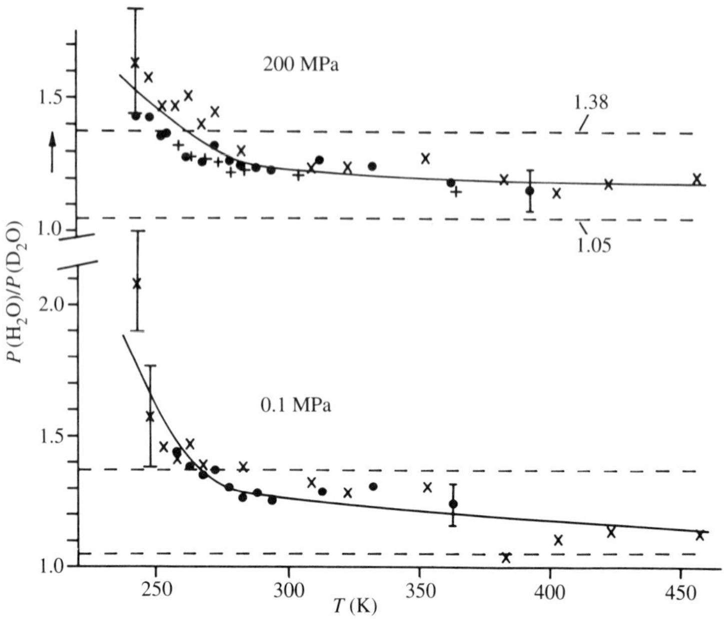

Fig. 5.17 The dynamic isotope effect for water as measured by: (×) spin-lattice relaxation, $1/\tau_2$; (•) self-diffusion, D, and (+) viscosity, η.

application of pressure is much more pronounced for rotation when monitored by deuteron or oxygen-17 T_1 measurements than for the translational self-diffusion. A more detailed discussion of this effect is given in the original papers.[18,19]

Figure 5.17 describes the dynamic isotope effect as described by the ratio:

$$\left(\frac{P(H_2O)}{P(D_2O)}\right)_{p,T}, \quad \text{where } P = \tau_2, D, \eta \tag{5.3.7}$$

Two unusual features can be observed in these plots (and, at least for the self-diffusion coefficient, this behaviour is common to all hydrogen-bonded liquids). This ratio is a function of temperature. At constant temperature and pressure, rotation and translation reveal the same isotope effect. From simple sphere dynamics one would expect the rotation to scale as the square root of the ratio of the moments of inertia (=1.38) while for translational mobility the square root of the ratio of the molecular masses (=1.05) should be found. This is clearly not the case, indicating that the dynamics of liquid water are really the dynamics of the hydrogen-bond network. The hydrogen bonds in D_2O are stronger than those in H_2O and thus the mobility in the D_2O network decreases more rapidly as the temperature decreases.

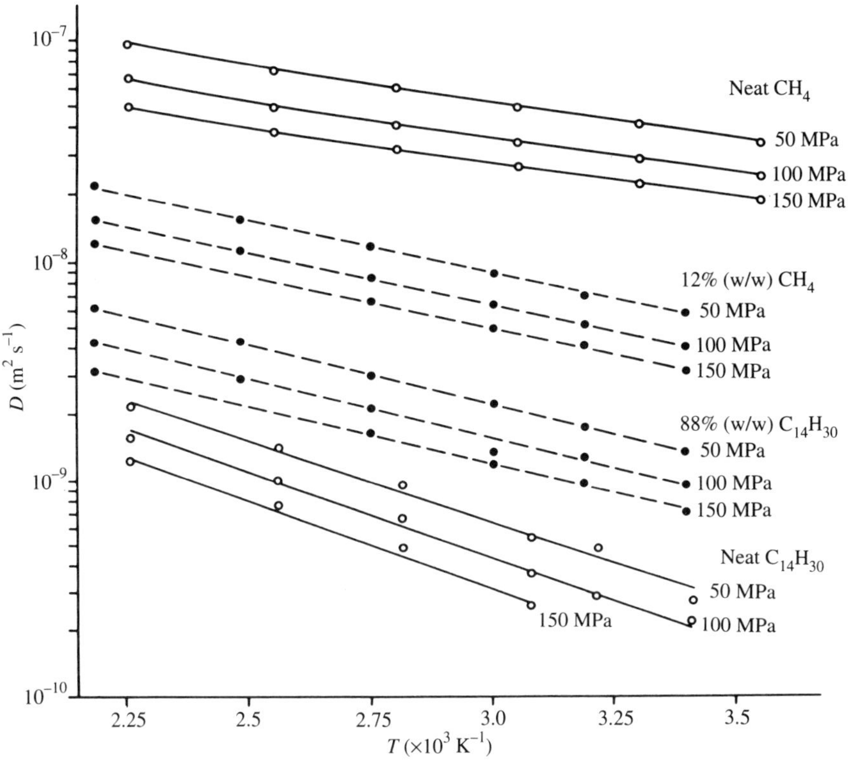

Fig. 5.18 Isobars for the intradiffusion of the components of a mixture of methane (12% (w/w) CH_4) and tetradecane ($C_{14}H_{30}$) together with data for the neat liquids.

In Fig. 5.18, isobars of the intradiffusion of the two compounds of a methane–tetradecane mixture containing 12% (w/w) methane are compared to the self-diffusion in the neat liquids.[50] The behaviour of this mixture appears to be typical for all mixtures consisting of nonpolar molecules: for neat substances the isobaric-temperature and isothermal-pressure dependence are significantly different. The larger molecules show a larger change of D with temperature and pressure. In all mixtures of nonpolar compounds studied to date both substances show at least a very similar (p,T)-dependence for D. In the binary mixture given in Fig. 5.18, $\Delta V^{\ddagger}$ and ΔE_a are identical within the limits of experimental accuracy. In Fig. 5.19, data obtained on a hexane–ethanol mixture are compiled.[28] Here the pressure dependence is also very similar for both compounds; however, the self-diffusion of ethanol decreases with falling temperature much faster than the self-diffusion of hexane (an alkane), indicating that even in fairly dilute ethanol–alkane mixtures two-dimensional hydrogen-bonded aggregates are formed by the ethanol molecules and this retards the translational mobility of the ethanol fraction of the mixture.

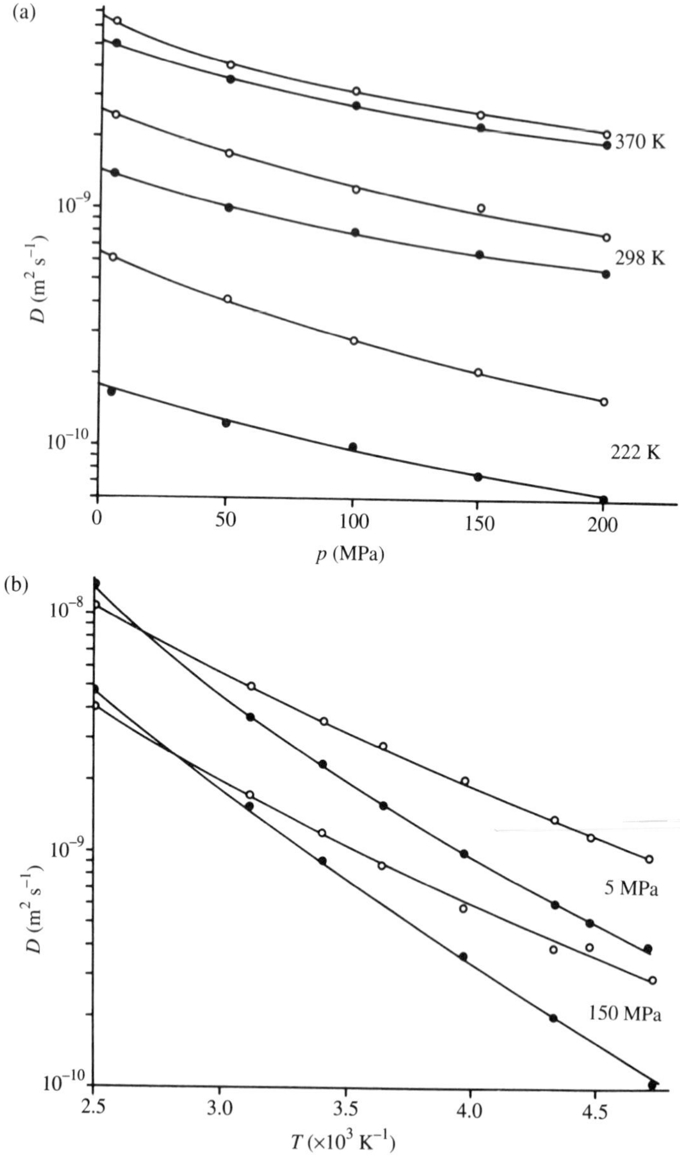

Fig. 5.19 Isobars (a) and isotherms (b) for intradiffusion in a binary mixture of hexane and ethanol (90:10 mol%): (○) hexane, and (●) ethanol.

In Figure 5.20, the rotational mobility of deuterons in neat D_2O is compared with the behaviour observed in solutions of some alkali and alkaline earth halides. In order to obtain these data, the aqueous liquids had to be emulsified in a cycloalkane mixture, which permits reliable undercooling of

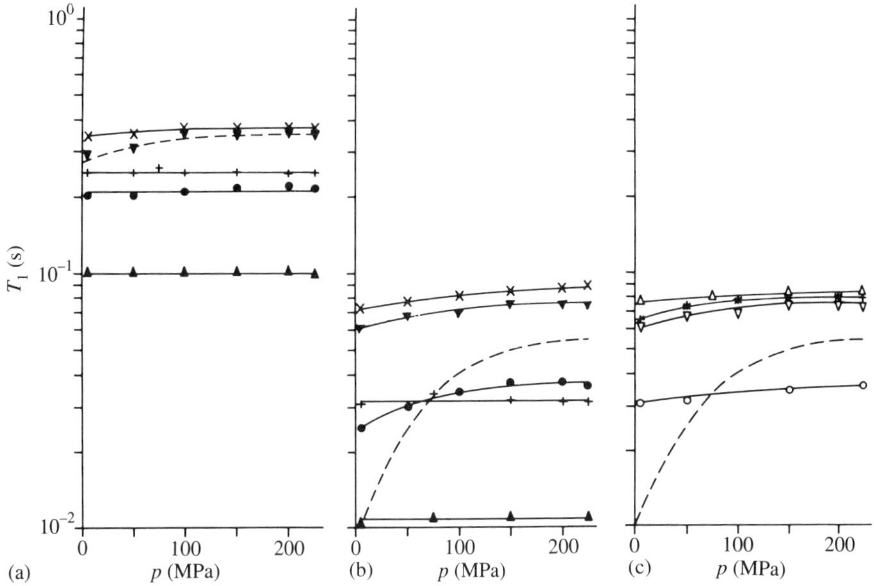

Fig. 5.20 The rotational mobility of deuterons in D_2O and solutions of alkali and alkaline-earth halides at temperatures of: (a) 283 K, for (– – – –) neat D_2O, (×) 3M CsCl, (▼) 3M KCl, (+) 3M NaCl, (●) 3M LiCl, and (▲) $MgCl_2$; (b) 238 K for (– – – –) neat D_2O, (×) 3M CsCl, (▼) 3M KCl, (+) 3M NaCl, (●) 3M LiCl, and (▲) 3M $MgCl_2$; and (c) 238 K, for (– – – –) neat D_2O, (△ 3M KI, (*) 3M KBr, (▽) 3M KCl, and (○) 3M KF.

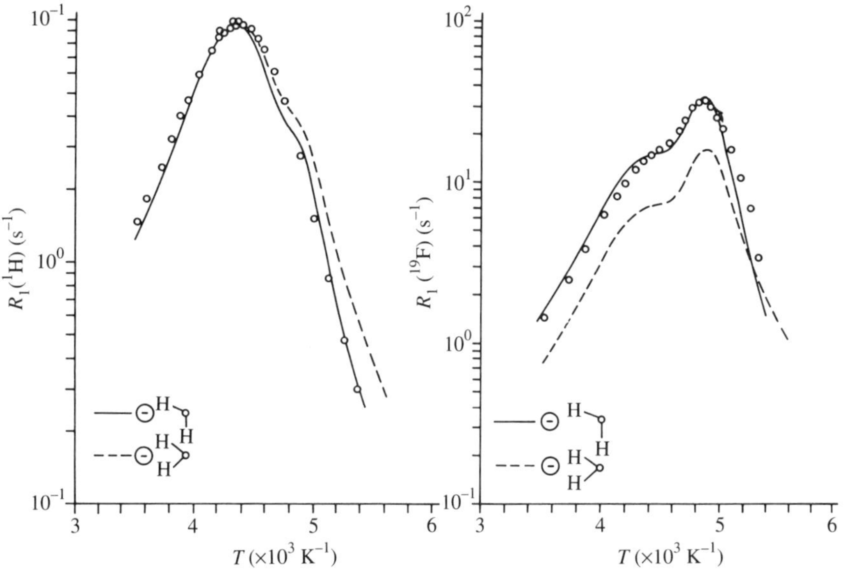

Fig. 5.21 Relaxation data for fluoride and protons in an aqueous 12M potassium fluoride solution in the supercooled region at frequencies of: (a) 300.1 MHz, and (b) 282.4 MHz.

the solutions. The data collected here show that, with the exceptions of the fluoride anion, the change in the rotational mobility caused by the three halide ions is small compared to the influence of the cations. Also the influence of the different cations becomes more pronounced in the metastable supercooled range. In Fig. 5.21 the temperature dependence of the fluoride anion and the proton of the water molecules are given far into the supercooled range at 200 MPa. The R_1 values in both Fig. 5.21(a) and (b) on undercooling pass through a maximum and extend into the temperature range beyond the extreme-narrowing limit, where the relaxation becomes frequency dependent. With these data and with data for heavy- and light-water solutions of KF as function of concentration, Lang *et al.* succeeded in distinguishing unambiguously between the two hydration structures proposed for the hydration shell of the fluoride ion.[51] The data can only be explained by an unsymmetric anion–water arrangement, in which the water molecules form one almost linear hydrogen bond with the fluoride. The symmetric arrangement, in which the positive end of the water dipole points into the direction of the centre of the anion can be safely excluded.

4. Protocols

Protocol 1.
Setting up a Hahn spin-echo pulse sequence and calibrating the gradient-coil constant, *k*

Caution! For safety procedures see Appendix 2.

The precision which is obtainable for self-diffusion coefficients obtained by NMR depends critically on exact determination of the two pulse widths in the Hahn spin-echo experiment and on the coil constant, *k*.

For this calibration a liquid probe with only one strong and narrow NMR signal is recommended. Water, benzene, and cyclo-octane are the preferred substances. It is essential that the geometry of the samples used in the calibrations is identical to that of the sample which is to be investigated.

1. Set the well shimmed NMR instrument for a single-pulse experiment, and determine the pulse width by systematic variation; that is, determine the time necessary for a 360° pulse—after obtaining a 360° pulse the spin system is returned to its starting condition and no NMR signal should be found in the Fourier-transformed spectrum. (This condition is usually the easiest to observe.) Then find the 180° and 90° pulses by dividing the time for the 360° pulse by 2 and 4, respectively.
2. (The Hahn spin-echo sequence consists of a 90° pulse followed after a time τ by a 180° pulse. The second pulse refocuses the magnetization, and the spin echo is formed after a time 2τ.) Vary τ to allow T_2 to be determined by

inserting the intensity $A(2\tau)$ in eqn (5.1.1). Measure the intensity of the echo for at least ten different values of τ.

3. Determine the self-diffusion coefficients of probes with more complicated NMR spectra by collecting the echo intensity $A(2\tau)$ as a function of g $(= kI)$ while keeping all the other variables in eqn (5.1.1) constant. Use this procedure for the calibration also. (The time τ between the 90° and the 180° pulse should be around $\frac{1}{4}T_2$.) Set up the pulse sequence for $\tau = \frac{1}{4}T_2$, and determine the intensity $A(2\tau)$ for at least ten different pulse currents I; then calculate k from eqn (5.1.1).
4. Control the determination of k by repeating step 3 with a different substance whose self-diffusion coefficient is known.

Protocol 2.
Filling of an oxygen-free sample of cyclo-octane into the Yamada glass cell

Caution! For safety procedures see Appendix 2.

Equipment

- As given in Figs 5.1 and 5.6
- The Young system is recommended for the joints and valves of the high-vacuum apparatus. This eliminates the use of vacuum grease, a possible source of contamination

Method

Note: cyclo-octane is one of the liquids recommended for calibrating the gradient coil. The filling procedure is representative of all volatile-liquid samples.

Caution! Wear safety glasses whenever working close to a high-vacuum apparatus.

1. Clean all glassware and the Yamada cell thoroughly.
2. Assemble the filling apparatus depicted in Fig. 5.6.
3. Connect the apparatus to a high-vacuum system capable of producing a final pressure of 10^{-4} Pa.
4. Put approximately 3 g of cyclo-octane which has been dried over a molecular sieve (0.3 nm) in the sample container, and then close valve of the container.
5. Evacuate the apparatus and close the valve leading to the part of the apparatus containing the high-pressure cell.
6. Degas the cyclo-octane for at least four freeze–pump–thaw cycles; control the degassing process using the vacuum gauge. Finally, close the valve of the sample container.

Protocol 2. *Continued*

7. Open the valve connecting the container and the vacuum line and pump for at least 24 h, in order to remove residual gas from the Yamada cell.
8. Close the valve between vacuum system and the filling apparatus. Open the valve of the sample container, and cool the lower tip of the cell container with liquid nitrogen. Continue until approximately 1 mL of cyclo-octane has sublimated into the tip. Close the connecting valve.
9. Allow the cyclo-octane to warm up, and lower the Yamada cell into the liquid by turning the spindle.
10. Open the connecting valve, and open the valve leading to the argon inlet. The argon will force the cyclo-octane into the glass cell.
11. Remove the cell from the container and flame seal the end of the glass capillary. Store the sealed cell in a refrigerator until use.

Protocol 3.
Filling of an aqueous salt solution into the Yamada glass cell

This procedure should also be used for mixtures if they contain compounds with very low vapour pressures.

Caution! For safety procedures see Appendix 2.

Equipment

- As given in protocol 2 and Figs 5.1 and 5.6, except that the sample container is not used in this process
- A portable magnetic stirrer and a Teflon-coated magnet

Method

1. Fill approximately 1 mL of solution and the magnet into the lower tip of the cell container. Use a long pipette for this operation in order to avoid contact between the liquid and the upper sections of the filling container. Then assemble the filling apparatus in the way described in Protocol 2.
2. Slowly cool the lower tip with liquid nitrogen.
3. Open the valve connecting the filling apparatus to a vacuum line and evacuate the apparatus.
4. Perform at least four freeze–pump–thaw cycles until the solution is completely degassed. Perform the warming up and melting of the solution very slowly and carefully in order to avoid splashing any of the solution into the upper sections of the cell container.
5. Freeze the solution with liquid nitrogen, and evacuate the cell container with the Yamada glass cell for 24 h in order to remove any residual gas

from the glass cell. (Make sure that there is always sufficient liquid nitrogen in the cooling Dewar flask.)

6. Close the connecting valve. Allow the solution to warm up. Stir the solution thoroughly with the magnet after it has melted. (During the freeze–pump–thaw cycles, part of the dissolved salt tends to crystallize, and it is obviously essential that the solution should be homogenized before filling the glass cell.)
7. Sink the glass cell into the homogenized solution by turning the spindle. Flush the apparatus with argon until the solution has filled the cell completely.
8. Remove the cell from the container and flame seal the upper glass capillary. Store in a refrigerator until use.

Protocol 4. Conducting an intradiffusion experiment by the diaphragm-cell method

For example intra(tracer) diffusion of tritiated water ^3HHO in water at 25 °C.

Caution! For safety procedures see Appendix 2.

Equipment

- A Stokes-type diaphragm cell with glass sinter
- Two Teflon stoppers with an aperture large enough to allow opening and closing when in position (this enables air to be expelled whilst the stopper is being inserted to ensure that no small air bubbles are left in the cell, which could interfere with the experiment)
- Two magnetic stirrer bars to sweep the sinter (one to float, one to sink)
- A thermostated water bath
- A submersible stirrer mechanism with a cell stand
- A submerged cell-holding device
- Scintillation counting equipment (for β/γ emissions)
- Two 50 mL bulb pipettes with the bottom aperture *on the side of tube*
- 50 μL syringe and needle

Materials

- 150 mL distilled, deionized water (18 MΩ)
- Tritiated water **caution, radioactive**
- A standard β/γ scintillation liquid (for example, Biofluor)

Method

1. Fill the bottom half of the diaphragm cell with the liquid of interest and insert the stopper. Half fill the other compartment.
2. Apply a vacuum (from a water aspirator) to the top compartment of the cell for a few seconds, producing bubbles from the sinter. Repeat this process several times. (This ensures the sinter is free of air and that it is filled with the liquid.)

Protocol 4. *Continued*

3. Completely fill the top volume of the cell with liquid, loosely stopper the top compartment (do not seal) and place the cell in the water bath to equilibrate. Begin stirring of the cell contents.
4. Add 10–20 μL of ^{3}HHO to the top compartment after equilibration. Avoid approaching the sinter too closely with the needle. Begin timing the experiment.
5. Replace the stopper, ensuring that the volume of the top compartment is completely filled. Seal the stopper.
6. Leave the cell for a specific time. Estimate the optimum time from t (in seconds) $= 1.1 \times 10^{-4}/D\beta$.
7. Remove the top stopper, and carefully sample the liquid in the top compartment using one of the special pipettes (collect approximately 25 mL).
8. Replace the top stopper in the cell but leave the cell unsealed.
9. Invert the cell in the submerged cell-holding device. (This keeps the temperature of the cell equilibrated whilst sampling the bottom compartment. This minimizes errors from the bulk movement of material in/out of the sinter.)
10. Sample the bottom compartment (25 mL).
11. Dilute some of the liquid from the top compartment, using unlabelled solvent, by a factor of two, *accurately by weight*. (If the optimal time for the experiment was used then the ^{3}HHO activity in both compartments will now be the same—if the estimate for the diffusion coefficient, D, which was used was accurate. This dilution method helps minimize errors in the ratio of the concentrations of the top and bottom solutions from the scintillation counting.
12. Measure the activity of both the top and the bottom solutions by scintillation-counting techniques using a suitable scintillation liquid. Calculate the total activity of the ^{3}HHO added to the top compartment initially by mass balance, taking into account the volume of the sinter, V_D.
13. Calculate the intradiffusion coefficient using eqns (5.2.1–5.2.4).

Acknowledgements

The authors wish to thank the Deutsche Forschungsgemeinschaft and the Fonds der Chemischen Industrie who financially supported some of the experimental developments and results described here. WEP thanks the Alexander von Humbolt Foundation for a Fellowship.

References

1. Harris, R. K. *Nuclear Magnetic Resonance Spectroscopy*; Pitman: London, **1983**.
2. Ernst, R. R.; Bodenhausen, G.; Wokaun, A. *Principles of Nuclear Magnetic Resonance in One and Two Dimensions*; Clarendon Press: Oxford, **1987**.
3. Diehl, P.; Fluck, E.; Günther, H.; Kosfeld, R.; Seelig, J. *NMR Basic Princip. Prog.* (J. Jonas, guest ed.) **1991**, *24*, 1–263.
4. Tyrrell, H. J. V.; Harris, K. R. *Diffusion in Liquids*; Butterworth: London, **1984**.
5. Stejskal, E. O.; Tanner, J. E. *J. Chem. Phys.* **1965**, *42*, 288–292.
6. Holz, M.; Weingärtner, H. *J. Magn. Res.* **1991**, *92*, 115–125.
7. Derome, A. E. *Modern NMR Techniques for Chemistry Research*; Pergamon: Oxford, 1987.
8. Sanders, J. K. M.; Hunter, B. K. *Modern NMR Spectroscopy*: Oxford University Press: Oxford, **1987**.
9. Yamada, H.; Kazuoka, T.; Sera, A. *J. Am. Chem. Soc.* **1988**, *110*, 7552–7553.
10. Gaarz, U.; Lüdemann, H.-D. *Ber. Bunsenges. Phys. Chem.* **1976**, *80*, 607–614.
11. Merbach, A. E.; Akitt, J. W. In reference 3; pp. 191–232.
12. Yamada, M. *Rev. Sci. Instr.* **1974**, *45*, 540–542.
13. Lang, E. W.; Lüdemann, H.-D. In reference 3, pp. 129–187.
14. Lang, E. W.; Lüdemann, H.-D. *Prog. NMR Spectros.* **1993**, *25*, 507–637.
15. Lang, E. W.; Lüdemann, H.-D. *Ber. Bunsenges. Phys. Chem.* **1980**, *84*, 462–470.
16. Angell, C. A. In *Water a Comprehensive Treatise*; Franks, F., ed.; **1985**, *7*, 1–81.
17. Lang, E. W.; Lüdemann, H.-D. *Angew. Chem.* (*Int. Edn*) **1982**, *21*, 315–329.
18. Prielmeier, F. X.; Lang, E. W.; Speedy, R. J.; Lüdemann, H.-D. *Phys. Rev. Lett.* **1987**, *59*, 1128–1131.
19. Prielmeier, F. X.; Lang, E. W.; Speedy, R. J.; Lüdemann, H.-D. *Ber. Bunsenges. Phys. Chem.* **1988**, *92*, 1111–1117.
20. Rasmussen, D. H.; McKenzie, A. D. In *Water Structure and the Water Polymer Interface;* Jellinek, H. H. G., ed. Plenum: New York, **1972**; pp. 126–149.
21. Prielmeier, F. X.; Lang, E. W.; Lüdemann, H.-D. *Mol. Phys.* **1984**, *52*, 1105–1113.
22. Jonas, J. In Reference 3; pp. 85–128.
23. Bachl, F.; Lüdemann, H.-D. *High Press. Res.* **1990**, *6*, 91–105.
24. Vardag, T.; Karger, N.; Lüdemann, H.-D. *Ber. Bunsenges. Phys. Chem.* **1991**, *95*, 859–865.
25. Hauer, J.; Lang, E. W.; Lüdemann, H.-D. *Ber. Bunsenges. Phys. Chem.* **1979**, *83*, 1262–1269.
26. Greiner-Schmid, A.; Wappmann, S.; Has, M., Lüdemann, H.-D. *J. Chem. Phys.* **1991**, *94*, 5643–5649.
27. Wappmann, S.; Tarassov, I.; Lüdemann, H.-D. *Z Naturforsch.* **1993**, *48a*, 613–618.
28. Heinrich-Schramm, A. Diploma thesis, Universität Regensburg, **1993**.
29. Stokes, R. H. *J. Am. Chem. Soc.* **1950**, *72*, 763–767.
30. Mills, R.; Woolf, L. A. *The Diaphragm Cell*; ANU Press: Canberra, **1968**.
31. Barnes, C. *Physics* **1934**, *5*, 4–13.
32. McCool, M. A.; Woolf, L. A. *High Temp. High Press.* **1972**, *6*, 85–95.
33. Woolf, L. A. *J. Chem. Soc. Faraday Trans.* I, **1975**, *71*, 784–796.
34. Woolf, L. A. In *Transport Property Measurements in Fluids*, Wakaham, W. A., Nagashima, A., Senger, J. V. eds.; Blackwells: Oxford, 1991; Vol. 3, Chapter 9.

35. Mills, R. J. *Phys. Chem.* **1973**, *77*, 685–688.
36. McCool, M. A. Ph.D. thesis, Australian National University, Canberra, **1971**.
37. Easteal, A. J.; Price, W. E.; Woolf, L. A. *J. Chem. Soc. Faraday Trans.* I, **1989**, *85*, 1091–1097.
38. Stilbs, P., *Prog. NMR Spectros.* **1987**, *14*, 1–45.
39. Jackman, L. M.; Cotton, F. A., eds, *Dynamic Nuclear Magnetic Resonance Spectroscopy*; Academic Press: New York, **1975**.
40. Isaacs, N. S. *Liquid Phase High Pressure Chemistry*; Wiley: Chichester, ***1981***.
41. van Eldik, R.; Asano, T.; le Noble, W. J. *Chem. Rev.* **1989**, *89*, 549–688.
42. Jennings, W. B.; Toley, M. S. *Tetrahedron Lett.* **1976**, *9*, 695–699.
43. Hauer, J.; Völkel, G.; Lüdemann, H.-D. *J. Chem. Res.* **1980**, S, 16–17; M, 426–455.
44. Hauer, J.; Treml, E.; Lüdemann, H.-D.; Mannschreck, A. *J. Chem. Res.* **1982**, S, 14–15; M, 253–270.
45. Spiess, H. W. *Rotation of Molecules and Nuclear Spin Relaxation in NMR, Basic Principles and Progress*; Diehl, P., Fluck, E., Kosfeld, R., eds; **1982**, *15*, 55–214.
46. Speedy, R. J.; Prielmeier, F. X.; Vardag, T.; Lang, E. W.; Lüdemann, H.-D. *Molec. Phys.* **1989**, *66*, 577–590.
47. Bachl, F.; Vardag, T.; Wappmann, S.; Lüdemann, H.-D. *J. Mol. Liquids*, **1992**, *54*, 193–203.
48. Vardag, T.; Karger, N.; Lüdemann, H.-D. *Ber. Bunsenges. Phys. Chem.*, **1991**, *95*, 859–865.
49. Karger, N.; Vardag, T.; Lüdemann, H.-D. *J. Chem. Phys.* **1990**, *93*, 3437–3444.
50. Vardag, T.; Bachl, F.; Wappmann, S.; Lüdemann, H.-D. *Ber. Bunsenges. Phys. Chem.* **1990**, *94*, 336–342.
51. Lang, E. W.; Fink, W.; Radkowitsch, H.; Girlich, D. *Ber. Bunsenges. Phys. Chem.* **1990**, *94*, 342–348.

6

Kinetic and thermodynamic measurements

DOUGLAS MAGDE and RUDI VAN ELDIK

This chapter describes the use of high pressures in the elucidation of the reaction mechanisms that govern chemical and biochemical processes near ambient conditions. In general, it is preferable that such investigations do not change the internal properties of molecules, but only perturb the molecules and their interactions by the minimum amount needed to measure the effect on their kinetic and thermodynamic behaviour. Since liquids are not very compressible, and solids are even less so, this modest aim still requires pressures up to 200 or 300 MPa. Over the past two decades, methods have been developed which allow virtually any kinetic or thermodynamic measurement to be performed at such pressures. The goal has been to devise instruments and procedures in which the rates or concentrations are measured in exactly the same manner as they are at ambient conditions. Protocols for such standard procedures are not included here. In order to permit coverage of the wide range of methods now available, we emphasize those aspects unique to operation at high pressures, including the selection or construction of a suitable apparatus, the choice of reaction conditions, the display of the pressure dependence of the data, and the interpretation of such data.

1. Thermodynamic and kinetic theory

When a component A is added to a solvent, there is a volume change; usually the volume increases but occasionally it decreases. The volume change on adding 1 mol of A is termed the partial molar volume, and it is denoted by $V_{m,A}$. It is dependent on all the thermodynamic conditions: the temperature, the pressure, the nature of the solvent, the concentration of A, the concentration of all the other solutes, and any other pertinent thermodynamic variables. Measuring $V_{m,A}$ is equivalent to measuring the solution density at specified conditions, a measurement that can be made quite accurately. For stable solutes, partial molar volumes are frequently quoted to a precision of 0.01 $cm^3\ mol^{-1}$. It often suffices to consider only the partial molar volume at infinite dilution, $V^0_{m,A}$, particularly in the study of reaction mechanisms, which usually treat dilute solutions. Under such conditions, for a reaction

$$aA + bB + \cdots \rightleftharpoons pP + qQ + \cdots,$$

the volume change associated with passage from reactants to products is

$$\Delta V^0 = pV^0_{m,P} + qV^0_{m,Q} - aV^0_{m,A} - bV^0_{m,B} + \cdots.$$

Le Chatelier's principle assures us that if external pressure is applied to a solution, the chemical equilibrium, characterized by the equilibrium constant K, will change in the direction that reduces the solution volume, which depends on ΔV^0, so that

$$\Delta V^0 = -RT\left(\frac{\partial(\ln K)}{\partial P}\right)_T.$$

It is apparent that ΔV^0 can be determined either by measuring the partial molar volumes of the products and reactants and subtracting the volume of the reactants from that of the products or by measuring the pressure dependence of K. There is also a third method: this involves the pressure dependence of the rates for the forward and backward reactions, as discussed below.

By analogy with the thermodynamic treatment of the solute equilibrium, transition-state theory describes rates of chemical change by postulating that a transition state lies somewhere on the pathway between the reactants and the products and that this transition state can be characterized by its own thermodynamic parameters, including its partial molar volume. The difference between the partial molar volume of the transition state and that of the reactants is the activation volume, $\Delta V^{\ddagger}$. The activation volume cannot be measured by a direct density measurement, because the transition state is not a chemical species, not even a short-lived one. It can be measured only by the effect of the pressure on some rate constant k that characterizes the chemical process:

$$\Delta V^{\ddagger} = -RT\left(\frac{\partial(\ln k)}{\partial P}\right)_T.$$

The goal of a pressure analysis of a chemical reaction is to construct the *volume profile*, which specifies at least the relative partial molar volumes of the reactants, products, and transition state. Figure 6.1 shows examples of several profiles. Note that the transition state may lie above, below, or in between the reactant and the product states; and note that the difference between the forward and backward value of $\Delta V^{\ddagger}$ must be equal to ΔV^0.

The complete characterization of a reaction requires that the temperature dependence of the equilibrium constant and all the pertinent rates are also measured, so that the enthalpies, entropies, and free energies for the reaction and for the transition state are also known. Such temperature effects are familiar to most chemists, since they are accustomed to thinking about

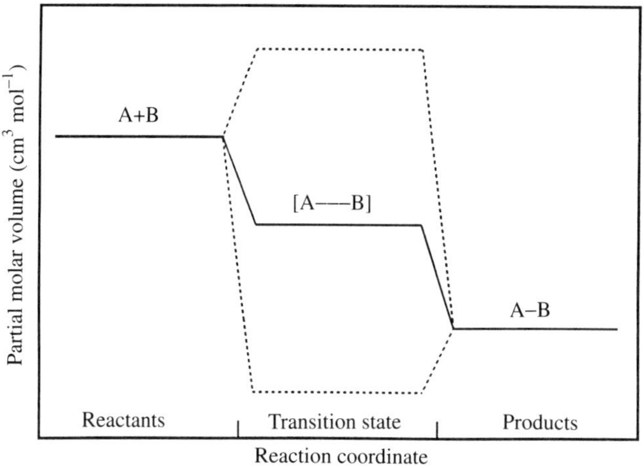

Fig. 6.1 A volume profile for the reaction A + B $\rightleftharpoons$ [A- - -B]‡ $\rightarrow$ A–B. Three possible cases are demonstrated for which the partial molar volume of the transition state is either higher, in between, or lower than the partial molar volumes of the reactant and product species, respectively,
$\Delta V^‡ = V^‡ - V_A - V_B$,
$\Delta V^0 = V_{AB} - V_A - V_B$.

energy profiles. Now that suitable procedures for making kinetic measurements at high pressure are becoming available, volume profiles are being recognized as an essential adjunct to energy profiles for clarifying reaction mechanisms.[1–5] One of the reasons why they are helpful is that volume changes are fairly easy to visualize.

2. Overall considerations for instrument design

2.1 Performance goals

Even when a laboratory is concerned initially with a single type of measurement, it is inevitable that additional capabilities are soon required. In order to prepare for a broad range of kinetic and equilibrium studies, it is best to plan from the beginning for a full set of instruments; and these instruments should, ideally, be designed to:

(1) determine concentrations *in situ*, by any of the following spectroscopic or electrical methods,
  (i) ultra violet/visible absorption spectrophotometry,
  (ii) conductivity,
  (iii) potentiometric and amperometric methods,
  (iv) vibrational spectroscopy (infrared or Raman),
  (v) ultra violet/visible luminescence,
  (vi) circular dichroism and related methods,

(vii) nuclear magnetic resonance (NMR) spectroscopy,
(viii) electron paramagnetic resonance (EPR) spectroscopy;

(2) use all traditional methods for initiating a reaction for kinetic studies, including
(i) stopped flow,
(ii) temperature perturbation,
(iii) pressure perturbation,
(iv) pulse radiolysis,
(v) flash photolysis;

(3) work on any time scale, from subpicoseconds to hours;

(4) facilitate batch reactions, including mixing and extracting sample aliquots;

(5) measure the density to determine partial molar volumes;

(6) give excellent precision, so that the initial slope of $\ln K$ (or $\ln k$) versus P can be determined from a small pressure range and so that curvature can be treated rigorously;

(7) offer precise, accurate, and versatile control of pressure, with
(i) absolute hydrostatic conditions,
(ii) convenient measurement of the pressure with a precision and accuracy of about 2 per cent and a calibration that is indefinitely stable and immune to accidents;

(8) offer precise, accurate, and versatile control of temperature, which is
(i) variable near the ambient value over about $\pm 50\,°C$,
(ii) uniform throughout the sample,
(iii) capable of measurement accuracy to at least $0.1\,°C$,
(iv) extendible to lower or higher temperatures, when desired;

(9) impose few constraints on the sample volume and geometry, offering
(i) the smallest possible volume for rare samples,
(ii) an easy change to larger volumes when needed,
(iii) variable pathlengths for transmission measurements,
(iv) the ability to flow or replace samples to combat degradation;

(10) provide for efficient data collection, while minimizing systematic errors by
(i) facilitating measurement at many pressures and temperatures in one run,
(ii) permitting measurements of different types on the same sample,
(iii) allowing an easy change of the samples without disturbing the windows, the electrical feedthroughs, the optical alignment of spectrometers, etc.;

(11) use conservative specifications, well below the material limits, ensuring
(i) no downtime, expense, loss of the sample, or damage to other equipment,

(ii) an indefinite lifetime for the components, whether in regular use or left unused,

(iii) an unlimited number of pressure cycles, with minimum maintenance;

(12) have low cost for acquisition and upkeep;

(13) be simple to operate after a brief, but thorough, training period;

(14) ensure a large safety margin for personnel, with

(i) no chance of bursting lines or vessels,

(ii) the minimum stored energy to dissipate if there is a break,

(iii) no toxic or corrosive fluids.

2.2 Design strategy derived from the performance goals

Not all the performance goals listed above are equally important in every measurement; some of the goals are even contradictory, so that compromises are necessary. Design simplicity and safety considerations require that compressed gases should be avoided. A modular approach, such as the following, makes it possible to meet most of the goals simultaneously.

1. Use a common, semiportable unit to generate and measure pressures and to deliver them to different pressure chambers by a hydrostatic connection. For reliability, use oil for the hydraulic fluid in this unit.

2. Confine any liquid-sample solution in a small, interchangeable cell, which should be optimized for each different procedure. Use some type of moving piston or deformable membrane in this cell, so that the pressure is balanced inside and out.

3. Immerse the sample cell in a hydrostatic medium in a high-pressure chamber that is compatible with existing spectrometers and that is connected to the pressure-generating unit by semiflexible tubing. Since spectroscopic constraints often require a transparent fluid and preclude the use of oil, incorporate a phase separator in this line. Standardize the dimensions and specifications of the high-pressure chambers where possible, but allow flexibility to accommodate different measurements.

4. Use commercially available components, or even complete commercial systems.

2.3 Error budget

Good experimental design requires that the accuracy and precision should be considered, to avoid overdesigning one component while underdesigning others. The quantities desired, ΔV^0 and $\Delta V^{\ddagger}$, will usually have numerical values no larger in magnitude than 20–30 $cm^3\ mol^{-1}$; they are often closer to 5 $cm^3\ mol^{-1}$. This can be understood on general grounds. Under the conditions of a high-pressure experiment, atoms will neither be added to nor removed from the overall system; however, bonds might be formed or destroyed. Two bonded atoms typically have an internuclear separation that is only half the sum of their van der Waals radii. Consequently, creating or destroying a bond changes the molar volumes by roughly half the molar volume of an atom, which could range from 10 to 20 $cm^3\ mol^{-1}$. (Recall that the molar volume of water is about 18 $cm^3\ mol^{-1}$.) On the other hand, reactions (and even transition states) often involve moving an atom from one bonded situation to another, nearly equivalent, situation, which should be accompanied by almost no change in molar volume. Also important is the effect of charge separation, in which two ions are created from a neutral species, or of charge neutralization. This results in electrostriction of the solvent, which is significant and can be as large as 30 $cm^3\ mol^{-1}$.[1-5] In order to distinguish trends based on small differences among similar reactions, we need conservatively estimated worst-case errors no larger than ± 0.5 $cm^3\ mol^{-1}$. All these considerations apply equally well to both ΔV^0 and $\Delta V^\ddagger$.

For measurements under hydrostatic conditions, with a gauge accurate to one per cent of the full scale (perhaps ± 0.2 MPa), the pressure may be assumed to be an independent variable with a negligible random error. One need merely guard against systematic errors. If we estimate the uncertainty in $\Delta V^\ddagger$ from a calculation of the slope of ln k (determined by only two extreme points), if we assume that each rate measurement has the same fractional error, $\delta k/k$, and if we assume that there is no error in T or P, we find

$$\delta(\Delta V^\ddagger) = \frac{\sqrt{2}RT(\delta k/k)}{P_2 - P_1}.$$

If $P_2 - P_1 = 200$ MPa and $T = 298$ K, we need a precision of 3 per cent in the rate measurements to achieve $\delta(\Delta V^\ddagger) = 0.5$ $cm^3\ mol^{-1}$. It is some help that the slope is derived not from just two points but from a least squares fit to several points, but the slope is heavily influenced by the extreme values. Overall, a 3 per cent precision is not an easy goal if it is to represent a 95 or 99 per cent confidence interval and imply conscientious tests for systematic errors at that level. This error analysis shows why the pressure range cannot be reduced to much less than 200 MPa and still yield the desired precision. One would prefer a bigger range, were it not for the fact that the curvature, or uncertainty about the curvature, in the plots makes a large range less useful.

So far we have assumed no error in T. In fact, errors in T are a major concern. The explicit T dependence in the definition of $\Delta V^\ddagger$ is not a significant problem; but k is strongly affected by T, depending on $\Delta H^\ddagger$. Consider a large, but not unreasonable, value for $\Delta H^\ddagger$ of 80 $kJ\ mol^{-1}$. Near room temperature, a difference of only 0.2°C in the temperature between measurements at 0 and 200 MPa would contribute an error of 0.3 $cm^3\ mol^{-1}$ to $\Delta V^\ddagger$. It is important to ensure that temperature drifts are not systematically correlated with pressure settings. This is one reason why it is customary to cycle the pressure up and down, while measuring rates during both parts of the cycle. The reaction volumes ΔV^0 will usually have much smaller values of ΔH; otherwise the reaction would go to completion in one direction and one would not be measuring K. (Some electrochemical measurements could be exceptions.) Consequently, small errors in T are not so important. However, it is often difficult, using methods which are practical at high pressures, to measure K to a precision of 3 per cent at a 95 per cent confidence limit, with full attention to all the possible systematic errors.

2.4 Pressure range for the measurements

Not only is an upper limit to the pressure range of about 200–300 MPa both necessary and sufficient for measuring ΔV^0 and $\Delta V^‡$, it is also desirable from the point of view of goals 7–10 listed in Section 2.1 calling for instrumental flexibility, and from practical considerations (goals 11–14, listed in Section 2.1). Although most of the analytic procedures could be adapted to pressures as high as 2000 MPa, many experimental changes would be needed at such a pressure. Multiple radial apertures with large windows could not be maintained. The sample volumes would be smaller, or the chambers would have to be very massive. The seals would be more complex. One would not separate the measurement chamber from pressure generation and measurement, and connect them by narrow bore, semiflexible tubing: there would be too much concern about pressure differentials and slow relaxation of pressure gradients. There would be very little reserve in the material parameters, and one would expect frequent failures. Off-the-shelf commercial components are not rated for 2000 MPa. All these concerns are relaxed at pressures in the range 700–1000 MPa, where a variety of commercial apparatuses are available. It is possible to work routinely in this range, but compromises would still have to be accepted. At the pressure range recommended here, with an upper limit of 200–300 MPa, a greater variety of less-expensive commercial apparatus is available; and it is easier to design, construct, and use any other components which may be needed.

3. Details of pressure generation and measurement

3.1 Pressure generation and measurement

A possible design for a transportable high-pressure unit is shown in Fig. 6.2, which, while intended as a schematic, is almost a scale drawing of a real implementation.[6] Except for the phase separator, all the components are available from commercial sources. A hand-operated, mechanical screw pump generates pressures up to 200–400 MPa without the need for a secondary intensifier. A lever pump could also be used, or a pump powered by compressed air or electricity; but screw pumps are compact and portable. The tubing has an outer diameter of 6.4 mm (0.25 in). A one-way check valve allows the pump to be refilled without any loss of pressure; this is occasionally necessary for large chamber volumes, and is always useful for the initial set-up. Two pressure gauges are useful, partly to give an increased resolution,

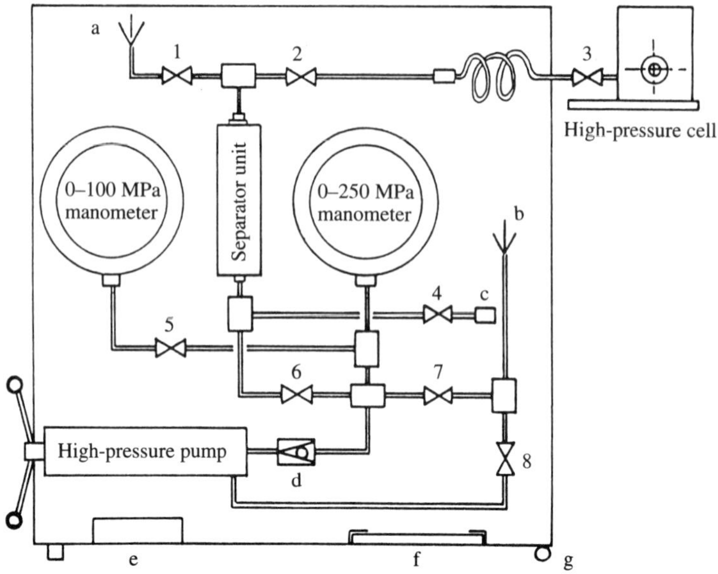

Fig. 6.2 A schematic diagram of a transportable high-pressure unit complete with a high-pressure cell: (a) pressure-medium reservoir, (b) oil reservoir, (c) to vacuum line, (d) one-way valve, (e) holder for pump handles, (f) holder for a high-pressure cell, (g) wheels for easy transportation and (1–8) high-pressure valves.[6]

but more importantly to show that both these gauges are working properly. Mechanical Bourdon gauges are relatively inexpensive and they are generally rugged, while offering excellent precision and accuracy, namely 1 per cent of the full scale to 400 MPa and 2 per cent to 1000 MPa. The pump and gauges use oil as the pressure medium to achieve good sealing and good lubrication, particularly of the sliding seal in the pump. Brake fluid has been recommended for pressures somewhat above 1000 MPa. Below 400 MPa, hydraulic oil is often used. Two procedural issues arise in filling the system with oil. The first is to eliminate all air from the lines. To purge air, which can be trapped at numerous points, the fittings are loosened one at a time, and then re-tightened. The presence of air is detected by a lag or 'softness' in the gauge reading when the pressurization is initiated. Air should be eliminated at the lowest possible pressures. The second issue is to find and correct oil leaks, which can be identified by a slow decrease in the gauge reading. The leak can be located by looking for oil droplets near fittings. Fittings should be checked to be sure that they have been tightened, but they should not require excessive force. They are designed to work with modest forces on precisely machined, mating surfaces. An apparent need for excessive force indicates that the fittings are deforming to accommodate some defect, such as a nick or a piece of dirt. Consequently, the fitting should be removed and cleaned or replaced.

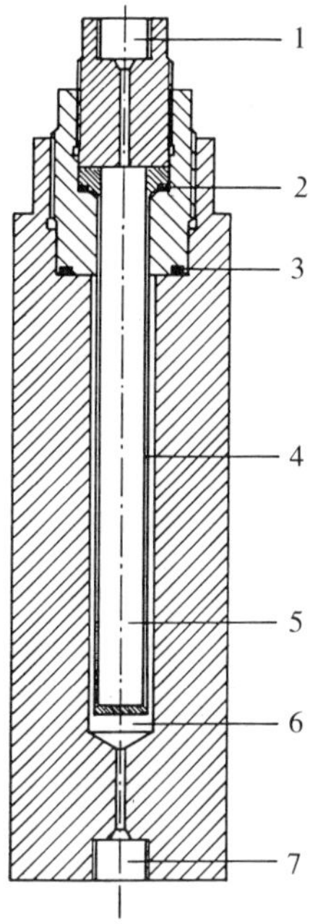

Fig. 6.3 A schematic diagram of a separator unit complete with a Teflon finger: (1) connection to high-pressure optical cell, (2, 3) O-rings, (4) Teflon finger, (5) pressure liquid for the optical cell, (6) pressure liquid for the high-pressure pump, and (7) connection to the high-pressure pump.[6]

The phase separator between the oil and a secondary pressurizing fluid is included with the pump unit. The construction details are shown in Fig. 6.3. The secondary fluid is usually either water for ultraviolet/visible spectrophotometry,[†] unless it would freeze at the temperature needed, or *n*-heptane for electrical measurements. The phase separator is a flexible, Teflon finger, which is compressed by the oil. It is less rugged than the other components, and it is important to return it to its original full expansion after each pressure run; otherwise it will eventually break. For this purpose, a vacuum pump with a trap is used to extract oil from outside the finger. When the phase separator is omitted for any reason, *n*-heptane is the preferred hydraulic fluid; it is acceptable for use in the pump and gauges, and it also gives adequate ultraviolet transparency for most purposes. Use of the single-isomer *n*-heptane gives better ultraviolet transmission than mixed isomers. Although most high-pressure tubing is very rigid the smallest diameters (for example, an outer diameter of 3.2 mm and an inner diameter of 0.5 mm) can be coiled, and then they are surprisingly flexible while still being able to support pressures greater than 1000 MPa. Consequently, a semiflexible connection can be

[†] Ethanol may also be used, Editor.

made to an experimental chamber. The design in Fig. 6.2 was optimized for easy transport by air freight. In a laboratory environment it is a little too compact, with too many short, rigid connections. In a laboratory situation, the components can be mounted on a rolling frame, with bends in the interconnecting tubing, so that the fittings can be loosened more conveniently.

Protocol 1.
Procedures for operating the high-pressure unit shown in Fig. 6.2

Caution! For safety procedures see Appendix 2.

1. Open valves 2, 3, 5, and 6, and close valves 1, 4, 7, and 8 (also close valve 5 for $p > 100$ MPa) to pressurize the high-pressure cell.
2. Open valve 7 and then valve 1 to release pressure from the cell.
3. Open valves 1 and 4 and close valves 2 and 6 to refill the separator unit.
4. Open valve 8 to refill the high-pressure pump.

3.2 High-pressure housing

A chamber that is small enough to fit in many standard spectrometers but large enough to accommodate a variety of cells and procedures is shown in Fig. 6.4. It is a vertical cylinder, with an outer diameter of about 10 cm and an internal diameter of about 3 cm. It can be constructed in different heights to accommodate different experiments, without affecting the other aspects of the design. There can be four radial ports 90° apart to accommodate optical windows or other points of access, and another radial port can be provided for the pressure line. There is a large port on the top, through which samples can

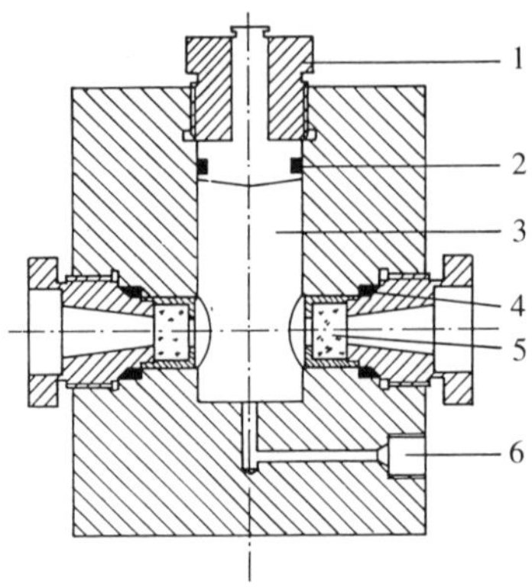

Fig. 6.4 A schematic diagram of high-pressure cell for ultraviolet/visible optical measurements: (1) pressure plug, (2) O-ring, (3) sample holder with a pillbox cell, (4) Δ-ring and O-ring system, (5) sapphire window, and (6) connection to pressure line.[6]

be introduced without disturbing anything else. It is often desirable that the bottom surface is flat and unobstructed, although there are occasions when feedthroughs might be introduced along the axis. The entire cylinder is surrounded by a copper shield with coils for circulating a temperature-regulated fluid (not shown). The large thermal mass makes temperature changes somewhat slow, taking about 20 min to reach equilibrium, but this also means that the temperature is not perturbed by minor changes in the environment and it should be predictable once careful, initial calibration has been completed. This pressure housing is not available commercially. Its design requires careful calculation, and its construction requires excellent machine work. The considerations involved are treated elsewhere in this volume or can be found in other texts.[7–9] Less versatile chambers are commercially available for selected applications, especially for optical-transmission measurements.

3.3 Sample cells

A liquid sample should be contained in a separate cuvette immersed in a pressurizing medium in the high-pressure chamber. The cuvette will require either a frictionless piston or a deformable membrane, in order to balance the internal and external pressures. Some examples are shown in Fig. 6.5. Electrodes or optical windows may need to be incorporated, and provision for mixing or circulating the solution may be needed. A sample volume of around 1 cm^3 is most common, but volumes which are ten to one hundred times smaller or larger may be required at times. The sample should contact only the most inert of materials: electrodes made of platinum or gold; windows made of quartz or sapphire; syringes and the other components made of glass, Teflon, or Kel–F*. The last is easy to fabricate into custom-made shapes. An O-ring made of Viton or another material may be needed; it should be selected for its compatibility with a particular solvent

4. Details of concentration measurements

4.1 Ultraviolet/visible spectrophotometry

Absorbance measurements in the ultraviolet/visible are convenient and familiar, especially for transition-metal chemistry and biochemistry. Fortunately, window materials with good mechanical properties at high pressures, such as sapphire and fused silica have excellent ultraviolet transparencies. Because sophisticated ultraviolet/visible spectrophotometers are likely to be available, it is usual to adapt a high-pressure apparatus to fit an existing instrument. The sample chamber of the spectrophotometer is often large enough to hold the high-pressure housing. If it is not, optical fibres can be used to transmit the beams to and from the high-pressure housing. Determination of the equilib-

* 'Kel-F' is the trade name of poly(chlorotrifluoroethene). It is more rigid than PTFE and chemically resistant (Ed.).

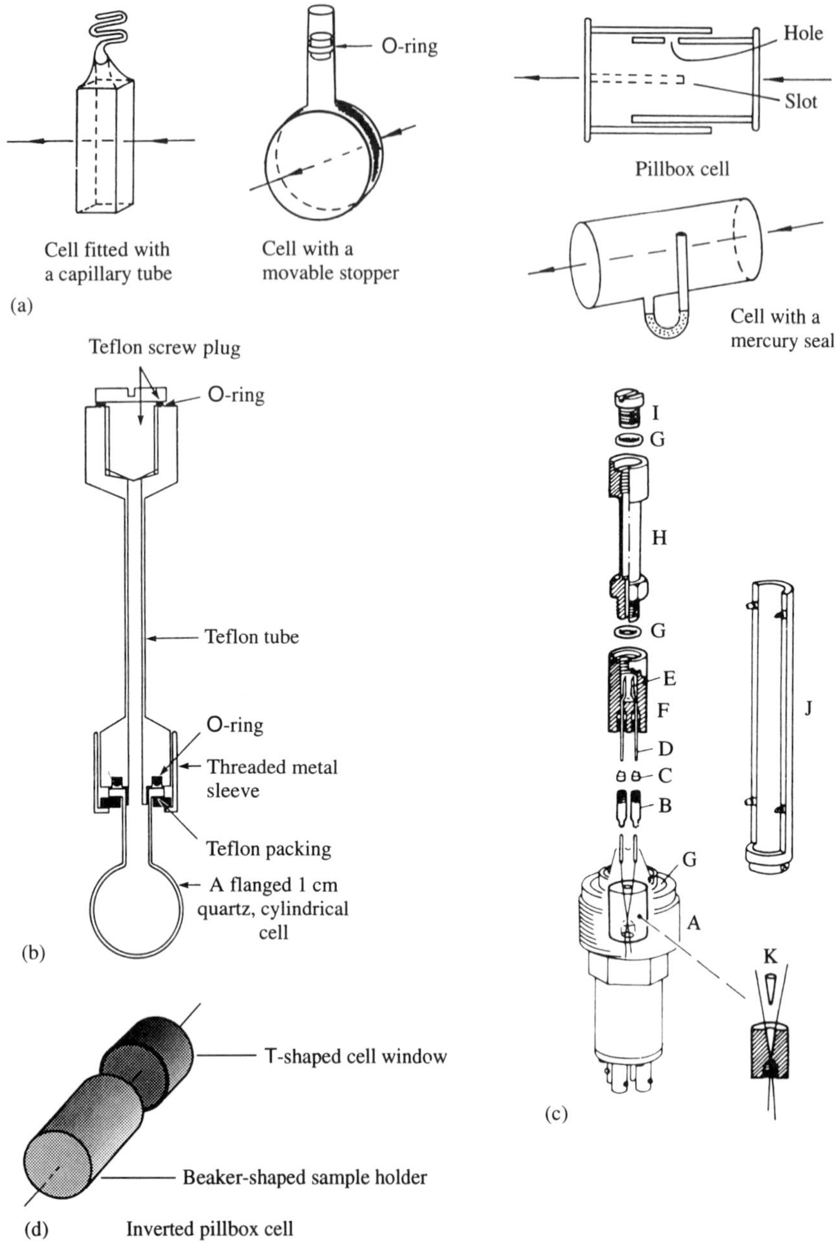

Fig. 6.5 Some typical sample cells. (a) Various types of optical cell. (b) A completely sealed optical cell. (c) An exploded view of a high-pressure conductivity cell: (A) the plug of the high-pressure cell, (B) a hollow cylindrical Kel–F screw, (C) a Teflon cone, (D) a platinum wire lead, (E) a platinum electrode, (F) a Kel–F body, (G) a Viton O-ring, (H) a flexible Teflon tube, (I) a Teflon screw, (J) an aluminium holder, and (K) a steel cone. (d) An inverted pillbox cell for infrared spectroscopy.

rium concentrations requires accurate and precise concentration measurements. Beer's Law is assumed for each component (but it should be verified that linearity actually applies for each component at the conditions required):

$$A = \varepsilon Cl,$$

where A is the absorbance, ε is the molar absorptivity at a particular wavelength, C is the molar concentration, and l is the optical pathlength. Absorbance can only be measured over a limited range, and it can only be accurately measured over an even narrower range, roughly between 0.3 and 1.6. To measure the equilibrium constants, one cannot arbitrarily change C; and the choice of wavelength is often restricted as well. Consequently, it is important that l can be varied. This is one reason for using interchangeable cells, as discussed in Section 3.3. Ideally, one wants to know the value of ε as a function of pressure separately for each component. The variation in question is only the change (if any) per molecule, and not the change associated with the overall compression of the solvent. Consequently, one should measure ε as a function of pressure using the pillbox cell shown in Fig. 6.5(a). In this design, compression occurs exclusively in the direction parallel to the optical beam, and the number of molecules in the beam remains constant. Pillbox cells[10] (in different optical pathlengths) are commercially available from Hellma. A small, but frustrating, point is that they can be difficult to fill without leaving unwanted gas bubbles. A simple device to help with this problem has been described by Wieland and van Eldik.[11]

Protocol 2.
Procedures for recording ultraviolet/visible spectra as a function of pressure

Caution! For safety procedures see Appendix 2.

1. Mount the high-pressure vessel shown in Fig. 6.4 into the cell compartment of an ultraviolet/visible spectrophotometer. Place the cell in such a way that maximum light goes through the cell, with the cell causing the lowest possible absorbance.
2. Record the baseline of the spectrophotometer with the pressurizing liquid in the cell; this is usually water or *n*-heptane.
3. Put the sample into the pillbox cell (Fig. 6.5(a)), and place the pillbox cell with a special cell holder into the high-pressure vessel.
4. Fill the high-pressure vessel with the pressurizing liquid, and close the cell with the pressure plug. Open the valve to the pressurizing system (Fig. 6.2) during this procedure so that the excess-pressure medium can flow into its reservoir (a in Fig. 6.2).
5. Follow the procedures outlined in Protocol 1 to pressurize the cell to 5 MPa.

Protocol 2. *Continued*

Allow 20 min for temperature equilibration, and record the spectrum of the sample.

6. Repeat the measurement at pressure intervals of 25 MPa up to 150 MPa by allowing 20 min for temperature equilibration after each change in pressure.
7. Repeat the measurements on releasing the pressure at intervals of 25 MPa. (These spectra should coincide with those measured during the increase in pressure; any deviations may indicate a slow decomposition or a subsequent reaction induced by pressure.)
8. Use spectra recorded as a function of pressure to estimate the equilibrium constant as a function of pressure, and subsequently to estimate the reaction volume with the help of the equation $\Delta V^0 = -RT[\partial(\ln K)/\partial P]_T$.

Data are shown in Fig. 6.6 for the change in the absorption as a function of the wavelength for various pressures for a recent study of O_2 binding to the protein myoglobin (Mb).[12] The spectra show an excellent isosbestic point, giving evidence for the single equilibrium process: $Mb + O_2 \rightleftharpoons MbO_2$. At the wavelength of the maximum change, the absorbance increases by a little more than 0.1 out of 0.7 absorbance units over a pressure range of 150 MPa. This corresponds to a doubling of K between the ambient pressure and 100 MPa; and this, in turn, implies rather large value for ΔV^0 of $-19.2\ cm^3\ mol^{-1}$,

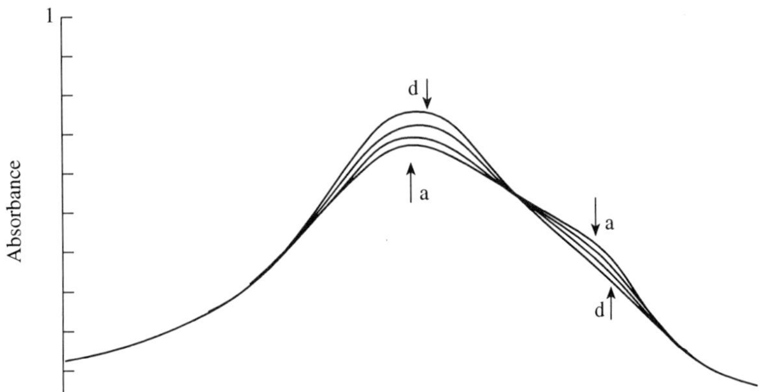

Fig. 6.6 The effect of the following pressure on the absorbance spectrum of Mb/MbO_2 (Mb = myoglobin) for the conditions, [total Mb] = 0.6–1.5 μM, [Tris buffer] = 5 mM, T = 25 °C, pH = 8.0–8.5, optical pathlength = 1 cm, and the ionic strength = 0.1 M (NaCl):[12] (a) 5 MPa, (b) 50 MPa, (c) 100 MPa, and (d) 150 MPa.

with an uncertainty of about 1.5 $cm^3 mol^{-1}$. Twelve measurement cycles over the pressure range were averaged to achieve this precision. This system is reasonably favourable for a determination of ΔV^0 by spectrophotometry. The full report of this study[12] is instructive because it compares different methods of deriving the desired thermodynamic and kinetic parameters. The equilibrium constant, K, was also calculated from forward and backward rate constants measured at each pressure; and ΔV^0 was calculated both from the plot of ln K versus pressure and from the difference between the forward and backward $\Delta V^{\ddagger}$.

Changes in concentration for kinetic studies may be measured on any time scale using ultraviolet/visible transmission. The techniques vary for different time ranges; some aspects are treated below.

4.2 Electrical conductivity

Conductometric analysis is well suited to high-pressure measurements; phenomena such as dissociations of weak acids were studied very early.[13] Conductivity electrodes, which may be commercially available, can be mounted in small cells with moving pistons or with deformable walls, and the electrical leads may be passed through one commercially available feed-through (see Fig. 6.5(c)). The electrodes themselves should be arranged so that they are not subject to deformation or vibration. It is relatively easy to make differential measurements by using two cells, both in the pressure bomb, as the arms of a Wheatstone bridge, one of which contains the reaction mixture and the other is a reference (such as pure solvent) that is designed to cancel out factors which are not of interest. An alternating-current bridge is usually employed. The conductivity is not a useful variable in situations where the ionic strength is maintained at a high value or where the pH is well buffered; and the conductivity is not selective when reactions involve charge transfer among similar types of ion. However, it complements spectrophotometry, in that it is good at detecting changes in the concentration of small, colourless ions, such as protons and hydroxyl anions in organic solvents. Large conductivity changes arise when ions are formed or consumed, or when charge is transferred to produce ions with greatly differing mobilities. For equilibrium studies, the proportionality constant between the concentration and the conductivity must be established. For kinetic studies, the conductivity at each pressure can be linear in concentration over a wide dynamic range, and it does not matter if the proportionality constant changes with pressure. For kinetic studies, the frequency used for an alternating-current bridge must be several times greater than the fastest observed rate; however, a frequency of 10–20 kHz is sufficient for stopped-flow measurements, and the higher the frequency that is used, the more concern must be given to unwanted capacitive coupling. Today's digital oscilloscopes, with their large memories, allow the entire alternating-current signal to be digitized, and all rectification can be performed numerically. One usually expects better signal-to-noise levels in a conductivity measurement than in a spectrophotometry measurement.

4.3 Potentiometric and related methods

A variety of electrochemical methods are useful in reduction–oxidation (redox) chemistry, and while they give more detailed information than conductivity methods they have not yet been fully developed for kinetic measurements. One should be aware that heterogeneous chemistry at surfaces is ultimately involved, and many processes could be responsible for a particular observed pressure effect. Furthermore, one is always comparing a working electrode with a reference electrode, and it is embarrassing to overlook the possibility that an observation is due to pressure effects on the reference electrode! Significant advances were made recently in high-pressure electrochemical methods. Details of two recently published apparatuses[14,15] both employ cyclic voltammetry, although one[14] apparatus interfaces the high-pressure cell to a commercial potentiostat, while the other[15] uses home-made electronics. The electrochemical cells are quite similar; one is shown in Fig. 6.7. It is designed to isolate the cell from the hydraulic fluid by using a free piston sealed by O-rings. Note the extra piston, with a porous Vycor disc, in the reference electrode to equalize the pressures there. Both instruments use platinum wire for the counter electrode and a 1 mm disc is used for the working electrode. Both groups of experimenters verified that either platinum or gold work equally well for the disc electrode. One group[15] used Ag/Ag^+ for the reference electrodes while the other group[14] used Ag/AgCl/KCl.

Since ln K is proportional to the negative of the standard free-energy change in a reaction, which, in turn, is proportional to the negative of the standard potential, E^0, for a reaction, the volume change in a reaction is given by

$$\Delta V^0 = -nF\frac{\partial E^0}{\partial P},$$

where F is Faraday's constant and n is the number of electrons transferred in the reaction. Measuring the reaction volume to a precision of 0.5 $cm^3\ mol^{-1}$, requires the voltage difference to be measured to within 1 mV over a 200 MPa range in the pressure. This assumes that there is no significant error in the temperature and that the temperature is near ambient value. The reversible potential, E, for a cell obeys the Nernst equation

$$E = E^0 + \frac{RT}{nF}\ln Q,$$

where Q is the reaction quotient. Measurements are made at a constant ionic strength, I, typically between 0.1 and 1 mol kg^{-1}; and the concentration of the reactant is varied to verify that the slope RT/nF is equal to $57/n$ mV, and that

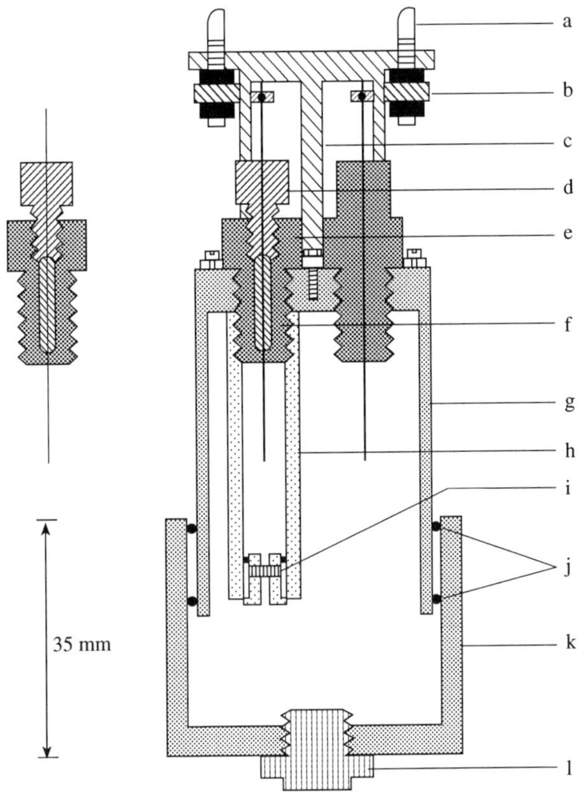

Fig. 6.7 A high-pressure electrochemical cell with an Ag/Ag$^+$ reference electrode and an Au disc working electrode; a platinum wire auxiliary electrode is shown on the left: (a) bomb electrical terminals, (b) brass connectors, (c) mounting plate, (d) Kel–F backing screw, (e) Kel–F electrode housing, (f) Teflon sleeve, (g) upper cell body, (h) reference-electrode compartment, (i) internal piston with a Vycor glass frit, (j) O-rings, (k) Teflon piston, and (l) screw plug.[15]

it does not change with pressure. Many aspects of the operation of a cell over the pressure range can be tested by operating it as a concentration cell.[15]

Cyclic voltammograms for $Fe(CN)_6^{3-}$ are shown in Fig. 6.8. Under reversible conditions, the peaks corresponding to oxidation and reduction would occur at the same applied potential. In order to estimate the reversible potential E from the measurements, the potentials of the two peaks are averaged. The peak position is taken to be the point at which the (smoothed) curves have a zero slope. What is actually considered in measuring the effect of pressure is $E(P) - E(0.1\ \text{MPa})$. These instruments have unexplained drifts in the value of $E(0.1\ \text{MPa})$ from day to day, but they appear to be stable over the time required for a set of measurements. The precision in ΔV^0 achieved

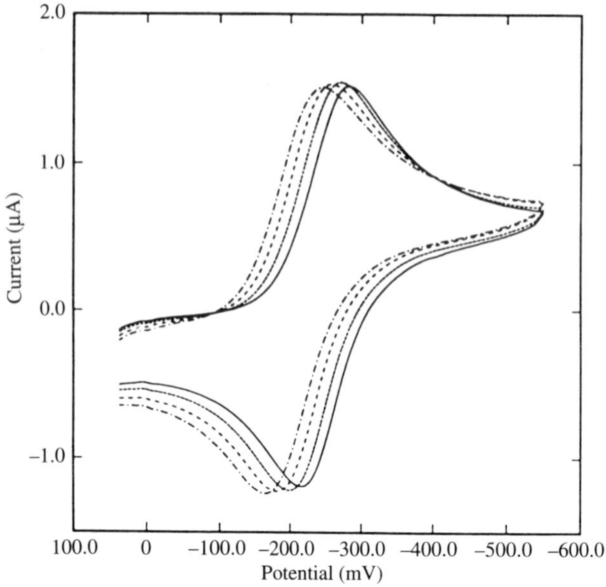

Fig. 6.8 Cyclic voltammograms of $Fe(CN)_6^{3-}$ (0.001 M) in 0.1 M KNO_3 versus Ag/$AgNO_3$ (0.01 M) in 0.1 M KNO_3 at: (——) 1 bar, (· · · ·) 351 bar, (– – – –) 688 bar, and (– · · –) 1025 bar. A 1 mm gold disc was used as the working electrode, and the scan rate was 0.060 V s^{-1} at 25 °C.[15]

by both instruments appears to be better than ± 1 cm^3 mol^{-1}. The dependence of the reaction volume on the ionic strength is significant compared with the precision obtained. It seems to be accepted that partial molar volumes should be quoted at $I = 0.1$ mol kg^{-1}. There are many issues involved in interpreting partial molar volumes for ions. The review by Millero[16] is essential reading.

4.4 Infrared and Raman methods

Vibrational spectroscopy gives direct information about internal bonding in molecules. Spectral shifts can measure a change in the equilibrium, and line broadening can reveal very fast exchanges or collisions, but generally become significant only at higher pressures than we consider. With some exceptions in the near-infrared, which can exploit ultraviolet/visible methods, little use has been made of either infrared or Raman spectroscopy simply to measure concentrations for mechanistic studies, even at ambient pressures, despite the good selectivity that they offer. This situation is changing because of the availability of improved lasers and detectors. However, both lasers and detectors may need to operate at liquid-nitrogen temperatures. Some traditional solvents, such as CCl_4, freeze at inconveniently low pressures. Water absorbs throughout the infrared; but work in aqueous solution has been accomplished using very-short-pathlength cells.[17] Infrared studies often use very concentrated solutions and short pathlengths. It is worth noting that pill-box cells can be constructed from single-crystal silicon in the inverted design shown in Fig. 6.5(d). Special strategies are available to make cells with path-lengths of a few micrometres.[17] Even with thin cells, the hydraulic fluid is still a problem. Windows also become an issue, especially if they must function over a wide spectral range.

Raman spectroscopy covers the complete vibrational-energy range using windows and solvents for the visible region. Low sensitivity is the major problem. Resonance Raman spectroscopy combines stronger scattering with selectivity for a particular chromophore. Because of the selectivity and the fact that water is a very good solvent for Raman studies, resonance Raman spectroscopy is attractive for biophysical studies. Kinetic studies have now been demonstrated on all time ranges down to subpicoseconds. The key issues raised by high pressures are the restrictions on the solid angle of the detector and the extra surfaces, causing scattering and polarization artefacts, that are introduced by strain in the window materials. Laboratories experienced in Raman studies should be able to add high-pressure capabilities, if the signal strengths are sufficient.

4.5 Luminescence

Luminescence appears in mechanistic chemistry in two distinct contexts. The details of the emission process itself are important in photochemistry, which is considered in Section 8. In other applications, fluorescence is simply an assay for qualitative and quantitative analyses. Good selectivity and sensitivity have earned it a role in commercial apparatus. However, the sensitivity is compromised by the small windows of a high-pressure cell, which limit the solid angle available. Laser sources help, but they restrict luminescence methods to specialized facilities. It is important to understand the polarization artefacts which may be introduced by thick windows and to appreciate that they may change with pressure. Many optical elements, including monochromators, have transmission properties that depend on the polarization.

4.6 Circular dichroism and optical rotatory dispersion

Circular dichroism and optical rotatory dispersion and other variations on the measurement of transmitted light may be carried out with essentially the same strategies as are used for absorbance measurements. The main concern is that any measurements involving polarized light must be standardized at each pressure. Sapphire is inherently birefringent, and isotropic glasses show strain-induced birefringence, and these factors change as the pressure increases.

4.7 Nuclear magnetic resonance (NMR)

With constantly improving capabilities, NMR has become a powerful tool for measuring rates and concentrations while providing detailed structural information. Designing a high-pressure NMR probe is a much more difficult challenge than anything considered above.[18] For both these reasons, NMR is considered in a separate chapter, Chapter 4.2.

4.8 Electron paramagnetic resonance (EPR)

Electron paramagnetic resonance (EPR) is useful for characterizing transition-metal compounds, and it is of particular interest for metalloproteins. Aqueous liquid solutions are important, but such systems present special challenges for high-pressure studies. Problems are posed by the limited space within a magnet, by line broadening because of severe dissipation due to the high dielectric constant of water, and by the challenge of constructing a high-pressure feedthrough for microwaves when limited to nonmagnetic materials (a problem not only for metals but also for the pressurizing fluid). Solutions have been demonstrated[19] for an X-band spectrometer. The initial efforts were limited to qualitative observations in haemoglobin. Quantitative extraction of reaction volumes or activation volumes was not attempted. Electron paramagnetic resonance is particularly effective in studies of solid (often crystalline) samples at low temperatures in the liquid-helium range. These have also been adapted to high pressures.[20]

5. Initiating reactions for kinetic studies

5.1 The stopped-flow method

The stopped-flow method is a method of mixing reagents in order to initiate a reaction for kinetic study. Concentration changes are usually monitored by ultraviolet/visible spectrophotometry, but conductivity is also commonly used, and almost any method is possible. The stopped-flow method has become the most common method of investigating chemical kinetics, and the invention of high-pressure stopped-flow instruments has made variable pressure a useful parameter for mechanistic studies. Manufacturers are beginning to support high-pressure studies. Figure 6.9 shows a recently described apparatus.[21] The entire device fits within a high-pressure chamber of the type described in Section 3.2, although the cylindrical housing is taller than usual, about 33 cm high. The reagents are contained in two syringes and then expelled quickly through a mixing jet into an observation chamber, where the spectrophotometry is performed. A receiving syringe allows the process to be repeated several times. The entire device is in a hydrostatic medium, in which the syringe plungers move to balance the internal and external pressures. The drive for the syringes is electrical, using a rotary solenoid, which is a stepper motor that moves a fraction of a full revolution and then returns to its initial position. A screw fitted with a ratchet turns in one direction. Only a small displacement occurs for each electrical impulse. The observation cell must be correspondingly small, but as many as forty kinetic measurements can be

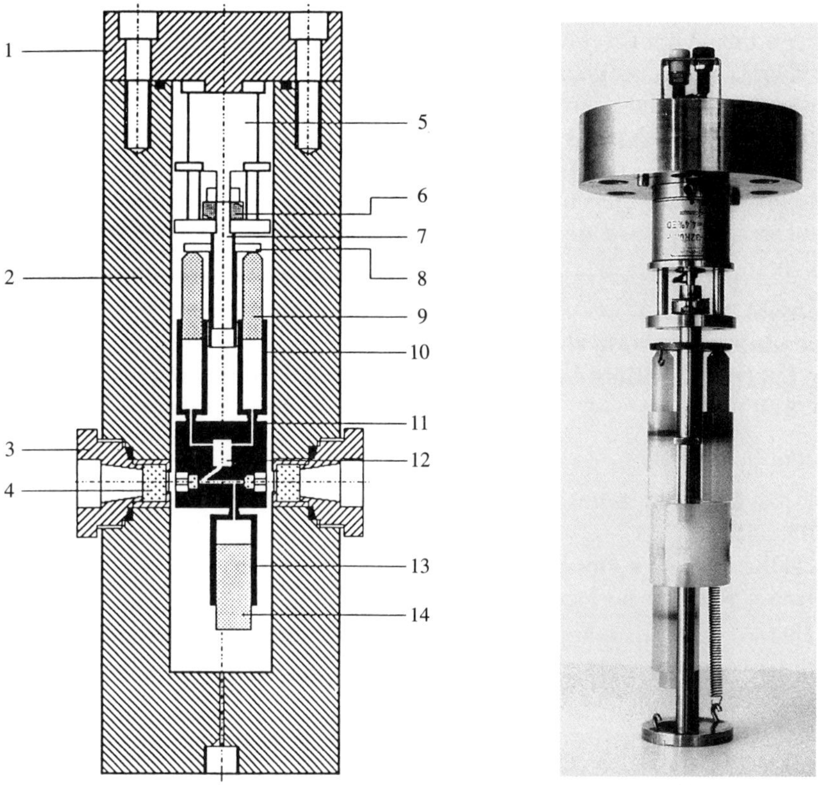

(a) (b)

Fig. 6.9 (a) A cross section of a high-pressure vessel and stopped-flow apparatus: (1) pressure-vessel lid, (2) pressure vessel, (3) high-pressure window-seal assembly, (4) high-pressure sapphire window, (5) stepper motor, (6) screw-drive mechanism, (7) drive shaft, (8) syringe thrust plate, (9) reactant syringe pistons, (10) reactant syringes, (11) Kel–F cell block, (12) mixing jet, (13) receiver syringe, and (14) receiver-syringe piston. (b) A photograph of the stopped-flow apparatus.

made with a single filling of the syringes. The electrical motor cannot give quite as large an impulse as a pneumatic drive,[22] but the small volumes which are used tend to keep the dead time short, that is, near or slightly below 10 ms. The mixer is critical; it must give thorough mixing with minimum obstruction of the flow. Naturally, the viscosity of the solvent affects syringe motion. This is not usually a problem, but it should be considered when a new solvent is used. In practice, the most troublesome components are the seals in the syringes, which must slide freely without leaking and which must be chemically compatible with both the hydrostatic medium and the solvent used for the reaction. The optical pathlength is fixed; a sliding pillbox design is not required for kinetic measurements. The small sample volume poses a challenge for the design of the optical system. An obvious concern is having enough throughput to give an adequate signal-to-noise ratio. A more subtle problem is that movement of the observation chamber may affect light transmission. The impulse given to the syringes must not introduce oscillations. Usually, kinetic changes are monitored at one wavelength only; however, recently developed rapid-scan spectrophotometers make it possible to collect complete spectra. Measuring 100 spectra per second is common; and much faster instruments are appearing.

Protocol 3.
Procedures for filling the high-pressure stopped-flow instrument in Fig. 6.9 (the numbers in parentheses are the numbered parts in Fig. 6.9)

Caution! For safety procedures see Appendix 2.

1. Wash all syringes and the mixing block with the solvent used in the experiments.
2. Fill the receiver syringe (13, 14), with solvent and wash the mixing cell from below. Push the syringe until it is almost empty.
3. Fill the reactant syringes (9, 10) with the test solutions and mount them on the mixing block. Be sure that there are no air bubbles in any part of the stopped flow and that the reactant syringes are exactly the same height.
4. Return the syringe thrust plate (8) to its top position and mount the mixing block complete with syringes.
5. Connect the springs which assist the movement of the syringes when the stepper motor is activated.
6. Insert the stopped-flow unit into the high-pressure vessel, fill it with the pressurizing liquid, and tighten the bolts.
7. Pressurize the system to 5 MPa and allow 20 min for temperature equilibration before performing a number of kinetic runs.
8. Increase the pressure in steps of 25 MPa, and allow 20 min for temperature equilibration before performing the kinetic runs at each pressure,
9. Dismantle the stopped-flow unit by reversing the procedure above after the pressure has been released at the end of the experiment, when the reactant syringes will be empty, the receiver syringe will be full.

A typical kinetic trace from a test reaction,[21] the formation and heterolysis of $(H_2O)_5Cr—C(CH_3)_2OH^{2+}$, is displayed in Fig. 6.10. The rising portion of the curve describes the formation; the slower decay monitors the phosphate-catalysed heterolysis. The first-order rate constant for heterolysis decreases with increasing pressure, as shown in Fig. 6.11. The negative slope implies a positive volume of activation, $\Delta V^{\ddagger} = 14.2 \pm 0.3\ cm^3\ mol^{-1}$.

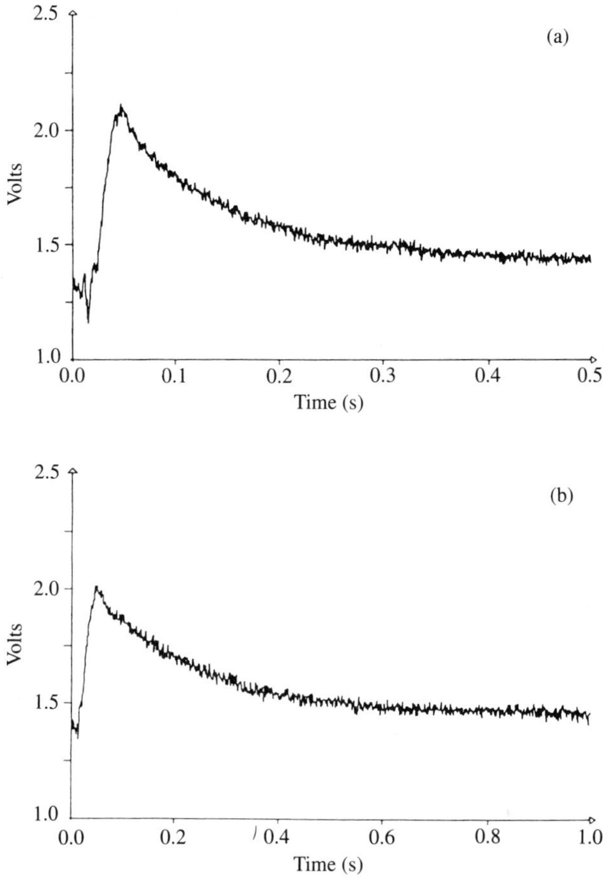

Fig. 6.10 Typical absorbance-time traces for the formation and phosphate-catalysed heterolysis of aq_5–Cr—$C(CH_3)_2OH^{2+}$ at pressures of: (a) 10 MPa, and (b) 180 MPa. The experimental conditions were: [total Cr(II)] = 3×10^{-3} M, [H_2O_2] = 3×10^{-4} M, [i-PrOH] = 0.5 M, [$H_2PO_4^-$] = 0.32 M, ionic strength = 0.5 M, pH = 3.2, and temperature = 25°C. On the voltage axis, 10 V = 1 absorbance unit.

5.2 Temperature jump

It is impossible to mix liquids in much less than a millisecond. Faster reactions must be studied by perturbation methods. The reaction of interest is allowed to equilibrate with the reactants and the products both present in the solution volume. A sudden change is made uniformly throughout the entire sample in one thermodynamic variable, most commonly temperature. This causes the equilibrium constant to change. The concentrations will then *relax* to the new equilibrium condition. The time resolution typically achieved in a *T*-jump (discontinuous temperature change) is 100 to 1000 times shorter than

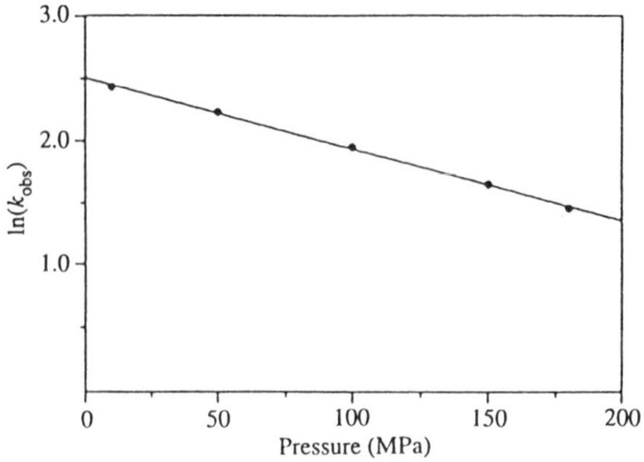

Fig. 6.11 A plot of ln(k_{obs}) against pressure for the phosphate-catalysed heterolysis of Cr—C$(CH_3)_2OH^{2+}$. (The experimental conditions are listed in Fig. 6.10.)

the stopped-flow dead times, although even faster instruments have been described. The applications are restricted to reactions for which the equilibrium does not lie too far in either direction. An advantage of reversible reactions is that the system may be returned to the original condition after each measurement, and the process can be repeated several times.

The magnitude of the temperature change required for a *T*-jump depends on the enthalpy of the reaction, but 5 °C is typical. This implies an energy deposition of about 20 J cm^{-3}, which could be supplied by a beam of radiation, such as microwaves or visible light. A high-pressure laser *T*-jump apparatus has been reported.[23]

More commonly, energy is deposited by resistive Joule heating in a conducting solution. To achieve fast heating, a capacitor is charged to a high voltage, about 15 to 20 kV, and then discharged through the sample solution, which is usually an aqueous salt solution with a resistance of 10–50 Ω. Conductivity measurements cannot be used with Joule heating, so ultraviolet/visible spectrophotometry dominates. These restrictions on the solvent and on the detection method make the *T*-jump method most suitable for inorganic and biochemical reactions. It is possible to average multiple measurements on the same solution. However, the return to the pre-jump temperature must be complete between jumps. Furthermore, the kinetic data refer to the temperature after the jump, which should be uniformly reproducible and accurately known. Normally, the slowest rate measurable is 0.1 s^{-1}, because the temperature begins to fall. The high electric field will cause hydrolysis, so the pH should be buffered. The larger electric fields can affect polymers, but they have little effect on small molecules.

Figure 6.12 shows a *T*-jump apparatus.[24] The cell body is machined from Kel–F. The electrodes are platinum-coated steel. The top electrode mates with a socket in the housing cover, which is connected to the electrical

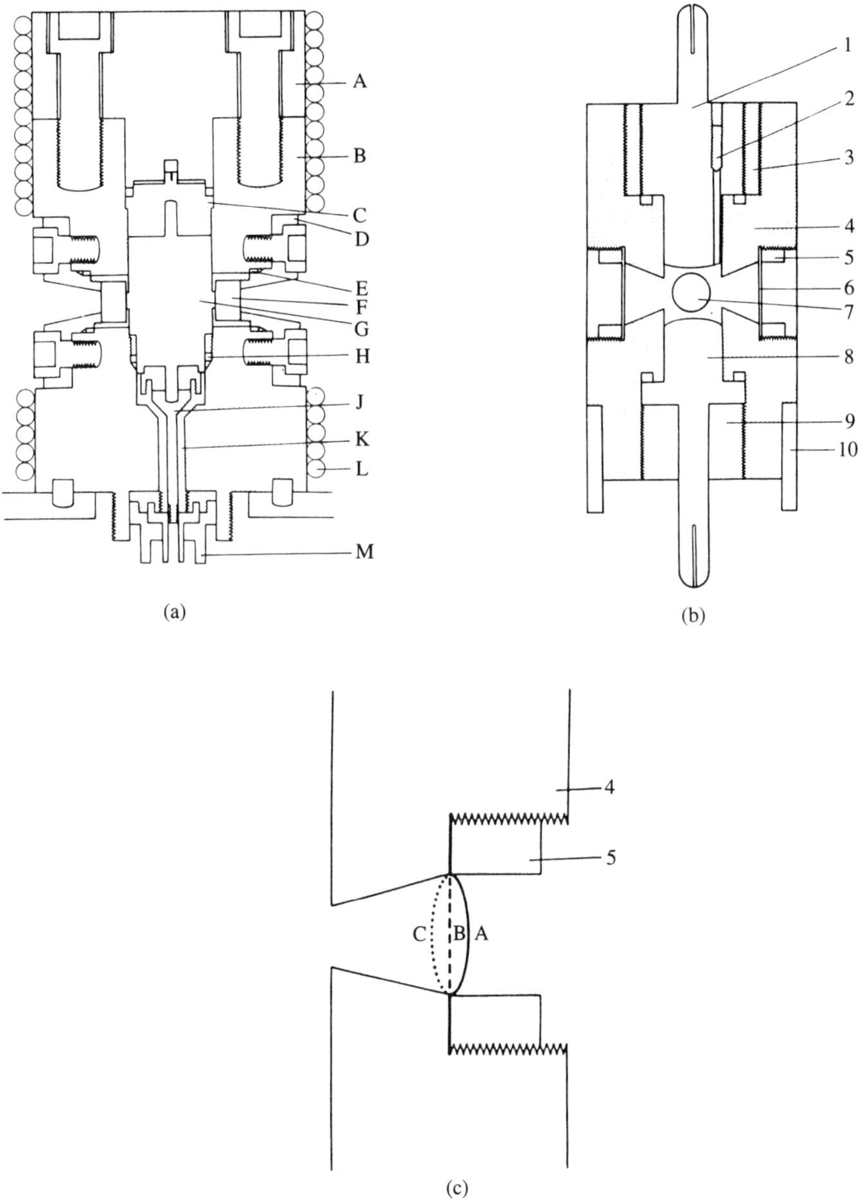

Fig. 6.12 (a) A high-pressure chamber: (A) vessel lid, (B) vessel body, (C) steel piston, (D) window support, (E) Δ- and O-ring, (F) sapphire window, (G) space for the *T*-jump cell, (H) Δ-ring and O-ring, (J) steel high-voltage connector, (K) insulation material, (L) circulation coil, and (M) high-voltage plug. (b) A *T*-jump cell: (1) upper electrode, (2) deaeration hole, (3) electrode thread support, (4) Kel–F cell body, (5) membrane support, (6) Teflon membrane, (7) optical window, (8) lower electrode, (9) threaded bolt, and (10) support ring. (c) A membrane system: (A) $P = 0.1$ MPa, (B) $P = 100$ MPa, (C) $P = 200$ MPa.[24]

ground. The lower electrode is inserted into a plug which is connected to the high-voltage supply and which is electrically isolated from the pressure housing by Kel–F. This is a case in which the construction is made easier if a port is provided on the bottom of the high-pressure housing. Figure 6.12(c) shows the method for transmitting pressure to the liquid inside the cell. Two deformable membranes made from PTFE are sealed in place with threaded support rings. Each is a circle 8 mm in diameter, with a dimple in the centre which is 4 mm in diameter and 1 mm deep. When the cell is filled at the ambient pressure, the dimple faces outwards. At 200 MPa, the dimple faces inwards. If the cell volume is small enough, 1.1 cm^3 in the original design, the small dimple can accommodate the compressibility of water. It is always important to eliminate gas bubbles from any inner sample cell, but it is especially important with these Teflon membranes. The optical windows are positioned so that the light beam is kept a short distance away from the electrodes, where the disturbing effects are at their severest. The hydraulic fluid is *n*-heptane, a transparent, nonpolar electrical insulator.

For sufficiently small changes in absorbance, as are often encountered with *T*-jumps, one may show that the change in concentration is a linear function of the voltage change. Hence, kinetics have sometimes been calculated directly from the voltage trace. This is a risky procedure, since a systematic error will be introduced when one is blessed with unusually large signals. With the widespread availability of computers there is no reason to use such a shortcut. When a time resolution below 10 μs is required, the design of the optical probe and electronics requires extra care. We cannot describe the construction here, but every user should verify two aspects of any data set: (i) be careful that any rate is a chemical rate and not the *RC* time constant of the electronic detection circuit or the heating time or the decay time of the high-voltage field, which is twice as long as the heating time; and (ii) be sure that the optical detection system is working in a range in which the voltage measured is proportional to the light intensity. If the signal drops by exactly 50 per cent when a calibrated 50 per cent attenuator is substituted for the sample, then the small changes induced by the equilibrium shift will certainly be reproduced accurately.

5.3 Pressure jumps (*P*-jumps) and ultrasonics

The rates of reactions may be measured when the concentrations adjust to sudden changes in pressure. This is often combined with conductivity measurements. The value of ΔP needed depends on ΔV^0, but typically a value of 10 MPa is used. Normally, *P*-jumps are generated by forcing water into a fixed volume until a membrane (made of brass foil, 0.1 mm thick) ruptures, permitting the pressure to drop to the ambient value in about 50 μs. There is no upper limit for *P*-jumps on an acceptable time scale. A variable-pressure version has been demonstrated.[25] The key feature was a closed system with a large reservoir pressurized to the desired final pressure and with a smaller sample compartment which is further pressurized until a brass membrane bursts allowing the two chambers to equilibrate. The final pressure could be varied from 0.1 to 98 MPa, and it could be controlled to within 0.5 MPa. The pressure drop was always about 13 MPa. The instrument was tested by measuring the kinetics of formation of a nickel–malonate complex. The solution conductivity was measured using an alternating-current bridge at 200 kHz. The activation volume was measured to be $8.7 \pm 2.1\ cm^3\ mol^{-1}$. The precision was only fair.

Now, *P*-jumps are not as common as *T*-jumps, even at ambient conditions, even though the apparatus for *P*-jumps is probably easier to construct. The signals are small, and the ruptured-membrane principle limits signal averaging. The high-pressure version is even less attractive. An alternative strategy for *P*-jumps has been demonstrated by Clegg and Maxfield.[26] They generated a repetitive pressure perturbation up to 0.5 MPa by driving a piezoelectric stack with an electrical square wave. With modern signal-averaging methods, which can measure a full decay curve for each half-cycle and accumulate many repetitions, both conductivity and photometry methods can work with such small perturbations quite well. Clegg and Maxwell demonstrated capabilities over a broad time range, from 20 μs to 100 s. This instrument design did not displace conventional *P*-jumps, perhaps because of the small ΔP, but its features are well suited for use at variable high pressures. A recent improvement uses a smaller volume to extend ΔP to 7 MPa, and this makes the design more attractive.[26a]

Ultrasonic sound waves propagating in a liquid solution create repetitive pressure/density variations. Some versions measure the absorption of sound energy and others involve the scattering of a light beam. They are useful for very fast reactions, but each instrument tends to be limited to a small time range. Although operation at high pressures is possible, such an application would be very specialized.

5.4 Pulse radiolysis

Jumps in temperature and pressure (*T*-jumps and *P*-jumps) are restricted to reactions for which the equilibrium does not lie too close to the reactant or the product side. Radiolysis and photolysis are also perturbation methods, but they can be applied to systems for which thermal equilibrium heavily favours one direction. Both these techniques are effective at times well below 1 ns. Pulse radiolysis initiates a chemical reaction with a beam of high-energy (MeV) electrons. The electrons are absorbed by the solvent, forming ions, which then undergo further reactions. The concentrations are usually monitored by ultraviolet/visible spectroscopy, using a three- or four-window pressure housing, with the electron beam at right angles to the optical-probe beam. Megavolt electrons can penetrate some distance through all materials, but there are no really transparent materials. Consequently, the experimental arrangement must minimize the amount of material traversed by the beam.

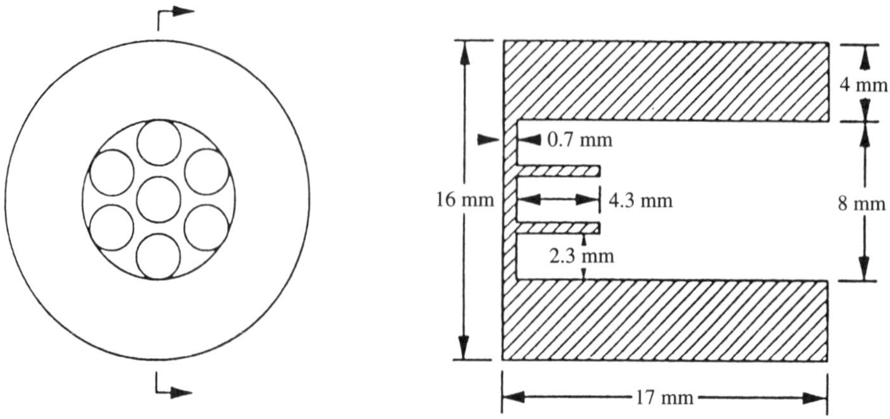

Fig. 6.13 A schematic diagram of an electron-beam window for high-pressure pulse radiolysis. The electron beam enters from the right-hand side.[27]

The optical measurement is made near the edge of a sample cuvette; the cuvette is close to the pressure window; and the high-pressure 'window' is itself reduced to a minimum thickness. Instead of using 1 cm of sapphire, 1 mm of steel makes a good 'window', if this 1 mm can be supported. An effective solution has been described recently.[27] A 17 mm thick steel disc has a honeycomb pattern of holes drilled to within 1 mm of the opposite face. This rigid structure increases the electron-beam intensity tenfold at the sample. This window design is shown in Fig. 6.13.

Electron transfer from ruthenium-(II)–amine complexes to cytochrome-*c* proteins has been characterized[28] using pulse radiolysis and the new window, the optical absorption being monitored at 550 nm. Large negative volumes of activation, $\Delta V^{\ddagger} \approx -16$ to -18 cm^3 mol^{-1}, were measured for different donor–acceptor combinations, with uncertainties quoted in the range 0.6–0.9 cm^3 mol^{-1} (see Fig. 6.14). The principal concern in pulse-radiolysis studies is verification of the potentially complex redox-reaction scheme involved. The electron beam creates hydroxyl radicals, which react with formate ions, which are present as part of the buffer system, to produce carbon-dioxide radical anions $CO_2^{\cdot -}$, that predominately reduce Ru(III) to Ru(II), transferring an electron to the protein, which reduces cytochrome(III) to cytochrome(II). However, it was necessary to rule out other pathways, including direct reduction of the cytochrome by the carbon-dioxide radical anion. Since the reduction of cytochrome is irreversible, it was necessary to stir the solution and correct for the changing concentration of cytochrome(III) (since the irradiation is nonuniform in the cell, and the stirring might not be completely effective). High pressures, as such, did not introduce new problems.

A major concern with both pulse radiolysis and photolysis is that the radiation or light is capable of initiating undesirable side reactions. Stirring works

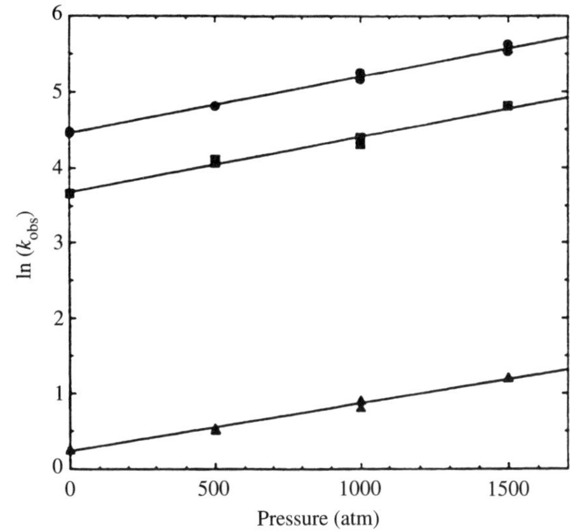

Fig.6.14 The pressure dependence of the observed rates of intramolecular electron transfer in: (●) pentaammineruthenium-modified *Candida krusei*, (■) horse-heart cytochrome *c*, and (▲) the observed rate of intermolecular reduction of horse-heart ferricytochrome *c* by $Ru(NH_3)_6^{2+}$.[28]

surprisingly well; one may use a magnetic bar and drive it with a magnetic stirrer located outside the high-pressure chamber. However, for very small cells or viscous solvents, it may be necessary to implement recirculating flow. This has not been done in the past, but it seems that one possibility would be something like a reversible version of the stopped-flow syringe system.

5.5 Flash photolysis

Photolysis creates excited electronic states; this is the subject of photochemistry, discussed in Section 8. The simpler applications of photodissociation and photoionization simply generate a supply of a reagent and distribute it throughout a volume. The subsequent reaction is usually followed by photometry or conductivity measurements. Dissociation and recombination of the same species can be conveniently studied by this method,[29] or the photogenerated reagent may react with another species, which is sometimes called a scavenger. If protons are released, there is a pH-jump. The small windows of a pressure chamber make conventional flash sources ineffective. Lasers can overcome this difficulty: *not* because they can provide a greater intensity of excitation light, but because they provide light in a collimated, monochromatic beam and in a much shorter duration pulse. It becomes easy to study rates as large as 10^8 s^{-1}. Taking full advantage of these features or working at still shorter times, for picosecond studies, requires sophisticated apparatus and some experience, to say nothing of the cost of lasers. Consequently, it is desirable that the high-pressure apparatus is portable so that it can be moved to the laser set-up.[6] To design a laser-flash-photolysis experiment, one should make a rough calculation of the number of transients that can be produced and then calculate the absorbance change that will result.

For a light pulse of energy J (in joules), with a wavelength λ (in nm), the number of photons N_{las} is

$$N_{\text{las}} = 5.0\,\lambda J \times 10^{15}.$$

The fraction, F, of the light of the excitation wavelength that will be absorbed by the sample can be calculated by assuming that Beer's law applies and from knowledge of the pathlength in the direction of the excitation beam. The concentration should usually be adjusted so that F is between 10 and 90 per cent. Estimate the volume, V, over which the light will be absorbed. Note the quantum yield, Q, per absorbed photon for producing the desired reagent. The initial concentration, C, of the photoinduced reagent will be given by

$$C = \frac{FN_{\text{las}}Q}{6.0V \times 10^{20}}.$$

For an optical pulse with an energy of 1 J, with $\lambda = 620$ nm, with a volume 1 cm^3, a 60 per cent absorption, and $Q = 1$, we have $C = 5$ μM. For a probe-beam pathlength of 1 cm, a molar absorptivity change of $\varepsilon = 10^4$ is required to have a transient absorbance of 0.05, a value which may be followed for many half-lives in a careful kinetics experiment, but which is not a really impressive value. Furthermore, a 1 J laser is very powerful and expensive. A more realistic energy value is 10–100 mJ. Consequently, the volume illuminated should be reduced; the ideal arrangement is a long, narrow volume which distributes transients over a long pathlength for the probe beam. One will never make more transients than there are precursors. As soon as some significant fraction is excited, the usual form of Beer's law fails because it assumes that the concentration of absorbers is never depleted. This effect starts to play a role at lower light intensities than might be expected.[30] Consequently, it is better to have a larger concentration of precursors which have only moderate absorptivity at the excitation wavelength. This runs counter to the intuition of some experimenters, who look for large values of ε at the pump wavelength. A better situation is a modest value of ε for the pump and the biggest possible $\Delta\varepsilon$ for the probe.

Figure 6.15 shows that the transmitted light is a function of the pressure for recombination of photodissociated ligands in a reaction of an iron porphyrin with CO.[29] The laser pulses had energies of about 3 mJ, and a few hundred laser shots were summed for each trace. These intensities have to be converted to absorbance signals. The data give $\Delta V^{\ddagger} = -19.3 \pm 0.4$ cm^3 mol^{-1}.

Although limited to photoreactions, lasers do make it easy to study fast reactions, and there are few special problems related to work at high pressures. One potential problem is the possibility that the reaction will change

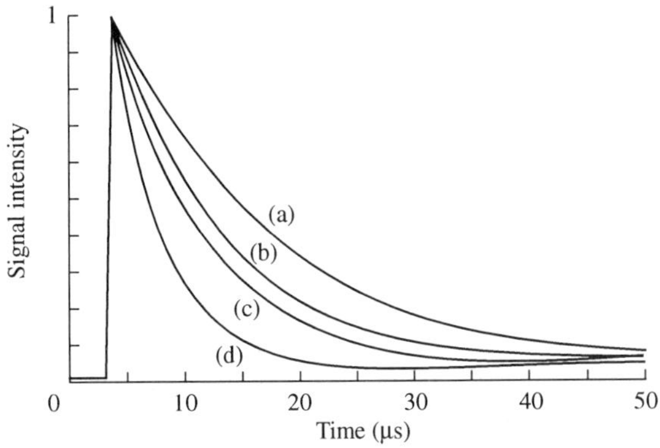

Fig. 6.15 A trace of the signal intensity against time for the following four pressures for the reaction MCPH—CO ⇌ MCPH + CO: (a) 5 MPa, (b) 50 MPa, (c) 100 MPa, and (d) 150 MPa. The experimental conditions were: the solvent was toluene, λ_{obs} = 440 nm, T = 25°C, [MCPH] = 15 μM, and [CO] = 6.7 mM.[29]

from being activation controlled to being diffusion controlled as a function of viscosity. Examples are rare. A nice illustration was reported recently by Traylor and co-workers.[31] Figure 6.16 shows the result: a biomolecular rate constant with a negative value of $\Delta V^{\ddagger}$ at low pressures (presumably due to the faster intrinsic reaction at each encounter under pressure) that changes to a positive value of $\Delta V^{\ddagger}$ at higher pressures (presumably due to slower diffusion when an encounter complex is formed). This effect was not an accidental discovery: it was predicted, and the conditions were chosen in advance to illustrate the point. This concern is a problem only with fast reactions, but the rates in Fig. 6.16 are no greater than 5×10^4 s^{-1}. On the other hand, a viscous solvent, mineral oil, was deliberately chosen.

6. Batch processing and remote sampling

Despite justified enthusiasm for *in situ* methods, there is still a place for mixing reagents, allowing the solution to approach equilibrium slowly, and extracting the products for detailed analysis, either after equilibrium is reached or several times during the approach to equilibrium. A large reaction vessel (with a volume of a few hundred cubic centimetres) of Teflon can be used for the inner reaction vessel, and a fine-control, high-pressure needle valve can be used to extract small aliquots, as shown in Fig. 6.17. Only a small pressure drop will occur at each extraction, and the pressure can be readjusted. Provided the reaction is sufficiently slow, a kinetic analysis can be accomplished by mixing the reagents at ambient conditions before pressurizing. However, compression of the hydraulic fluid and the reaction solvent generates heat, which will affect the reaction kinetics. Consequently, if kinetic studies are

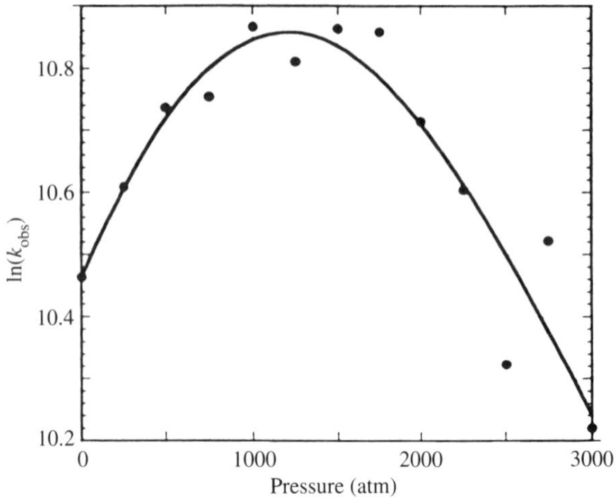

Fig. 6.16 A plot of $\ln(k_{obs})$ against hydrostatic pressure for data from six runs, investigating the effect of pressure on the reaction of CO with the intermediate BHm formed by nanosecond laser-flash photolysis of BHm—CO in a mineral oil/toluene solution 90/10 to 95/5 (v/v). The experimental conditions were [heme] = 1×10^{-5} M, [CO] = 7.5×10^{-3} M, and T = 25°C. The curve is a mathematical fit to these data based on the model described.[31]

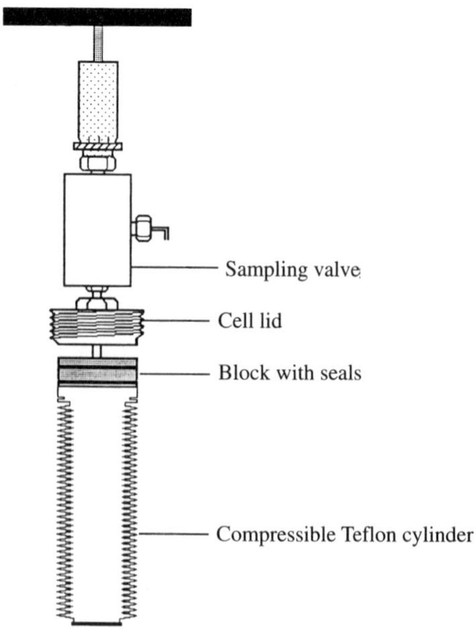

Fig. 6.17 A high-pressure reaction vessel with a sampling valve.

planned, only quite slow reactions should be used in which the temperature can stabilize at its regulated value before significant reaction occurs. Faster reactions require the mixing of the reagents under pressure. For sufficiently slow reactions, it is possible to extract aliquots of the sample and analyse them before the concentrations change. One should quench the reaction and stabilize the concentrations, either by lowering the temperature or by adding a reactive scavenger. This can be done quite quickly in some cases. Such sampling methods might be required in order to use titrations or chromatography to measure the concentrations, either because no other method is possible or in order to achieve greater precision. This might also be necessary in order to search for minor products and to verify the material balance.

7. Dilatometry and density measurements

The determination of ΔV^0 by subtracting partial molar volumes which have been measured separately for the products and the reactants is attractive in principle but difficult in practice. The volume change upon mixing, or equivalently the density, must be measured as a function of the concentration to permit extrapolation to infinite dilution, while all the other parameters that affect the density of the solvent are kept constant. In particular, millidegree temperature regulation is necessary. Isaacs has outlined various strategies;[7] and we concur that the most attractive, especially for work at high pressures, are those which allow the density to be directly measured as an intensive variable, notably when the resonance frequency of a tuning fork or crystal is measured. Various instruments have been designed to measure reaction volumes dilatometrically, and they have been reviewed elsewhere.[32]

8. Photochemistry and short-lived intermediates

By one definition, photochemistry is the study of the electronic excited states of molecules. A reactant A absorbs light and its wave function changes so it is in state A*. Now A* may react by a unimolecular process or by a bimolecular process to form a product P. If A* lasts long enough to reach thermal equilibrium, then the problem of deducing a mechanism for the reaction is the same as that for any other reaction: the concentrations of A* and P must be determined precisely as functions of time and as functions of temperature, pressure, solvent compositions and of the concentrations of the other reagents or of any catalysts used. (The issue of what to say about A* when it is not a stable, thermodynamic intermediate belongs to the realm of chemical physics.) The special problems of photochemistry relate to the fact that it is very difficult to determine the concentration of A* accurately as a function of time. For example, A* may be extremely reactive and short lived. Worse still, competing reactions must always be considered, and it may be difficult to determine a proportionality constant for A* for the relationship between concentration and some measurable. As with other reactions, such a proportionality constant must be re-examined at each pressure. The extent of formation of A* may vary with pressure. Once A* is formed, it is necessary to consider a minimum of three competing rate processes: the photochemistry of interest, k_{pc}; the radiative emission of light, k_r, and the nonradiative deactivation to A, k_{nr}. The lifetime of A*, τ, is given by the reciprocal of the sum of all the deactivation processes: $\tau = (k_{pc} + k_r + k_{nr})^{-1}$. If bimolecular reactions are involved in k_{pc}, pseudo-first-order conditions are assumed; the other processes are unimolecular. The integrated yield of any process is given by its rate multiplied by the lifetime; that is by $Q_x = k_x\tau$. Activation parameters such as $\Delta V^{\ddagger}$ should be defined only for k_x, and this means that τ must be known in addition to Q_x. It is apparent that the pressure dependence of k_{pc} can only be deduced if the pressure dependences of k_r and k_{nr} are also known. Usually, k_r is so small that emission of light does not change the concentrations, it just serves as a proportional measure. If the pressure dependence of k_r is known, then at least the relative concentrations of A* at different pressures can be determined. The dominant effect of pressure on k_r will usually be manifest through changes in the refractive index of the medium; and k_r should be proportional to the square of the refractive index. Any intrinsic change in k_r depends on the wave function, which should be reflected in spectral changes and should not be much perturbed by the pressure changes considered here. Luminescence is often selective for a single excited state. Luminescence kinetics are now measured routinely in the low picosecond range. When k_r is sufficiently large to change populations, the concentration of A* can still be extracted if k_r is known. When k_r is very small, transient absorption or conductivity measurements are usually used. Unfortunately, even when the time dependence of the concentration of A* has been established, one is still dealing with competitive reactions characterized by k_{pc} and k_{nr}. The role of k_{nr} is critical. It is usually the dominant process, controlling the concentration of A*. There is a fairly extensive literature on the effect of temperature on k_{nr} and there is a less extensive literature on the effect of pressure.[33,34] Unfortunately, it is impossible to make simple generalizations. The volumes of activation for k_{nr} are sometimes close to zero, but they can be as large as 10 $cm^3\ mol^{-1}$. There are dramatic changes with solvent composition and the selected temperature. The only safe strategy is to treat every reaction as though it involves competitive processes until proven otherwise.

Some new methods are available for measuring partial molar volumes which can only be applied to fast reactions.[35] Consider a reaction initiated locally in only part of the volume of a solution by some perturbation, such as laser-flash photolysis. The volume change associated with the rapid production of a product or intermediate generates a sudden pressure increase which initiates a sound wave. This propagates to the surface, where it couples to the surrounding gas and propagates again. A microphone can be surprisingly sensitive in detecting these photoacoustic changes, and this method can be developed into a procedure for determining volumes of reaction. If the intermediate is short lived, it will convert to another intermediate or product and generate yet another sound wave, which may also be detected. A second means of detection has also been developed. The volume change in a local region in the solution will change the refractive index (the refractive index usually decreases for a positive volume change). A region of locally diminished refractive index forms a negative lens, and such regions can be detected by a propagating laser beam. If the probe is arranged so that the decrease in the index is symmetric and centred, the beam will expand. If the beam is aligned to one side, the direction of propagation will be deflected away from the perturbation. These deflections can be used to measure very small perturbations. The principal difficulty with both these methods is that many variables affect the density. In particular, a temperature rise due to the enthalpy change in a chemical reaction also changes the density. However, with appropriate controls, it is sometimes possible to distinguish these contributions. In some cases, the reaction rates can be followed as can the integrated changes. If the rates are able to be followed, then the measurements could be carried out as a function of static pressure to determine the volumes of activation. Both the photoacoustic method and the laser-deflection technique are well established, and they are likely to persist as tools for special purposes.

9. Procedural guidelines

9.1 Solvents

Solvents raise two practical considerations related to high pressures, the viscosity change, and crystallization to a solid. Since viscosity is discussed in detail in Section 9 of Chapter 3, only the qualitative features pertinent to the study of a reaction rate are outlined here. Viscosity normally increases with pressure. As in so many of its other properties, water is an exception.[7] Increased pressure disrupts hydrogen bonding, so that at, and below, room temperature, the viscosity of water initially decreases, even though its density increases. Solvents which are already viscous at ambient pressure, especially those that are composed of relatively large molecules, often show large increases in their viscosity with increases in pressure, as though the molecules are locked together without adequate free volume to diffuse. Since these solvents are those which are most likely to be used in investigations of viscosity effects, their strong pressure dependence can be an advantage. Experimenters should be wary of large temperature effects in these systems. Solvents (such as glycerol) which are viscous because of multiple hydrogen bonding, but which are composed of small molecules, show significant, but lower, viscosity increases with pressure. Familiar, low-molecular-weight organic solvents change their viscosity in a fairly predictable way, showing an increase of 4–16 times over the pressure between the ambient value and 400 MPa. For the purpose of reaction rates, we are most interested in diffusion.

In general, one expects solute diffusion to parallel the behaviour of viscosity, and some data does support this assumption. However, there is not much data at the atomic level. Assumptions about diffusion should be checked, particularly in special circumstances, for example, where small translations or rotations are involved, for hydrogen-bonded systems, or for small solute molecules immersed in large solvent molecules. These examples will be of concern to physical chemists studying fundamental aspects of liquid behaviour, and they might affect other studies of fast reactions. However, for slow reactions, which are clearly activation controlled at ambient conditions, investigators need not be too worried that diffusion will be slowed so greatly that the reaction will become diffusion controlled at 200 MPa. Of greater concern is the fact that many common solvents freeze under modest pressures. Liquids may of course supercool. Larger molecules, solvent mixtures, and hydrogen-bonded systems, in particular, may form glasses with extremely high viscosities without crystallizing. Common solvents, however, can crystallize abruptly from a liquid of modest viscosity. At 298 K, some freezing pressures of possible concern include (in MPa): cyclohexane, 36; decane, 300; benzene, 72; *o*-xylene, 229; *p*-xylene, 36; bromoform, 72; tetrachloromethane, 146; 1,2-dibromomethane, 63; bromobenzene, 326; *t*-butanol, 0.1; acetic acid, 39; nitrobenzene, 84; 1,2-dimethoxybenzene, 14; aniline, 162; *N,N*-dimethylaniline, 93; dioxane, 120; and methyl benzoate, 65. A modest increase in temperature will extend the liquid range considerably, while a decrease has the opposite effect.

9.2 Concentrations

Concentrations are always quoted in pressure-independent units, which also means density-independent units. These could be the mole fraction or molality, but equally correct is the molarity measured at a specified temperature and pressure. Invariably, the pressure should be 0.1 MPa.[36] From this value, the mole fraction can be calculated if necessary, but for most purposes there is no reason to make this calculation. The ionic strength is quoted in units of the molality at 0.1 MPa.

9.3 Plots of ln *k* against *P*

In principle, one expects $\Delta V^{\ddagger}$ (or ΔV^0), which is the slope of the plot of ln *k* (or of ln *K*) against pressure, to vary with the pressure at which the slope is evaluated. It is mathematically undesirable to determine a slope at the end-point of a data set. Since we must do this, it would be best if the functional form for the curve and fit were known for a data set over a wide range of pressure.[1,4,7] At one time it was thought desirable to fit a quadratic function, but it is now generally agreed that there is no justification for such a fit since it is as likely as not to introduce additional systematic errors. Contemporary practice favours devoting one's efforts to making the best possible measurements over a limited range of pressure near the origin and fitting the data to a straight line. It is good to have enough precision and enough data points to verify that a straight line is a good fit over the range used, but one should also consider data accumulated on other, similar, reactions. If non-linearity is large the first suspicion should be that the reaction mechanism is changing, perhaps due to a change from activation control to diffusion control, as discussed in Section 8, or because of (unsuspected) competing reactions. There is no point in interpreting the volumes of activation until the elementary steps of the reaction scheme are established. If the non-linearity is slight and recognizable only because of very precise measurements, these factors could still be responsible, but one might also try to distinguish between effects due to the pressure dependence of the solvent compressibility and those due to the pressure dependence of the solute volumes, including that of the transition state. To begin such an analysis, one should re-plot ln k against the solution density, instead of against the pressure. It is clear that density is a preferable independent variable. If the density is not known as a function of pressure for the solvent system, there is no point in continuing, as one has no way to begin to interpret any nonlinearity. There is currently no interpretation for the compressibility of the transition state, but the availability of high-quality data might stimulate interest in this theoretical problem.

9.4 Activity coefficients

In principle, both K and k should be expressed in terms of activities, each of which is a concentration multiplied by an activity coefficient. However, standard states are chosen so that the activity coefficients are equal to unity at sufficiently low concentrations; and for uncharged small molecules, though not for polymers, this usually includes the types of studies considered above. For ions, however, the activity coefficients become significant in the low millimolar range, and then one should consider the pressure effects on the activity coefficients and distinguish them from effects related to the particular reaction being investigated. Experimental data show that the increase in the mean activity coefficients for ions with increasing pressure is significant and that this effect is more important at higher ionic strengths.[37]

9.5 Interpreting volume profiles

In the previous sections we outlined methods of measuring $\Delta V^{\ddagger}$ and ΔV^{0} which can now be used to construct a volume profile, as discussed in Section 1. The mechanistic interpretation of such profiles is in terms of intrinsic and solvational volume changes which occur along the reaction coordinate.[1–5] Such volume changes will be characteristic for different types of reactions, such as ligand exchange, ligand substitution, addition, elimination, isomerization, electron transfer. In a recent example,[38] complex-formation reactions of aquated Cr(II) with a series of aliphatic radicals were studied as a function of pressure in an effort to determine the nature of the solvent exchange and complex-formation mechanism of the metal ion. The volume profile obtained

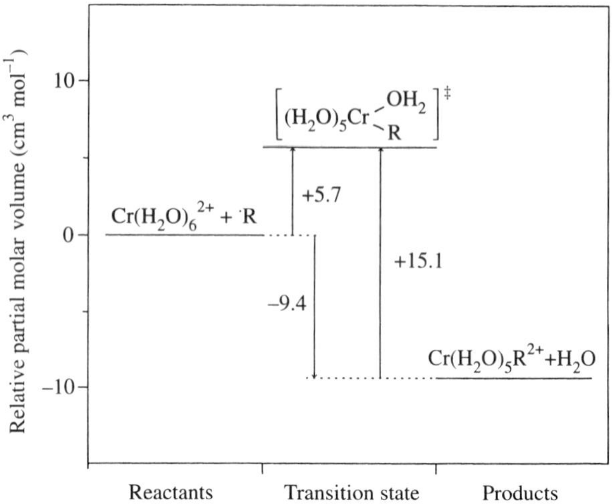

Fig. 6.18 A volume profile for the reaction $Cr(H_2O)_6^{2+} + \cdot R \rightleftharpoons Cr(H_2O)_5R^{2+} + H_2O$.

for a particular reaction is shown in Fig. 6.18, from which it follows that ligand substitution on $Cr(H_2O)_6^{2+}$ is controlled by the dissociation of a coordinated solvent molecule according to a dissociative interchange mechanism. The formation of the Cr—R bond is accompanied by a large volume collapse, partly because of contraction of the metal centre due to the process Cr^{II}— R → Cr^{III}—R^-.[38]

9.6 The thermodynamics of the transition state

One may speculate about correlations between transition-state parameters, such as whether large, positive, activation volumes correlate with large, positive, activation entropies. In general, researchers into volume profiles should understand that a full understanding requires knowledge of all the parameters.[5] However, complete characterizations are rare. Given the considerable effort required to measure a volume profile at one temperature, it is often appropriate to characterize the behaviour at one or two additional temperatures. Thus, in a short period of time, a database would accumulate which would become useful for considering different categories of reactions.

It is appropriate to distinguish between intrinsic, solvational, and steric contributions towards $\Delta V^{\ddagger}$. The solvational contribution can usually be obtained from a systematic solvent-dependence study. The steric contribution can only be obtained from a series of measurements in which the steric hindrance is changed systematically. It is not yet known whether steric hindrance will also affect the position of the transition state in terms of 'early' or 'late' with respect to the reactants and products along the reaction coordinate. Series of studies have shown that there is good correlation between the activation enthalpy and the location of the transition state in terms of volume changes. For a particular system, fast reactions will be characterized by a low activation enthalpy and an 'early' transition state will be characterized in terms of volume changes, whereas a slow reaction will have a high activation enthalpy and a 'late' transition state. This seems to be especially true for systems in which the reactivity is controlled by electronic factors.[39]

10. Summary

This chapter describes the construction of instruments and offers guidelines for protocols for a general laboratory dedicated to determining pressure effects on rates and equilibria in solution chemistry and biochemistry. We have tried to take advantage of two decades of accumulated experience, as reported by many investigators, and to explain why certain strategies have evolved in the design of apparatuses. However, our goal is not to convince readers that these are the only ways to proceed. We hope that by our explanation of the rationale for one choice, readers will recognize where their interests diverge from common assumptions and so be encouraged to strike out in new directions. We have also tried to bias our remarks somewhat in directions which we expect will attract increased attention in the next decade.

References

1. van Eldik, R., ed. *Inorganic High Pressure Chemistry: Kinetics and mechanisms*; Elsevier: Amsterdam, **1986**.
2. le Noble, W. J. ed. *Organic High Pressure Chemistry*; Elsevier: Amsterdam, **1989**.
3. Jurczak, J. Baranowski, B., eds *High Pressure Chemical Synthesis*; Elsevier: Amsterdam, **1989**.
4. van Eldik, R.; Asano, T.; le Noble, W. J. *Chem. Rev.* **1989**, *89*, 549–688.
5. van Eldik, R.; Merbach, A. E. *Comments Inorg. Chem.* **1992**, *12*, 341–378.
6. Spitzer, M.; Gartig, F.; van Eldik, R. *Rev. Sci. Instrum.* **1988**, *59*, 2092–2093.
7. Isaacs, N. S. *Liquid Phase High-Pressure Chemistry*; Wiley: Chichester, **1981**; Chapter 1.
8. Sherman, W. F.; Stadtmuller, A. A. *Experimental Techniques in High-Pressure Research*; Wiley: Chichester, **1987**.
9. Winter, R.; Jonas, J. *High Pressure Chemistry, Biochemistry and Materials Science*; NATO ASI Series C 401; Kluwer: Dordrecht, **1993**.
10. le Noble, W. J.; Schlott, R. *Rev. Sci. Instrum.* **1976**, *47*, 770–772.
11. Wieland, S.; van Eldik, R. *Rev. Sci. Instrum.* **1989**, *60*, 955–956.
12. Projahn, H.-D.; Dreher, C.; van Eldik, R. *J. Am. Chem. Soc.* **1990**, *112*, 17–22.
13. Baliga, B. T.; Whalley, E. *J. Phys. Chem.* **1963**, *73*, 654–666.
14. Doine, H.; Whitcombe, T. W.; Swaddle, T. W. *Can. J. Chem.* **1992**, *70*, 81–88.
15. Sachinidis, J.; Shalders, R. D.; Tregloan, P. A. *J. Electroanal. Chem.* **1992**, *327*, 219–234.
16. Millero, F. J. In *Water and Aqueous Solutions: Structure, thermodynamics and transport processes*; Hornes R. A., ed. Wiley–Interscience: London, **1972**; Chapter 13.

17. Buback, M. *Angew. Chem.* **1991**, *103*, 658–670.
18. Akitt, J. W.; Merbach, A. E. *NMR Basic Princip. Prog.* **1990**, *24*, 189–234.
19. Paul, J.; v. Goldammer, E.; Wenzel, H. R. *Z. Naturforsch.* C, **1988**, *43*, 162–166.
20. Hoffman, S. K.; Hilczer, W. *Inorg. Chem.* **1991**, *30*, 2210–2213.
21. van Eldik, R.; Gaede, W.; Wieland, S.; Kraft, J.; Spitzer, M.; Palmer, D. A. *Rev. Sci. Instrum.* **1993**, *64*, 1355–1357.
22. Ducommun, Y.; Nichols, P. J.; Helm, L.; Elding, L. I.; Merbach, A. E. *J. de Phys.* **1984**, *45, C8*, 221–224.
23. Caldin, E. F.; Grant, M. W.; Hasinoff, B. B.; Tregloan, P. A. *J. Phys.* E, **1973**, *6*, 349.
24. Doss, R.; van Eldik, R.; Kelm, H. *Rev. Sci. Instrum.* **1982**, *53*, 1592–1595.
25. Inoue, T.; Kojima, K. Shimozawa, R. *Chem. Lett.* **1981**, 259–262.
26. Clegg, R. M.; Maxfield, B. W. *Rev. Sci. Instrum.* **1976**, *47*, 1383–1393.

26a. Clegg, R. M. Private communication.

27. Wishart, J. F.; van Eldik, R. *Rev. Sci. Instrum.* **1992**, *63*, 3224–3225.
28. Wishart, J. F.; van Eldik, R.; Sun, J.; Su, C.; Isied, S. S. *Inorg. Chem.* **1992**, *31*, 3986–3989.
29. Taube, D. J.; Projahn H.-D.; van Eldik, R.; Magde, D.; Traylor, T. G. *J. Am. Chem. Soc.* **1990**, *112*, 6880–6886.
30. Magde, D. *J. Chem. Phys.* **1978**, *68*, 3717–3733.
31. Traylor, T. G.; Luo, J.; Simon, J. A.; Ford, P. C. *J. Am. Chem. Soc.* **1992**, *114*, 4340–4345.
32. Kotowski, M.; van Eldik, R. *Coord. Chem. Rev.* **1989**, *93*, 19–57.
33. Wieland, S.; van Eldik, R. *Coord. Chem. Rev.* **1990**, *97*, 155–165.
34. Ford, P. C.; Crane, D. R. *Coord. Chem. Rev.* **1991**, *101*, 153–160.
35. Braslavsky, S. E.; Heibel, G. E. *Chem. Rev.* **1992**, *92*, 1381–1410.
36. Hamann, S. D.; le Noble, W. J. *J. Chem. Educ.* **1984**, *61*, 658–660.
37. Hamann, S. D. *J. Chem. Soc., Faraday Trans.* I, **1984**, *80*, 2541–2548.
38. van Eldik, R.; Gaede, W.; Cohen, H.; Meyerstein, D. *Inorg. Chem.* **1992**, *31*, 3695–3696.
39. Pipoh, R.; van Eldik, R. *Organometallics* **1993**, *12*, 2668–2671.

7

Chemical transformations

1. Organic synthesis at high pressures

Neil S. Isaacs

1.1 How are reactions affected by pressure?

Only a brief outline of the effects of pressure on rates and equilibria is given here since the subject is amply documented elsewhere.[1–4] It has long been appreciated that the position of a chemical equilibrium may be shifted by the application of external pressure in reactions in both the liquid and the gaseous phase. This shift in equilibrium favours the direction of the reaction which results in the smaller volume; this is an application of the Le Chatelier's principle. In gas-phase reactions the term *volume* denotes the total volume of the system; in dilute solutions the term *volume* denotes the algebraic sum of the partial molar volumes of the individual reagents and products. The thermodynamic relationship which summarizes this effect is

$$\left(\frac{\mathrm{d}\,(\ln K)}{\mathrm{d}p}\right)_{\mathrm{p}\to 0} = \frac{-\Delta V}{RT}, \tag{7.1.1}$$

where ΔV is known as the volume of reaction, which is defined by

$$\Delta V = \Sigma V_{\text{products}} - \Sigma V_{\text{reagents}}. \tag{7.1.2}$$

For solutions, the V are partial molar volumes, that is, the volume change brought about by the dissolution of 1 mol of solute in an infinite volume of solvent.

Thus, the volumes of reaction may be measured by determining how the pressure affects the value of the equilibrium constant or by measuring the individual partial molar volumes of the reagents and the products and then applying an algebraic summation. A good agreement between both methods has been demonstrated. The rates of chemical reactions in solution may similarly be affected by the external pressure; they either increase or decrease, according to whether a decrease or increase of the volume, respectively, results from conversion of the products to the transition state. By analogy to eqn (7.1.1) we have

$$\left(\frac{\mathrm{d}\,(\ln k)}{\mathrm{d}p}\right)_{\mathrm{p}\to 0} = \frac{-\Delta V^{\ddagger}}{RT}, \tag{7.1.3}$$

where $\Delta V^{\ddagger}$ is the *volume of activation*, which is defined by

$$\Delta V^{\ddagger} = V_{ts} - \Sigma V_{reagents}, \tag{7.1.4}$$

and where V_{ts} is the partial molar volume of the transition state.

In eqns (7.1.1) and (7.1.2), the pressure is in bar (=0.1 MPa), the temperature is in Kelvin, and $R = 81\ cm^3\ bar\ K^{-1}\ mol^{-1}$ so ΔV has units of $cm^3\ mol^{-1}$.

Equation (7.1.3) indicates that a reaction with a negative volume of activation is accelerated by pressure, whereas a reaction with a positive value of $\Delta V^{\ddagger}$ is retarded. The magnitude and the sign of the volume of activation contains information about the mechanism of the reaction, as discussed in ref. 2, Chapter 4. The following generalizations can be made.

1. A negative contribution to the volume of activation results from associative processes, including bond formation in the rate-determining step, and especially from the formation of charged species from uncharged reagents. This last phenomenon is known as *electrostriction* and the volume change is

Table 7.1 Typical values for the volumes of activation[a]

Reaction	**$\Delta V^{\ddagger}$ ($cm^3\ mol^{-1}$)**
Homolysis	+5 to +20
Radical polymerization (propagation)	ca. −20
Cycloadditions	
Diels–Alder	−30 to −50
catalysed	−25 to −35
intramolecular	−25 to −30
dipolar	−40 to −50
(2+2)	−40 to −55
H_2 transfer	−25
Neutral S_N2	small
Solvolysis of RX	−15 to −25
Quaternization of RX	−30 to −50[a]
Ester hydrolysis	
basic	−10 to −15
acidic	>−10
Epoxide-ring opening	−15 to −20
Hydration of alkenes	ca. −20
Aldol reactions	−10 to −30?
Baylis–Hillman reactions	<−50
S_NAr reactions (ionogenic)	−30 to −40
Wittig reactions	−20 to −30

[a]Many reported volumes of activation appear to show a considerable dependence on the structure, on the solvent, and on other conditions.

due to solvation of the charges, which is another form of molecular association. Electrostriction is dependent on the solvent, and its magnitude tends to increase as the polarity of the solvent decreases.

2. Positive contributions to the volumes of activation result from dissociative processes, from bond breaking, and from the relaxation of solvation due to charge neutralization.

3. For multistep reactions, the measured volume of activation is the sum of the activation volume for the rate-determining step and the volumes of reaction for any pre-equilibria occurring prior to the slow step.

Table 7.1 sets out some reaction types and their characteristic volumes of activation. High-pressure conditions should be considered as a means of achieving products in reactions whose rates are much larger than the very low values obtained at atmospheric pressure.

1.2 How great a rate acceleration may be obtained?

We may use eqn (7.1.3) to calculate the change in the rate brought about by the application of external pressure to reactions with suitably negative activation volumes. For example, if the reaction in question is a Diels–Alder cycloaddition with $\Delta V^{\ddagger} = -35\ \mathrm{cm^3\ mol^{-1}}$, then, assuming a linear response, the increase in the rate at a pressure of 1000 bar and a temperature of 298 K relative to the rate at 1 bar is given by

$$\begin{aligned}\Delta(\ln k) &= -\frac{\Delta p\, \Delta V^{\ddagger}}{RT},\\ &= -\frac{1000 \times (-35)}{81 \times 298}\ \frac{[(\mathrm{bar})\ (\mathrm{cm^3\ mol^{-1}})]}{[(\mathrm{cm^3\ bar\ mol^{-1}\ K^{-1}})\ (\mathrm{K})]},\\ &= 1.45,\end{aligned}$$

or

$$k_{1000}/k_1 = 4.26.$$

A modest increase in the rate is observed at 1 kbar, but suppose the reaction is carried out at a pressure of 10 kbar then

$$\Delta(\ln k) = 14.5$$

and therefore $k_{10\,000}/k_1 = 1.98 \times 10^6$.

This logarithmic response means that one will only obtain rate increments that are useful in synthesis by going to quite high pressures (see Fig. 7.1). The pressure range 5–20 kbar is normally used for organic synthesis. However, eqns (7.1.1) and (7.1.3) are not linear over this wide pressure range. The measured volume of reaction and the volume of activation invariably decrease as the pressure increases; and a plot of ln k against pressure becomes markedly more curved towards the pressure axis. The acceleration of the rate which

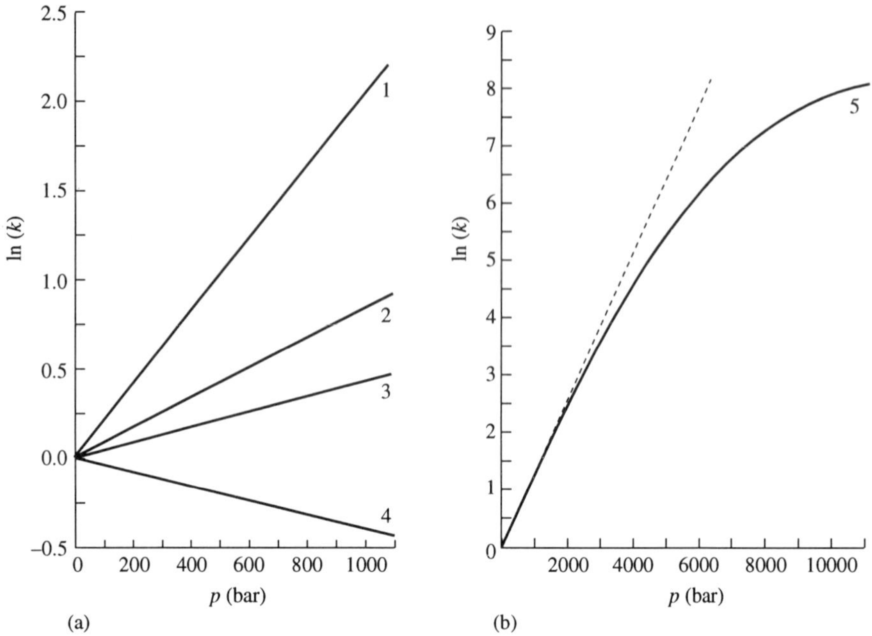

Fig. 7.1 (a) The kinetic response to pressure predicted from eqn (7.1.3) of reactions with the following volumes of activation: *curve 1,* 10 cm^3 mol^{-1}; *curve 2,* -10 cm^3 mol^{-1}; *curve 3,* -20 cm^3 mol^{-1}; and *curve 4,* -50 cm^3 mol^{-1}. (b) A schematic curve giving: (– – – –) the linear response, and (——) a more realistic (nonlinear) response, *curve 5,* for $\Delta V^{\ddagger} = -35$ cm^3 mol^{-1}.

can be obtained is therefore far below the predicted value (Fig. 7.1(b), curve 5). It is for this reason that volumes of reaction and of activation are conventionally quoted at the $p \rightarrow 0$ pressure limit. The curvature of these plots (which can be expressed by the second derivative $d^2(\ln k)/dp^2$ and which is related to the compressibility of the transition state) is not a quantity which is generally available. Nevertheless, very large rate increases can be obtained, and many syntheses have been successfully accomplished under high-pressure conditions which would otherwise not be possible (see Scheme 7.2).

1.3 For which reactions are high pressures appropriate?[5,6]

As will be seen, high-pressure equipment tends to be expensive, it is capable of handling only small or medium quantities of materials, and it is often not simple to operate (so the high-pressure method is most appropriate when more conventional approaches fail). A reaction for which the high-pressure approach might be tried would be one which gives poor or negligible yields under thermal conditions or which requires very long reaction times, while being a reaction with a large, negative volume of activation, say, greater than

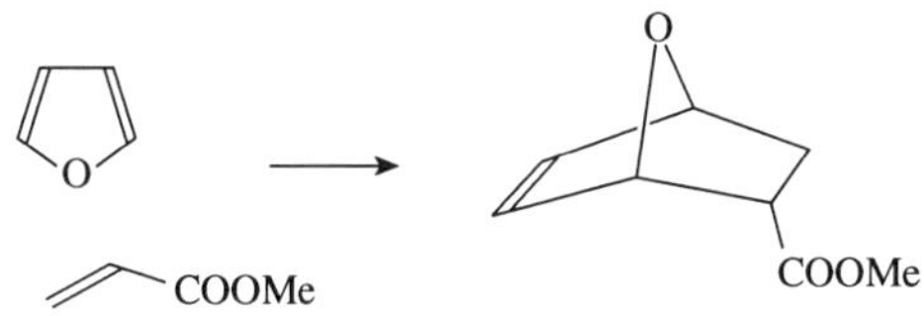

Scheme 7.1 The Diels–Alder cycloaddition of furan with methyl acrylate.

$-20\ cm^3\ mol^{-1}$. A typical example would be the Diels–Alder cycloaddition of furan with methyl acrylate given in Scheme 7.1. The yields under various conditions (with solvent) are as follows:

1 bar,	25 °C,	2–3 months		50%
1 bar,	75 °C,	4 days,		10%
1 bar,	40 °C,	2 days,	ZnI_2 catalyst	55%
15 kbar,	25 °C,	4 h,	CH_2Cl_2 solution	62%

High pressures are especially effective in furan cycloadditions since the desirable forward reaction is accelerated and the undesirable retro reaction, which is especially easy in furan cycloadducts, is also suppressed, although in this case the catalysed 1 bar reaction would probably be the reaction chosen for large-scale synthesis.

For the purposes of synthesis, one normally wishes to increase the rate of the reaction by as much as possible, since by this means reactions may be achieved which would not otherwise occur. The logarithmic relationship on the activation volume is shown in Fig. 7.1.

Scheme 7.2 illustrates some reactions which can only be successfully accomplished under high-pressure conditions.

1.4 What solvents may be used?

Normally, reactions are carried out in solution, and a liquid phase is necessary to permit intermolecular collisions. Pressure will, however, restrict this motion in two ways: the solvent may freeze under pressure, or the viscosity may increase. There is little information currently available on the effect of solubilities. Reactions may also occur in the solid state. The solvent can also have an effect on rates.

1.4.1 Freezing pressures

Liquids form solid phases on crystallization which are denser than the liquid (with the exception of water), and hence the freezing points are raised by increases in pressure. It is essential to avoid solidification because it can prevent a reaction from occurring and because possible damage to equipment may result. The freezing pressures of some common organic solvents at 25 °C are given in Table 7.2.

Water is anomalous since its freezing point initially falls with pressure;

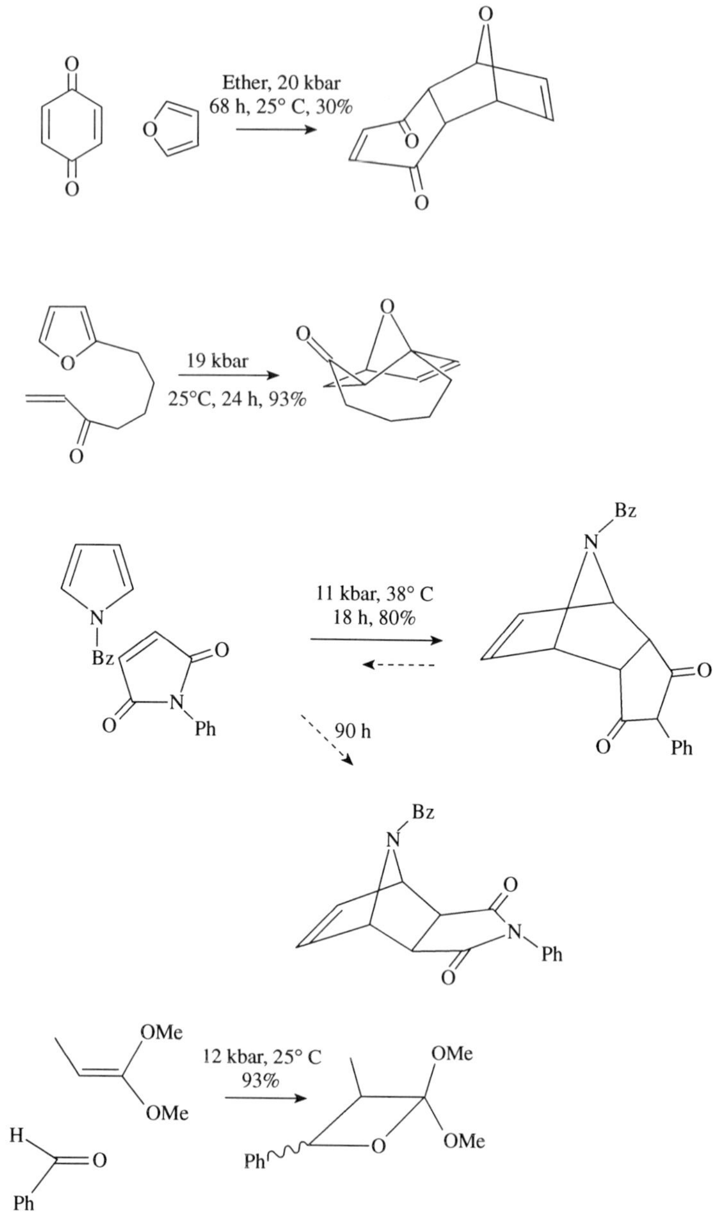

(a)

Scheme 7.2 Some reaction types which are characterized by large, negative volumes of activation: (a) Diels–Alder and other cycloaddition reactions, including heterocyclic and intramolecular examples; (b) ionogenic reactions, such as Menshutkin quaternizations; (c) additions to double bonds, for example, aldol reactions, Wittig reactions and, chiral-borane reductions (in the last type of reaction, pressure both accelerates the reduction and retards decomposition of the Alpine borane to achiral 9-BBN); (d) sequential

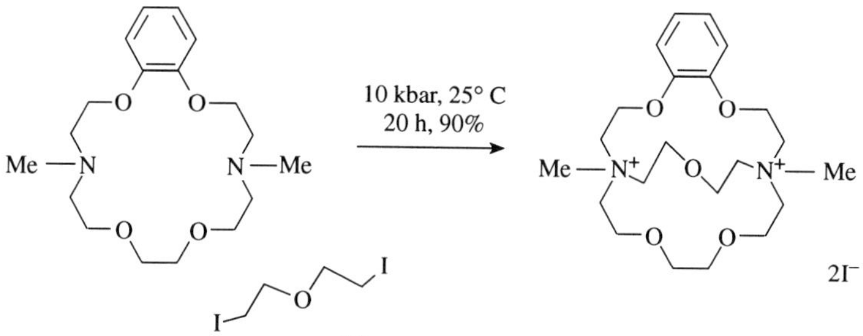

(b)

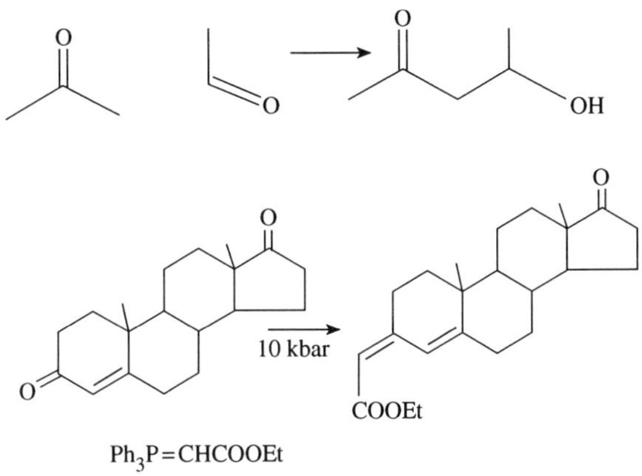

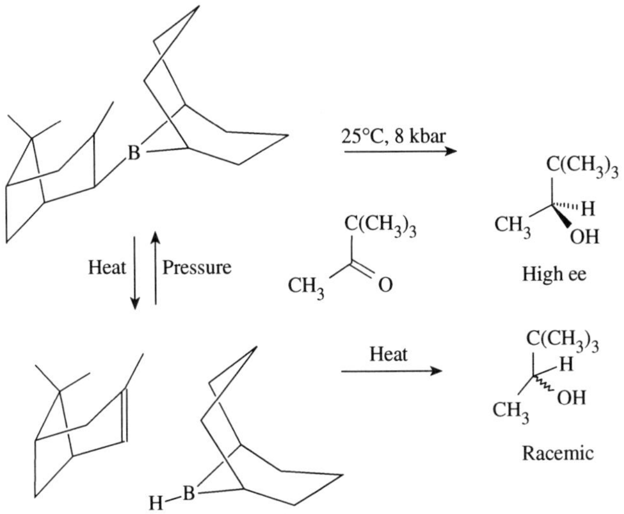

(c)

additions, which especially favoured by pressure, examples include the Baylis–Hillman reaction, formation of tetrathiafulvalenes and some organometal-promoted oligomerizations (polymerization is, in general, promoted by the application of pressure).

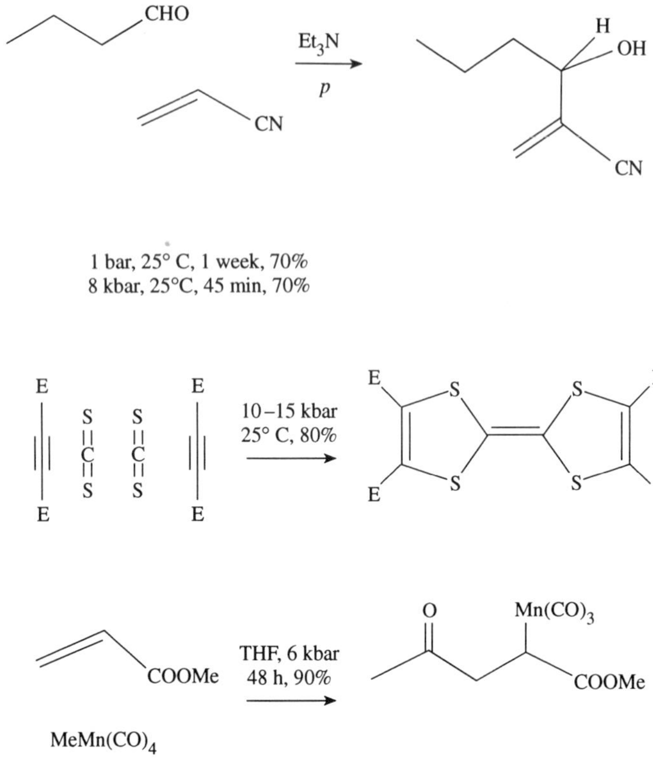

(d)

Scheme 7.2d

however, above 3 kbar it begins to rise again, and at 10 kbar it freezes at around room temperature. Solutions and mixtures of solvents will, in general, show a depression of the freezing point relative to that of the pure solvent and hence they may be the preferred media for reactions at very high pressures. Hexane isomers (petroleum ether), dichloromethane–ether or methanol–ethanol mixtures are often used where it is important to avoid solidification. It has been reported that the last mixture is still fluid at 100 kbar (though it is highly viscous).

1.4.2 Viscosity

Even if freezing does not occur, the viscosity of liquids increases with pressure (Fig. 7.2), and eventually the diffusion rates may limit the reaction. The viscosities of even mobile liquids (such as ethanol, ether and dichloromethane) are probably so high by 20 kbar that it is not profitable to increase further the pressure in an attempt to increase the reaction rates.

Table 7.2 The freezing pressures[7] of some organic solvents at 25°C

Solvent	**p_f (kbar)**
Cyclohexane	0.355
p-Xylene	0.36
Acetic acid	0.39
Benzene	0.72
Nitrobenzene	0.84
N,N-dimethylaniline	0.93
Carbon tetrachloride	1.462
Dioxan	1.2
o-Xylene	2.29
Decane	3.0
Mesitylene	3.5
Chlorobenzene	4.55
Octane	5.3
Chloroform	5.5
Heptane	11.1
Diethyl ether	ca. 11
Dichloromethane	>20
1-Butanol	11.7
Methylcyclohexane	13.8
Ethanol	30
Methanol	50

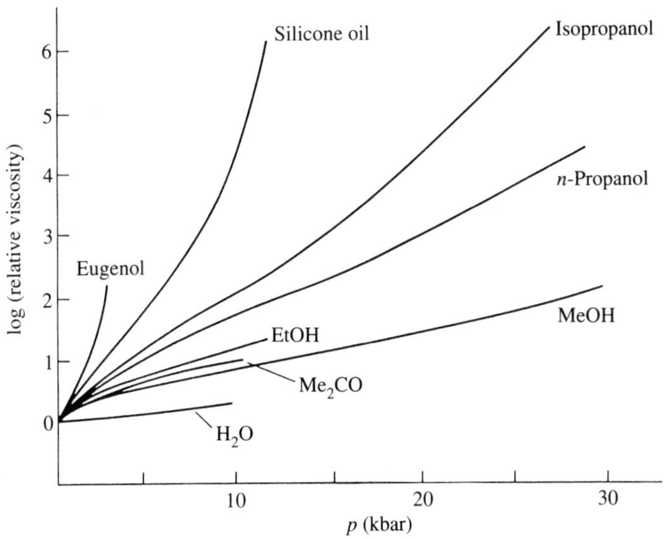

Fig. 7.2 The viscosities of several liquids as a function of pressure.[2]

1.4.3 Solubilities

Not much information is available on the effect of pressure on the solubilities of organic compounds in organic solvents. It is possible that the solubility could either increase or decrease, depending on the volume change which accompanies dissolution of the material, and this could be ascertained by measurements at atmospheric pressure.

1.4.4 Reactions in the solid state

Solid-state reactions can occur. They usually occur without a solvent, and they usually involve oligomerization reactions in which the requirements for molecular motion are minimal. In these cases there is no upper limit to the pressures which may be usefully employed (see Scheme 7.3).

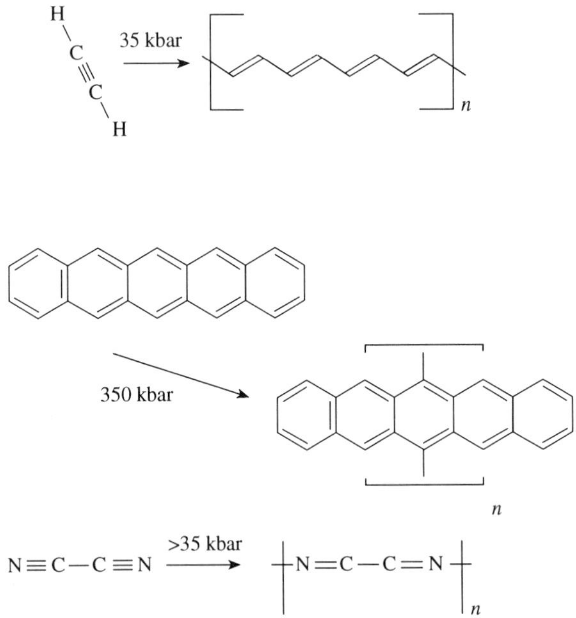

Scheme 7.3 Some high-pressure solid-state reactions.

1.4.5 Solvent effects

Ionogenic reactions (that is, reactions in which charges are generated in the rate-determining step) are accelerated by solvents capable of solvating charges—polar solvents—and this kinetic solvent effect can be very large; for example, there is an increase by a factor of 10^3–10^4 in the rate for hexane (nonpolar) and methanol (very polar) for a quaternization reaction. Now, if we are to use high pressures to assist a reaction, the activation volumes of the reactions of this type will be at their most negative when carried out in *non-polar solvents*. A situation may therefore arise in which the fastest reactions are

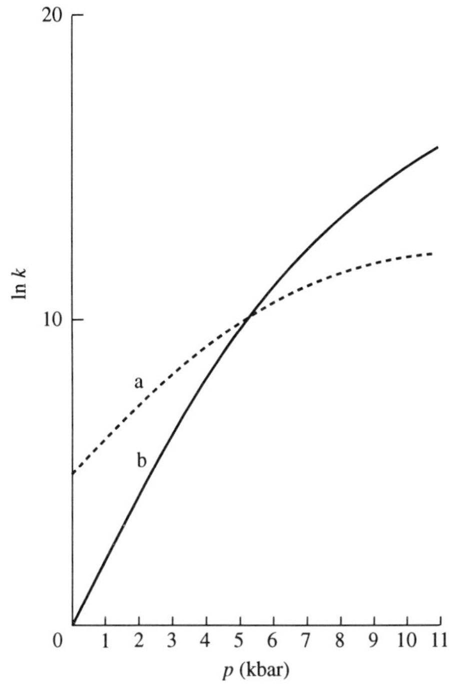

Fig. 7.3 A schematic rate-pressure profile for an ionogenic reaction such as a quaternization, for: (– – – –) a polar solvent, and (——) a nonpolar solvent.

found to occur in polar solvents, at relatively low pressures and in nonpolar media at high pressures (Fig. 7.3).

1.5 What equipment is required?

The design and production of pressure vessels with high capacities is an engineering problem for specialists. The initial considerations when setting out to make equipment for preparative chemistry would include:

(1) the maximum pressure (the design should be for a pressure considerably in excess of this 'maximum', because many pressurization cycles are usually required and vessels operating repeatedly near their stress limit are not long lived);
(2) the sample volume (cost is usually a consideration);
(3) the type of apparatus (this should be consistent with ease of operation and maintenance and it should be capable of many cycles of repetitive use).

The two basic elements are: (i) a container which is capable of withstanding 10–15 kbar or even higher pressures, and (ii) a means of reaching this pressure (two types of systems are in common use, piston–cylinder vessels and intensifier vessels; continuous-flow systems and other methods are also described below).

1.5.1 Piston–cylinder vessels

Figure 7.4 shows a cylindrical pressure vessel filled with fluid contents and contained in a strong pressure frame which holds a hydraulic ram which applies force to a piston. The vessel may be fabricated from a single piece of supertough steel (for example, maraging steel)—a *monoblock* design—or it may have an inner core of this material accurately fitted into a relatively elastic steel jacket. The piston and bottom plug are of hard tool steel. Both the piston and the cylinder need to be honed to a mirror finish with a very-close-tolerance fit. Seals are fitted to both the bottom plug and to the piston; they may take various forms such a Δ-ring of soft metal (bronze or brass) or a neoprene O-ring (Fig. 7.5). The cylinder may be fitted with an electric heater, and operation to at least 100 °C should be permitted. Pressurization is accomplished by means of a hydraulic pump (electrically or air operated, though hand pumping is feasible for small units) operating the ram. Depressurization is brought about by release of this hydraulic pressure and extraction of the piston, for which the hydraulic ram may operate in reverse.

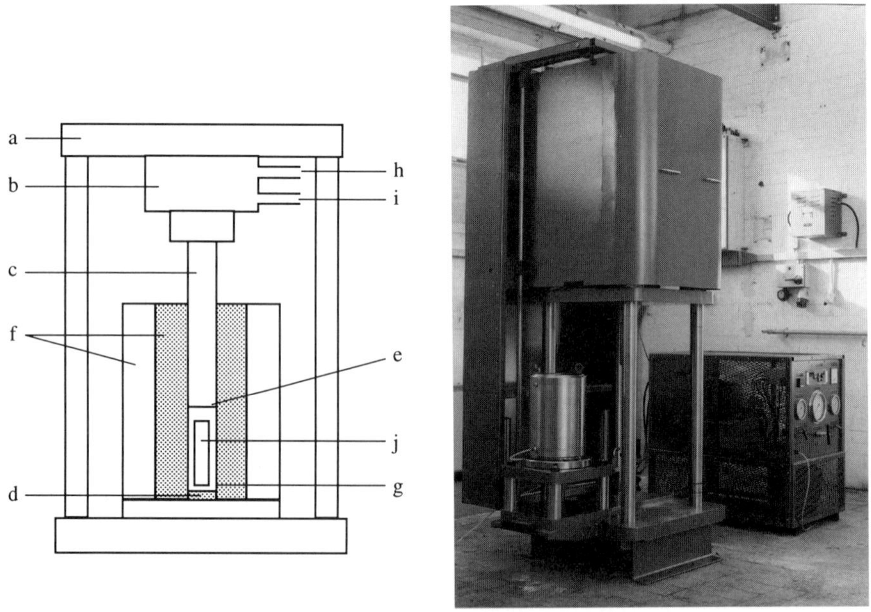

Fig. 7.4 A piston–cylinder high-pressure apparatus: (a) the pressure frame, (b) a 100 tonne hydraulic ram, (c) the piston, (d) the bottom plug, (e) high-pressure seals, (f) a double-walled pressure vessel, (g) the pressurizing fluid, (g) a 500 bar hydraulic fluid for pressurization, (i) hydraulic pressure used for piston extraction, (j) the sample. (Reproduced by permission from PSIKA Ltd).

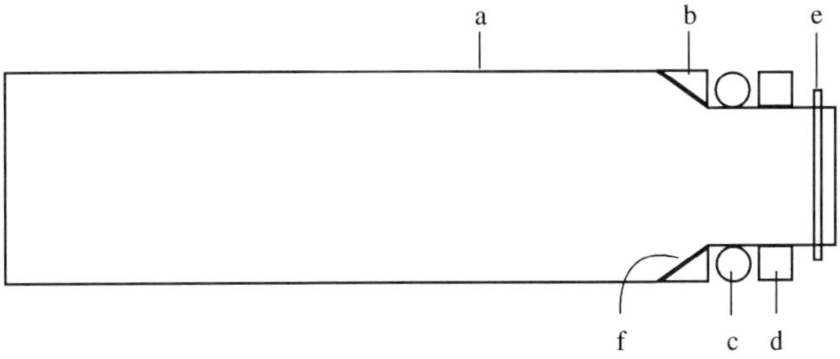

Fig. 7.5 Details of piston seals: (a) a piston honed to a mirror finish that is accurately cylindrical and machined to an interference fit to the cylinder, (b) a Δ ring of bronze or brass, fitting the chamfer of the piston and precisely fitting the diameter of the cylinder (this ring will deform somewhat to create the high-pressure seal above pressures of about 5 kbar), (c) a rubber O-ring which can seal pressures below 5 kbar (above this pressure elastomers tend to become brittle), (d) a backing ring of PTFE, (e) a circlip to prevent rings falling off when the piston is extracted, (f) an accurately machined 45° chamfer.

The advantages of this type of equipment are the simplicity and ease of operation and maintenance which it offers. The disadvantage is the relatively small sample volume available. The cross-sectional area of the bore of the cylinder cannot be increased without increasing the force on the piston which is necessary to reach a given pressure; this necessitates the use of larger hydraulic rams and a stronger pressure frame. Also, the length of the bore cannot be increased indefinitely because of the manufacturing difficulties of honing the interior and also because liquids, especially organic solvents, are compressible, and the volume increase concomitant with a longer bore necessitates a longer piston and a longer travel to reach a desired pressure.

Typical piston–cylinder vessels are 2–5 cm in diameter and 20–30 cm in length; thus, allowing for the space occupied by the piston; typical volumes are in the range 50–300 cm^3.

1.5.2 Intensifier vessels

If the pressure vessel is separated from the pressure-generating unit, it can be increased in size without limit (see Fig. 7.6). The pressure vessel can now be of any geometry; however, increasing the volume by increasing the diameter is relatively more costly than increasing the length because of the increased forces on the closure. The vessel containing the sample may be a monoblock or, more typically, it may have an outer fitted jacket or it may even be wire wound for strength. The closure consists of a steel plug fitted with a soft metal seal and it is held in place against the force of the pressurized liquid by a massive threaded collar. This collar may have a divided thread and it may permit engagement of the threads with a fraction of a turn (in the same manner as a gun breech). At the other end, a pipe connection (a *duplex* pipe) permits

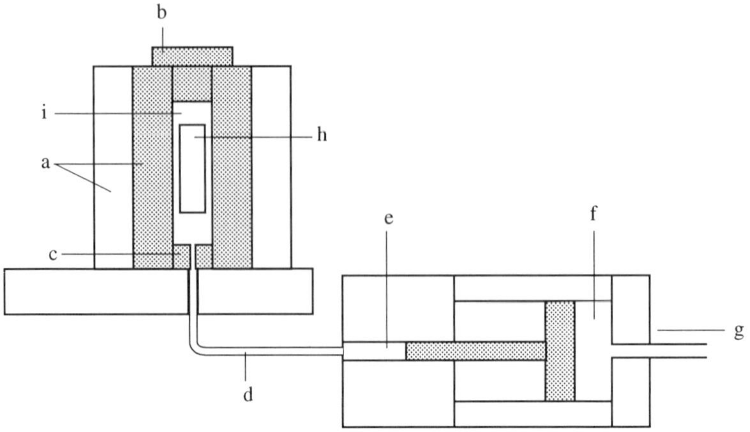

Fig. 7.6 An intensifier vessel: (a) a double-walled pressure vessel; (b) the top plug; (c) the bottom plug, which admits the high-pressure fluid; (d) duplex tubing; (e) the high-pressure fluid; (f) the low-pressure fluid (ca. 500 bar); (g) the hydraulic fluid; (h) the sample; (i) the pressurizing fluid.

pressurized fluid to be admitted to raise the pressure inside the vessel to the desired level. This pressurized fluid may be generated by an intensifier consisting of a pair of adjacent cylinders, one with a large diameter and one with a small diameter, and containing linked pistons. By applying pressure from a pump to the large piston, the force is transmitted to the small cylinders and the force is increased by the ratio of the areas of the two cylinders. Thus, if the large and small pistons have areas of 20 and 1 cm^2, respectively, a pressure of 500 bar applied to the former will generate a 10 kbar pressure in the small-bore cylinder, which should be filled with a suitable oil that is piped directly to the vessel. The intensifier may also be built directly into the end of the sample vessel, as in Fig. 7.7. An arrangement of nonreturn valves permits any number of strokes of the intensifier to be made to increment the high pressure gradually in the vessel. After each stroke, the two pistons must be returned to their starting position, either manually or by an automatic arrangement. Depressurization is accomplished by opening a valve in the high-pressure side (for example, a ball-bearing valve). The closure is then lifted off, and the sample removed. Electrically operated pumps are now available which can supply oil at 10 kbar or more continuously; this is more convenient than the intensifier method. This type of apparatus is a version of the well-established *isostatic presses*, which are used at pressures of around 3–6 kbar for powder-moulding technology. These presses can be constructed with volumes of 1000 L, and the pressure range could be adequate for food processing.

Intensifier systems tend to be more complex and expensive than piston–cylinder systems but they can deal with larger sample volumes.

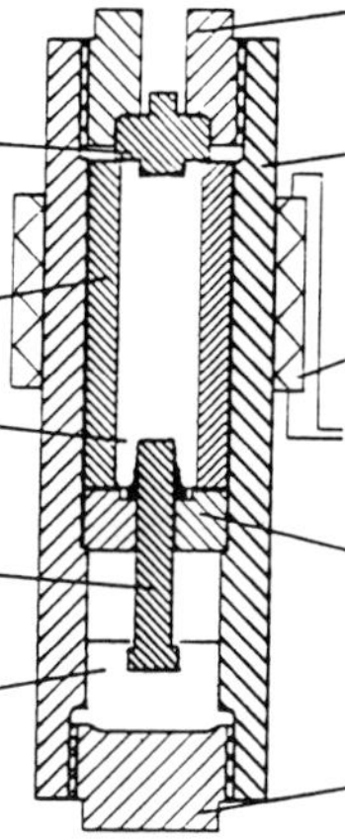

Fig. 7.7 An intensifier which is built directly into the end of the sample vessel.

1.5.3 Continuous-flow systems

The availability of high-pressure pumps raises the possibility of continuous pressure production, in which reagents are passed at the required pressure through a flow tube (with the flow rate and the length of the tube determining the residence time) then through a relief valve at which the pressure is dropped and the product is collected (see Fig. 7.8). Such an arrangement would be chosen for industrial production, and it has been shown to be feasible in the laboratory at a pressure of 4 kbar.[7] This apparatus was shown to be very effective in bringing about a number of acid-catalysed reactions by passing the reagents through a reaction tube packed with an acid resin. The reactions studied included the formation of esters from the alcohol and carboxylic acid and of bisphenol A from phenol and acetone. Further development, particularly in relief-valve design, is needed for these high pressures, however, before even higher pressures can be reached continuously.

1.5.4 Other methods

In principle, anvil-type pressure generators could be used for synthetic purposes if very high pressures are desired, if processes taking place in solid or glassy media are being studied, and if milligram quantities of reagents can be tolerated. Most applications of this type fall within the scope of inorganic and mineral chemistry.

1.6 How are the reagents contained?

High-pressure vessels are made of various steels, which are potentially subject to corrosion. Normally, the reagents are not permitted to come into contact with the vessel both to avoid corrosion and to maintain purity in the reaction mixture. Continuous-flow systems are an exception because the

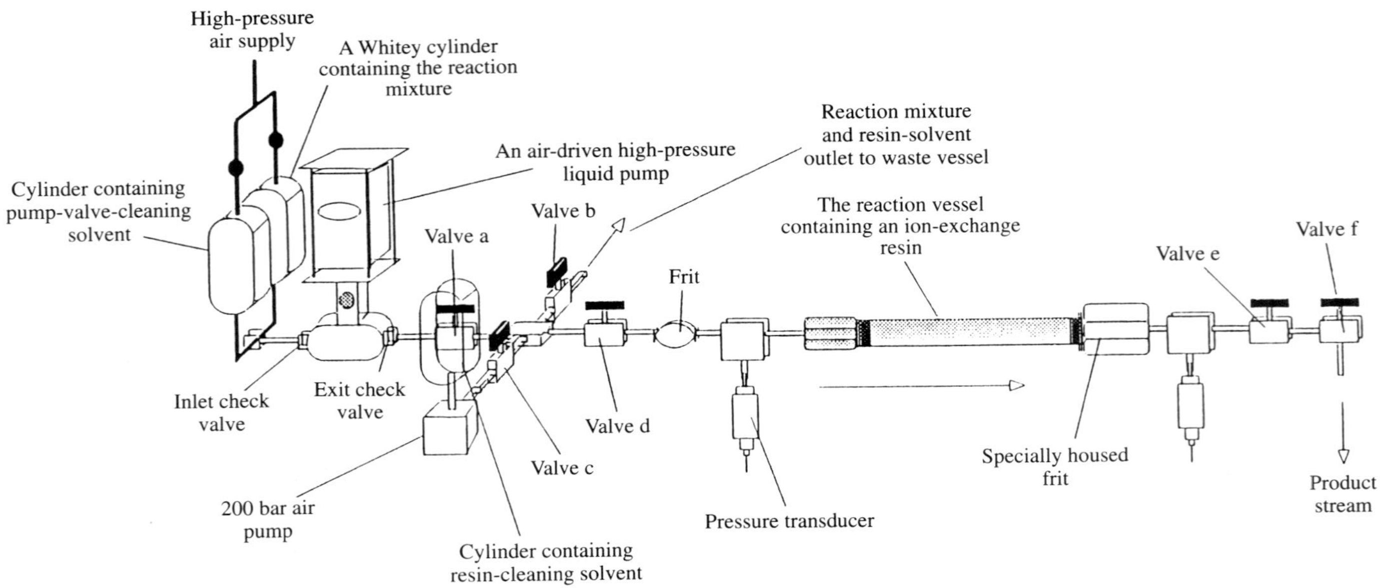

Fig. 7.8 A continuous-pressure-production system.

reactants *must* make contact with the pump and the reaction tubing. In other cases, reagents are introduced into the pressure vessel in some type of flexible package, and the remaining volume is occupied by a pressure-transmitting medium.

1.6.1 Pressure-transmitting media

Noncorrosive fluids which remain fluid at the highest pressures that are used include solvents such as hexane, ethanol, and some light hydrocarbon oils or an ester such as Plexol (dinonyl sebacate). Some vessels are designed to use oil–water emulsions.

1.6.2 Packaging of the reagents

The reagents must be packaged in a flexible container because the application of pressure causes compression of the contents (see Fig. 7.9). A heat-sealed plastic bag is the simplest container. Any residual air bubbles will immediately dissolve when pressure is applied. Screw-topped plastic ampoules which deform under pressure may be used. Re-usable cylindrical vessels made of PTFE and fitted with a tight-fitting piston have proved useful. Air may be expelled either through a simple seal in the piston or by using butane to fill the head space which dissolves in the organic solvent when the piston is pushed into place. A wire or rod reinforcement through the sides and across the top of the vessel may be necessary to contain volatile solvents if the sample which is to be placed in the vessel is heated above its boiling point while it is still at atmospheric pressure.

1.7 How may temperature and pressure be measured?

Heating may be accomplished by placing around the vessel an electrical band heater fitted with a thermostat, preferably of the fail-safe type, which cuts off if the sensor (a thermocouple or a thermistor) fails. If this is not the case, it is recommended—from bitter experience—that a thermal cutoff switch in contact with the vessel should be fitted, so that, if the temperature reaches, say, 100 °C, the heater will cut out. Overheating can damage the vessel and ruin a reaction.

The pressure on the high-pressure side may be measured by transducers. Strain gauges are available which measure pressures up to 14 kbar, while manganin gauges measure pressures above 20 kbar; manganin gauges consist of coils of manganin wire, whose electrical resistance changes linearly with pressure. Both types of pressure gauge can be supplied with a direct readout in pressure units.

An alternative procedure, which usually gives adequate precision for preparative chemistry ($\pm$ 5 per cent), is to measure the pressure on the low-pressure side with a Bourdon gauge and then to multiply this value by the intensification ratio, which is obtained from the geometry of the pistons. The frictional losses under these high pressures are insignificant.

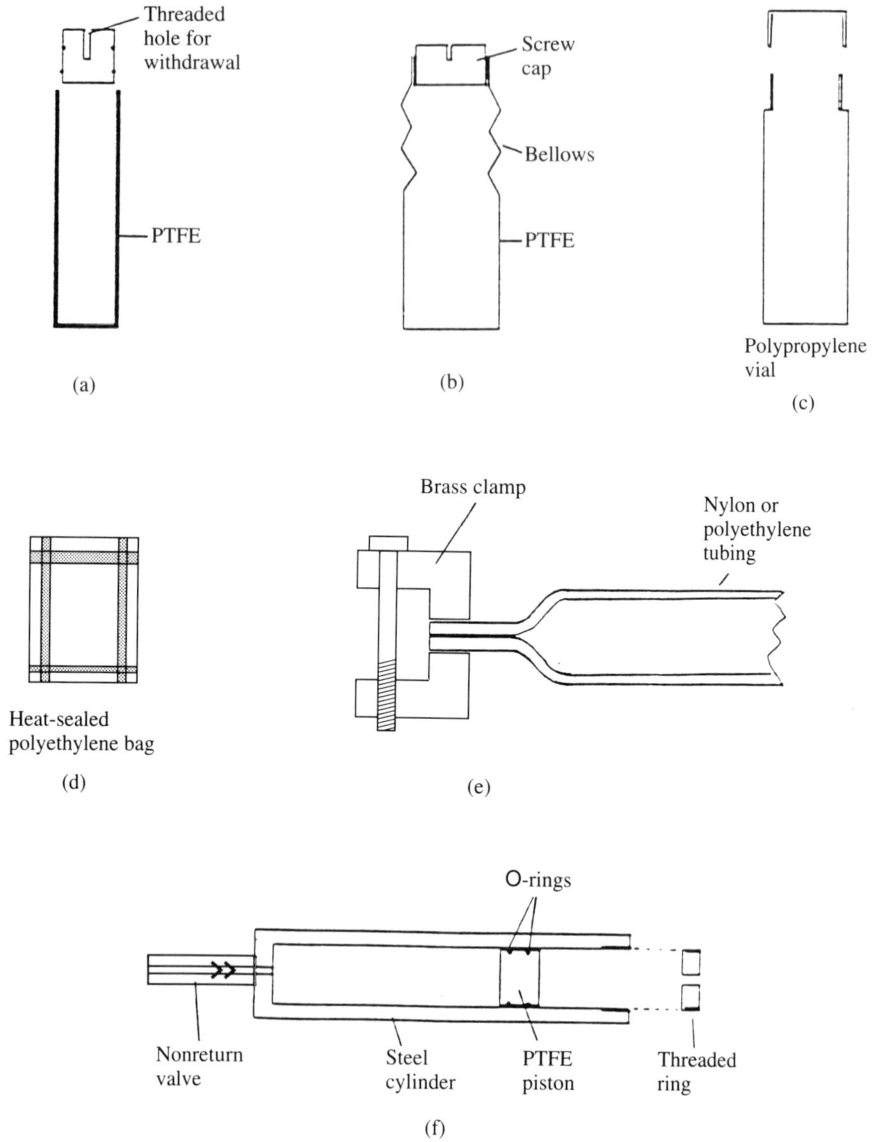

Fig. 7.9 Flexible sample containers used at high pressures: (a)–(e) containers which are suitable for liquids, (f) a container suitable for charging with a gaseous component such as liquid carbon dioxide.

1.8 Are high-pressure reactions hazardous?

While it is important to take proper precautions (see Appendix 2), properly built equipment offers few hazards if relatively small quantities of liquids are used under pressure. The energy stored in a compressed liquid is of the order of 0.2–0.5 kJ mol^{-1} $kbar^{-1}$; this is much less than in a gas. In the event of the catastrophic failure of a vessel, only around 10 per cent of the liquid would need to escape to reduce the pressure to the atmospheric value, and double-walled vessels are unlikely to suffer a failure of both cylinders. The author has experienced the failure of a piston–cylinder apparatus which led to nothing more than a sudden depressurization. On the other hand, hazards may arise from the low-pressure side, which is likely to be carrying oil at 500 bar or more; a pin-hole jet at this pressure can penetrate the human body. The valve threads may become worn and the stems may then be forcibly ejected. So hose connections should be inspected and renewed regularly; and, where possible, they should be shielded from personnel, the valves should be mounted with the stems pointing away from the operator, and the guidelines set out in the code of practice (see Appendix 2) should be followed.

1.8.1 Runaway reactions

A practical factor which *must* always be considered is the tendency of certain reactions to become uncontrolled as pressure is applied. Such reactions usually have a very large, negative volume of activation. The main warning sign of such an occurrence is a sudden increase in pressure on the gauge, with this increase continuing even when pumping stops. The result is that the reagents are converted largely to carbon and to a quantity of gases under extremely high pressure. Presumably, adiabatic heating of the sample during this pressurization initiates an exothermic process which then proceeds explosively. Scheme 7.4 indicates examples which are known to have exploded within pressure vessels when pressurized rapidly. The subsequent depressurization needs to be carried out with caution: the gases must be allowed to escape

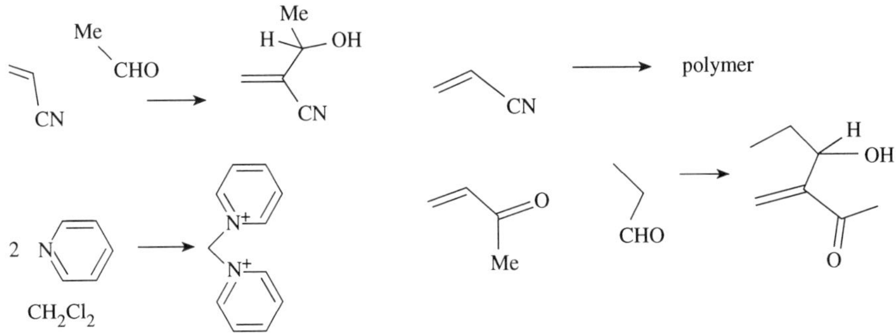

Scheme 7.4 Some reactions which have been found to cause explosions under rapid pressurization.

slowly, and a thorough *cleaning* of the vessel and *renewal* of the seals is invariably necessary. It may be mentioned that Bridgman has listed a number of solids which he found exploded upon pressurization.[9]

The remedy for successfully accomplishing a reaction showing this tendency is to pressurize in stages, say in 1 kbar increments at hourly intervals, and to start at or below room temperature and to increase the temperature, if necessary, only when the full pressure is reached, and to establish by trial and error the minimum pressure and temperature necessary for a successful reaction.

1.9 Applications to food processing

The high-pressure processing of foodstuffs is a potentially important method of preservation because bacteria, moulds, and yeasts can be killed, and spoilage enzymes can also be denatured.[8] Pressures in the region of 6–8 kbar are adequate for these purposes, and temperatures from 40 to 60 °C may be used. The methodology used on a laboratory scale is the same as the methodology used for chemical processes. Irregularly shaped materials (such as chunks of vegetables) are probably best pressurized when heat sealed into plastic bags in a surrounding medium of water (see Fig. 7.9(d)). This method may also be used to examine the pressure tolerance of bacteria and moulds; see *Protocol 4*.

1.10 Protocols

Protocol 1.
Synthesis of exo- and endo-10-benzoyl-4-phenyl-4,10-diazatricyclo[5,2,1,0^{2,6}]dec-8-en-3,5-dione

Caution! For safety procedures see Appendix 2.

1. Dissolve *N*-benzoylpyrrole (2 g, 11 mmol) and *N*-phenylmaleimide (2 g, 11.6 mmol) in dichloromethane (6 mL), and place them in a cylindrical vessel made of PTFE with a diameter of 2 cm and a height of 5 cm that is equipped with a tight-fitting piston with two Viton O-rings (see Fig. 7.9(a)).
2. Sweep most of the air out with a stream of butane and insert the piston to the liquid level; butane in the headspace dissolves in the organic phase.
3. Place the vessel in a high-pressure enclosure surrounded by the pressure-transmitting fluid (Plexol) at ambient temperature.
4. Apply pressure, going up to 11 kbar in 15 min.
5. Raise the temperature to 40 °C and leave the sample under these conditions for 18 h.
6. Cool to room temperature and depressurize then withdraw the piston using a tool which screws into a hole in the top.

7. Remove the sample from its container and work it up.

A yield of 3 g (78 per cent) of the endo-adduct was isolated with a melting point of 96–97 °C. When the reaction was left under the same conditions for 6 days, the product was found to contain 40 per cent exo-adduct and 60 per cent endo-adduct.

Protocol 2. Carrying out a high-pressure reaction in supercritical carbon dioxide

Caution! For safety procedures see Appendix 2.

Equipment

- A stainless steel tube (see Fig. 7.9(f)) fitted at one end with a check valve (for example, Swagelok 4c), and at the other end with an internally threaded section fitted with a steel ring and a PTFE piston, which is snugly fitted with two O-rings into the bore (the dimensions should allow a fit into the available high-pressure enclosure)

1. Remove the check valve, and place the reagents in the tube so that they fill the space between the piston and the valve connection without leaving an air space.
2. Insert the check valve and make it firm.
3. Screw the threaded ring fully into the other end of the tube, and connect the check valve to a source of carbon dioxide, such as a gas booster, and fill the cylinder with carbon dioxide to a pressure of about 65 bar (the vessel will then be full of liquid carbon dioxide).
4. Place the whole tube in the high-pressure enclosure in a pressure-transmitting medium, close the enclosure, and pressurize to the required pressure. Adjust the temperature also. (In this way, a medium of carbon dioxide at high temperature and pressure can be made available either as a solvent or as reagent.)
5. Cool the tube, and remove it. Depressurize the tube by carefully loosening the check valve, allowing the carbon dioxide to evaporate.

Protocol 3. Synthesis of 2-cyanohex-1-en-3-ol

Caution! For safety procedures see Appendix 2.

This is a Baylis-Hillman tandem reaction with $\Delta V^{\ddagger} = -60\ \text{cm}^3\ \text{mol}^{-1}$

1. Make up a mixture of acrylonitrile (8.0 g, 150 mmol), butyraldehyde (11.1 g, 150 mmol), and triethylamine (0.3 g, 3 mmol).
2. Place the solution in a PTFE cylinder of the type shown in Fig. 7.9(1) which

Protocol 3. *Continued*

is 25 mm in diameter and 100 mm long so that it is filled to within 10 mm of the top.

3. Insert the piston while sweeping out air with a stream of butane.
4. Cool the tube and the contents in ice (to minimize the risk of overheating during pressurization), and place it in the high-pressure apparatus at the ambient temperature.
5. Apply pressure up to 8 kbar and raise the temperature to 45 °C.
6. Allow it to stand at this temperature and pressure for 45 min.
7. Depressurize, and remove the sample containing the product, which may be purified by distillation.

This yield should be 11 g at 70 per cent purity.

Protocol 4.
Pressurization of a bacterial culture

Caution! For safety procedures see Appendix 2.

Escherichia coli is used as an example

1. Inoculate a sample of the bacterium from an agar slope into a medium of 9 mL Tryptone Soya Broth or Nutrient Broth and incubate overnight in an orbital shaker at 37 °C.
2. Subculture 1 mL of the overnight culture into 100 mL of fresh medium, and allow it to incubate at 37 °C in the orbital shaker. (If a sample at the midlog phase of growth is required, take samples at around 3 h and measure the absorbance at 680 nm using a 1 cm cuvette. Continue the incubation until the absorbance measures 0.5–0.6. Continue the incubation for 16 h if stationary-phase bacteria are required.)
3. Transfer the culture to a 500 mL centrifuge container and cool it in ice to inhibit further growth. Centrifuge at 6000 r.p.m. (revolutions per minute) for 15 min in a precooled rotor. Decant off the supernatant, and suspend the pellet of bacteria in 50–100 mL of 0.1 M Tris–HCl buffer at a pH of 7.5. Centrifuge again as before, and resuspend the pellet in sufficient buffer to give the required cell density or absorbance (for example, 0.5).
4. Transfer aliquots of this suspension into polythene sachets 1–2 mL capacity (see Fig. 1.9(d)), and heat seal each sachet. Several such sachets may then be placed in a PTFE container of the type shown in Fig. 7.6(a) which is filled with water, sealed, and placed in the high-pressure apparatus. (Pressures of 2–6 kbar for periods of 10–60 min will cause an observable diminution of the cell population, and, ultimately, complete sterilization.)

References

1. Le Noble, W. J. *Prof. Phys. Org. Chem.* **1967**, *5*, 207.
2. Isaacs, N. S. *Liquid Phase High Pressure Chemistry*; Wiley: Chichester, **1981**.
3. Le Noble, W. J. (ed.) *Organic High Pressure Chemistry*; Elsevier: Amsterdam, **1988**.
4. Kotowski M.; van Eldik, R. *Coord. Chem. Rev.* **1989**, *93*, 19. Wieland, S.; van Eldik, R. *Coord. Chem. Rev.* **1990**, *97*, 155.
5. Jurczak, J.; Baranowski, B., ed; *High Pressure Chemical Synthesis.*, Elsevier: Amsterdam, **1989**.
6. Matsumoto, K.; Acheson, R. M. *Synthesis at High Pressures*; Wiley: New York, **1991**.
7. Landolt-Börnstein, *High Pressure Properties of Matter*, IV/4; Springer-Verlag: Berlin, **1980**.
8. Balny, C.; Hayashi, R.; Heremans, K.; Masson, P., ed.; *High Pressure and Biotechnology*; John Libbey Eurotext: Paris, **1992**.
9. Bridgman, P. W., *Rev. Phys.*, **1935**, *48*, 825.

2. Inorganic reactions

Gérard Demazeau and Christian Cros

2.1 Introduction

The range of pressure in nature is very large in comparison with the range of temperature since it varies from approximately 10^{-30} bar (the partial pressure of hydrogen in intergalactic space) to 10^{30} bar (the pressure in the centre of a neutron star).[1]

Pressure is used for many different applications, including:

(1) the synthesis of new phases[2] or the stabilization of new structures;[3]

(2) the shaping of materials;[4,5]

(3) study of the effect of pressure on the physicochemical properties of materials (electric, magnetic, etc.).[6]

2.2 Equipment designed for the generation of high pressures

There are two main factors which should be taken into account in the design of a high-pressure apparatus: (i) the nature of the medium used, in particular its compressibility, governs the structure and the number of compression stages, and (ii) the use to which this equipment is put (for example, synthesis, crystal growth, or *in situ* characterizations) is the major factor in the design of the equipment. Pressure can be used jointly with high temperatures (in the synthesis of new compounds), or with a temperature gradient (in crystal growth), or with electromagnetic radiation such as X-rays, visible radiation, or infrared radiation (for *in situ* characterizations).

(a)

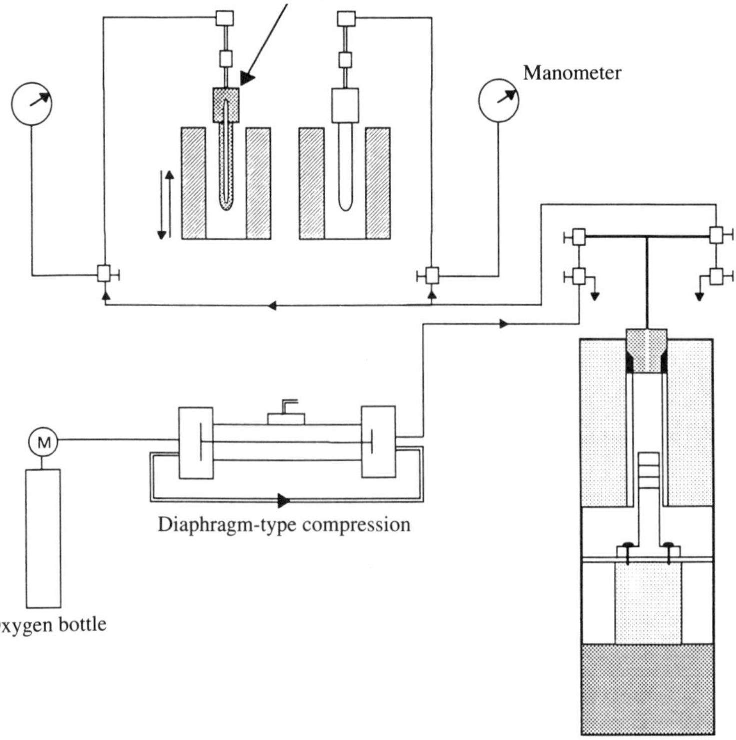

Fig. 7.10 High-oxygen-pressure equipment: (a) general view, and (b) a schematic view of the layout.

2.3 High pressures in gases

An apparatus which is able to reach pressures up to 5 kbar (500 MPa) needs to include several compression stages because gases have high compressibilities. Thus, due to the stored energy and the much smaller volumes, workers must be protected and safety codes must be observed.[7] (Safety procedures are given in Appendix 2.)

Two different approaches, using gases, are possible in solid-state chemistry:

(1) gases can be used as the reactive medium (the combination of the reagent and the medium);

(2) gases can be used as the transmitting medium.

2.3.1 The development of high pressures in reactive media

Oxides are the most common materials in our environment because of the oxygen partial pressure in air; and pressurized oxygen has been developed as a reactive medium to prepare new phases containing unusual oxidation states of transition metals.[8] Figure 7.10 shows a three-stage compression equipment built in Bordeaux to develop high oxygen pressures (up to 500 MPa at room temperature).[9]

Equipment using high pressures with reactive gases has been developed mainly for fundamental scientific research. The stabilization of unusual oxidation states of transition elements is one area of research involving such equipment.[8] Specific reaction vessels with external heating have been developed to synthesize oxides which require high temperatures (Fig. 7.11). Due to the large decrease of the mechanical properties of alloys with temperature, only a limited pressure–temperature domain is available (Fig. 7.12).

Fig. 7.11 A high-pressure reaction vessel.

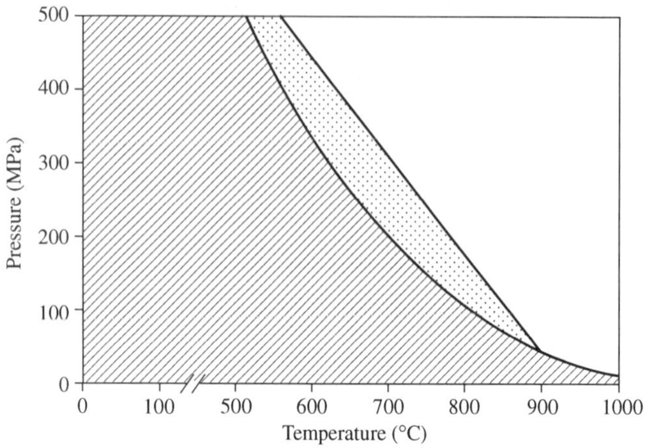

Fig. 7.12 Pressure and temperature domain for reaction vessels: [hatched box] working domain (the lifetime of the reaction vessel is acceptable), and [dotted box] limited domain (the reaction vessel has a short lifetime).

Protocol 5.
Example of a reaction protocol for compressed fluids (gases or liquids)

Caution! For safety procedures see Appendix 2. Experiments under high oxygen pressures require the absolute avoidance of any traces of combustible materials (such as grease or rubber), and all the parts of the equipment, particularly manometers, valve seats, and the tubing, etc., must be appropriately cleaned. For safety reasons the high-pressure unit, (that is, the compressors), the vessel, the furnace and any manometers must be placed behind protective shields, particularly when high oxygen pressures are used.

1. Place 1–5 g of the starting material (powdered) in a gold tube 0.3 mm thick, closed at one end, whose dimensions are suited to the inside diameter of the high-pressure vessel[a]. Reduce the tube aperture by finching (closing the end of the tube by pinching).
2. Close the vessel. Connect the upper part of the vessel to the high-pressure line through a flexible stainless-steel capillary (with an internal diameter of 0.5 mm and an outer diameter of 1.17 mm). The gas pressure is generated in special three-stage compression equipment, with a double-stage diaphragm-type compressor (up to ~100 MPa) and a pressure intensifier (100–500 MPa) in series.
3. Estimate the pressure to be generated in the reaction vessel at the ambient temperature, taking into account the required final pressure, the temperature conditions of the experiment, and the mechanical capabilities of the vessel (Fig. 7.12).

4. Increase the temperature of the high-pressure vessel using an external electrical furnace (see Fig. 7.10(a)). Measure the temperature with a thermocouple connected to a power controller. Adjust the reaction pressure with a valve when the required temperature is reached.
5. Maintain the experimental pressure and temperature conditions for a period of one to several days in order to get a complete reaction. End the experiment in one of two ways:
   (a) if the reacted oxide is metastable under standard pressure and temperature conditions then quench the reaction vessel in water, after removing the heating furnace, while maintaining the oxygen pressure;
   (b) if the resulting oxide is stable under normal pressure conditions, use slow cooling by switching the power supply of the furnace off.
6. Release the residual pressure slowly when the temperature of the high-pressure vessel is low enough, and recover the reacted materials for subsequent characterization.

[a] The main body of the vessel should be made from a refractory superalloy, such as PER 5 or PER 41 from Aubert et Duval Company in France (equivalent trade mark: UDIMET 500 and RENE 41). The vessel closure should be made from a refractory iron-based alloy, such as XN26TW from Aubert et Duval Company (Cr 15%, Ni 25%—equivalent trade mark, A286). Its design must fulfil the recommendations and safety rules available in specialized handbooks and in the literature. This vessel can be machined in a well-equipped laboratory workshop, or it can be supplied by high-pressure-equipment manufacturers. For example, the internal volume of a laboratory high-pressure vessel varies from 50 to 500 cm^3. A properly machined conical seal maintained by two flanges with four threaded rods and screws (as shown in Fig. 7.11), has proved to be the best closure system for high-temperature experiments. The conical surface of the two cones must be re-polished from time to time.

2.4 High pressures in liquids

Liquids are less compressible than gases, so their high-pressure equipment needs only a single-stage compression; this is usually provided by an air-operated actuator (Fig. 7.13).

The energy stored during liquid compression is less than that stored for compressed gases. Consequently, cells with large volume can be developed (Fig. 7.14). For fundamental scientific research, the design and the alloys used in reaction cells are the same than those used with gases, which were described in Section 2.3.

In inorganic chemistry, the liquid phase can be used as either a reactive or a transmitting medium.

2.4.1 High pressures in liquid phases used as reactive media

Oxidizing solutions (for example diluted solutions of NH_4ClO_4) have been used for preparing well-defined small particles of chromium dioxide[10,11] and for stabilizing thermally unstable oxidation states,[12,13] such as Hg^{2+}.

Nitriding solvents (such as liquid NH_3 or hydrazine NH_2NH_2) are now

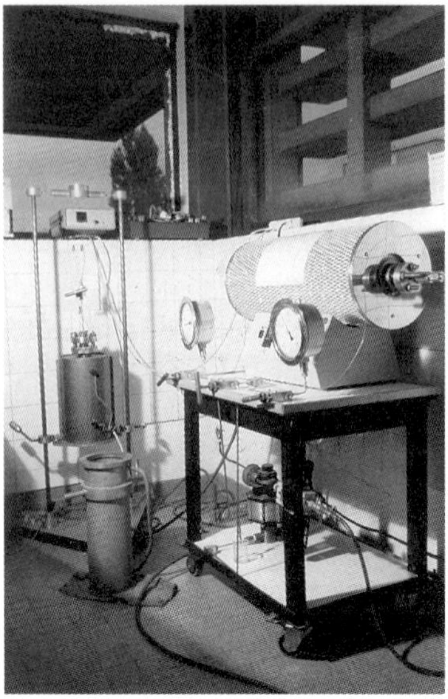

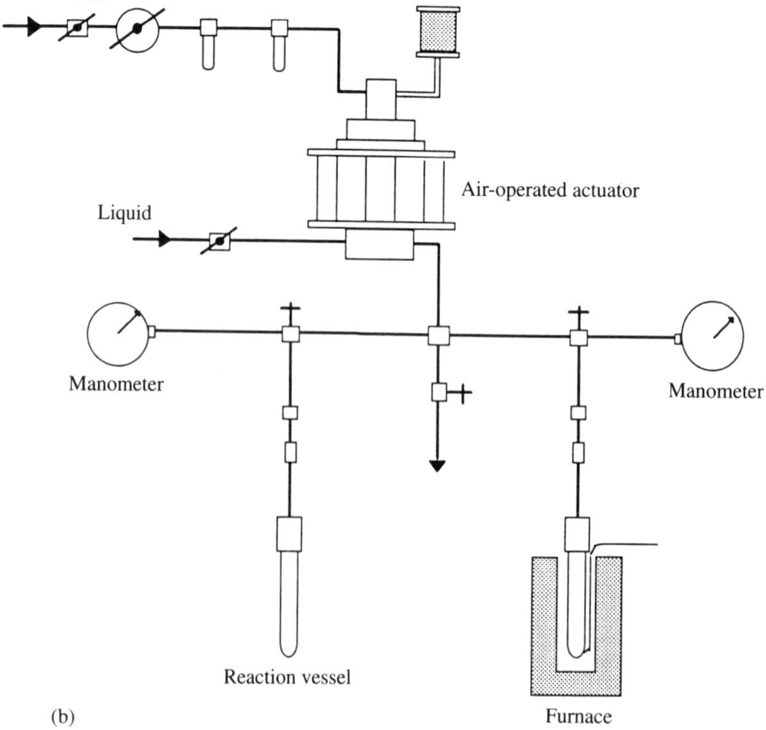

(b)

Fig. 7.13 High-pressure equipment used with liquid media: (a) general view, and (b) a schematic view of the layout.

Fig. 7.14 a 0.85 L pilot plant for high-pressure work in liquid medium.

being developed for the production of nitrides as fine particles or as single crystals.[14]

2.4.2 High pressures in liquid phases used for crystal growth

Crystal growth of low-temperature forms of materials requires its own specific processes. For examples, α-quartz is an interesting material because of its piezoelectric properties. The $\alpha \rightarrow \beta$ transition is close to 573 °C; a high-pressure hydrothermal, crystal-growth method has been developed because the application of pressure increases the solubility of the nutrient.[15,16] Such a process is shown in Fig. 7.15. The difference in temperature, ΔT, between the upper part of the reaction vessel (which contains the seeds) and the lower part (containing the nutrient) provides the driving force of the chemical transport for the crystal-growth process.

2.4.3 Pressure in liquids used as the pressure-transmitting medium

The two main applications which use liquids as the pressure-transmitting medium are:

(1) the shaping of materials (ceramics, metals, etc.) using cold isostatic pressing;[4]
(2) food processing using high-pressure conditions.[17,18]

In both applications, the highest attainable pressure valve is close to 500 MPa and the temperature is lower than 100 °C.

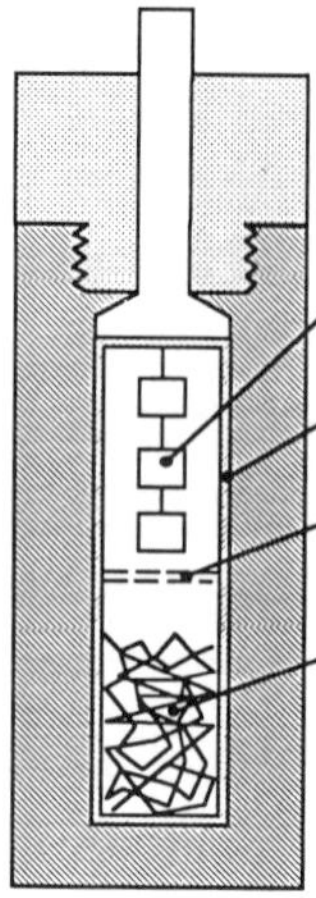

Fig. 7.15 A reaction vessel used for the low-temperature crystal growth of α-SiO_2.[15]

2.5 High pressures in the solid phase

The three main research areas in materials science which use high pressures are:

(1) the synthesis of new materials;[19]

(2) structural transformations, generally associated with the geosciences;[20,21]

(3) the investigation of pressure-dependent physicochemical properties (structural, electrical, and optical properties).[22]

2.5.1 The synthesis of materials under high pressures

The synthesis of materials under high-pressure conditions in the solid phase usually uses the *belt-type equipment* shown in Fig. 7.16. This type of apparatus was developed for the synthesis of diamond.[23] Different anvil shapes have been used.[24] One of the most appropriate for inorganic reactions at high pressures and high temperatures is the X-shaped annular die designed by M. Contré.[25]

The general design of the reaction cell is given in Fig. 7.17. The most commonly used transmitting medium is pyrophyllite, a natural aluminosilicate with the general formula $Al_2Si_4O_{10}(OH)_2$. The nature of the microfurnace depends on the type of reaction which is involved. Under oxidizing conditions a platinum heater is the most appropriate heater, but its temperature limit is close to 1400 °C.[26] Under nonoxidizing conditions, a graphite heater can be used, which allows higher temperatures (2000–2200 °C) to be reached.

The application of pressure may help in the synthesis of new materials. For example, in the preparation[28] of lanthanum monoxide (LnO) by the reaction

$$Ln_2O_3 + Ln \rightarrow 3LnO,$$

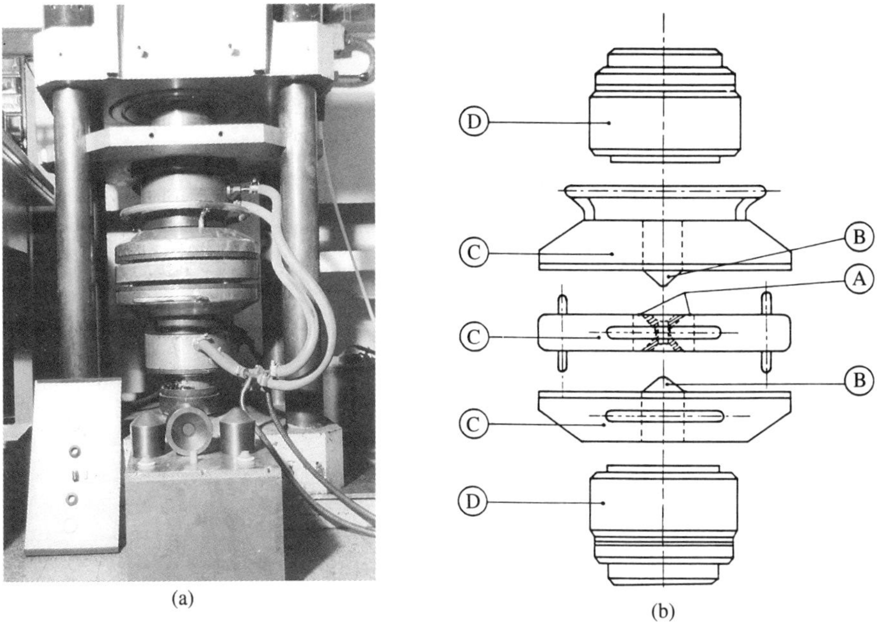

Fig. 7.16 A belt-type apparatus which is able to work at 10 GPa and 2000°C. (a) A general view. (b) A schematic view of the press: (A) a tungsten-carbide die, (B) tungsten-carbide anvils, (C) steel rings, and (D) backing parts.

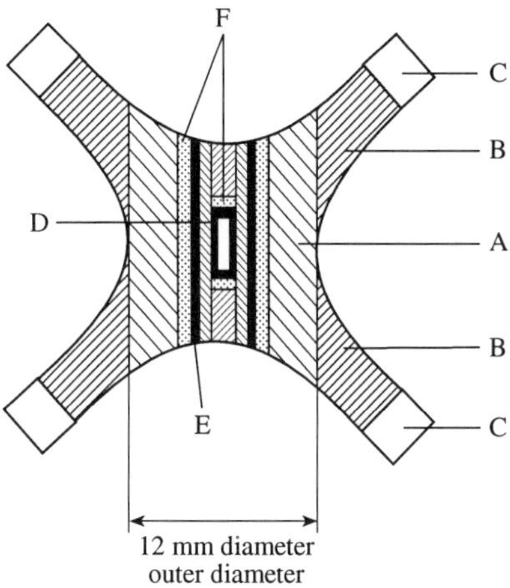

Fig. 7.17 The structure of the reaction cell used with a belt equipment: (A) a cylindrical body made of pyrophyllite, (B) a pyrophyllite gasket, (C) PTFE seals, (D) a capsule containing the reactants, (E) a microfurnace, and (F) a La_2O_3 layer.

application of pressure can stabilize the reactants (a recent illustration is given by the preparation[29] of new superconducting oxides containing Hg^{2+}).

High oxygen pressures are an important tool for stabilizing the high valencies of transition metals.[30] While oxygen pressures up to 4 kbar (0.4 GPa) can be obtained by gas compression, highest pressures can be reached (5 GPa or more) by using the *in situ* thermal decomposition of unstable oxides.

Example. The thermal decomposition of CrO_3 by

$$CrO_3 \rightarrow CrO_2 + \tfrac{1}{2}O_2\uparrow$$

has been used in the preparation[31] of $CaFeO_3$. In order to prevent the contamination of the resulting material by ferromagnetic CrO_2, a specific structure had to be adopted for the high-pressure cell.

Example. The thermal decomposition of $KClO_3$

$$KClO_3 \rightarrow KCl + \tfrac{3}{2}O_2\uparrow$$

was developed[7] mainly in order to stabilize unusual oxidation states of transition metals, for example, Fe(V), Cu(III), Ir(V), and Ir(VI).[32–35] Since KCl is soluble in water, it is easy to eliminate. Due to its diamagnetic properties, the presence of small amounts of KCl does not affect the magnetic characterization of resulting materials. By contrast, small amounts of ferromagnetic CrO_2 can drastically affect magnetic measurements.

Protocol 6.
An example of a reaction protocol using decomposition of $KClO_3$ as the oxygen source in belt-type equipment

Caution! For safety procedures see Appendix 2.

1. Introduce 50–200 mg of a mixture of the starting oxides and $KClO_3$ (molar fraction of $KClO_3$ between 0.17 and 0.33) in a platinum capsule (3 mm long, 0.1 mm thick, with an inner diameter of 3 mm, for a belt-type equipment characterized by a die with an inner diameter of 12 mm) which is subsequently sealed by means of an electric arc.
2. Place the sample and its protective capsule in a porous ceramic tube, which is itself introduced in a microfurnace (platinum or graphite; see Fig. 7.17). This microfurnace is later inserted in the pyrophyllite body, which is itself placed inside the die of the belt; see A in Fig. 7.16(b). The gaskets (B in Fig. 7.17) and the pyrophyllite body of the reaction cell are machined in advance from blocks of the raw pyrophyllite material, according to the shape of the anvils and the die.
3. Position the upper part of the belt (it is guided by four centring pins, as shown on Fig. 16(b)), and the whole belt inserted between the two jacks of a four-column hydraulic press, whose capabilities are, of course, suited to

the size of the inside die. Use a 300 tonne hydraulic press, for a 12 mm inner-diameter laboratory belt to reach 9 GPa. (The electrical supply for the microfurnace is provided by the contact of the upper and lower edges of the microfurnace and the tungsten anvils. A low-voltage and high-current direct-current power supply is necessary; for example, 0–10 V, 0–100 A.)

4. Increase the pressure gradually to the required value by adjusting the hydraulic pressure. A prior pressure calibration, based on abrupt changes in the resistivity of standard materials (Bi, Tl), is necessary. Increase the temperature gradually and maintain it at a constant value, which can be evaluated either using a thermocouple inserted in the high-pressure cell or using a power controller after a $T = f(P)$ calibration. Due to the kinetic effect of pressure, the reaction time is very short, varying between 5 and 20 min in most cases. The oxygen pressure generated from the *in situ* decomposition of $KClO_3$ is very difficult to evaluate. It can be assumed that it is of the same order of magnitude as the pressure applied to the high pressure cell, if the platinum capsule is sealed.
5. Quench the treated material when the reaction is completed, by switching off the electrical supply. Slowly decrease the pressure after ~5 min in order to release the internal stresses gradually in the die and the anvils.
6. Remove the belt-apparatus from the hydraulic press and open it by taking off the upper part. Recover the sample in its protective capsule, and characterize it by classical methods, after having removed the potassium chloride (resulting from the decomposition of $KClO_3$) by rapid dissolution in water. (A new high-pressure cell must be used for each experiment.)

This experimental protocol is also valid for most syntheses under high pressure, for example, the synthesis of diamond or cubic boron nitride in the laboratory or in industry. In the latter case, the high-pressure cell must be much larger (50 mm inner diameter) and it requires large belt-type equipment and a more powerful hydraulic press.

High-pressure reactions in specific dry conditions. In some cases, the synthesis of materials requires the elimination of all traces of water. High-pressure cells are usually made of pyrophyllite and the high-temperature treatment of such material leads to the *in situ* formation of water, which induces hydrolysis.

Now La_2O_3 can be used as a protective layer in order to trap water; the resulting lanthanum oxyhydroxide, LaOOH, is stable under pressure (8 GPa) up to 1000 °C.[7] A cell using this protective layer has been used to investigate the structural changes with pressure of K_2MO_3 phases. The experimental protocol is the same as Protocol 6.

2.5.2 Investigation of the physical properties of materials under high-pressure conditions

The study of *in situ* changes in the physical properties of materials under high-pressure conditions requires special designs for high-pressure cells. In the majority of cases, the temperature is lower than that developed in high-pressure synthesis. The main constraint is from the method which is used for the evaluation of the physical property. Two cases can be differentiated:

(1) the measurement does not require electromagnetic radiation (for example, electrical or magnetic measurements);

(2) the measurement uses electromagnetic radiation (X-rays, infrared, etc.).

The design and the structure of the equipment are directly related to which of these types of radiation are used. For example, using a belt apparatus with the specific high-pressure cell shown in Fig. 7.18, electrical measurements have been made at various temperatures and pressures.[37,38]

For X-ray diffraction (XRD) studies, due to the high-pressure range required, diamond-anvil cells seem to be the most appropriate type of equipment (see Section 3 of Chapter 3).[39,40] At medium pressures ($P < 10$ GPa) some specific multianvil apparatuses have been developed. In particular, the construction of a new hexahedral press is now in progress in Bordeaux (Fig. 7.19). Its design allows *in situ* XRD studies to be carried out at high temperatures ($T < 1500\,°C$), leading to new advances in materials science.[41]

Infrared investigations generally require a lower pressure range and

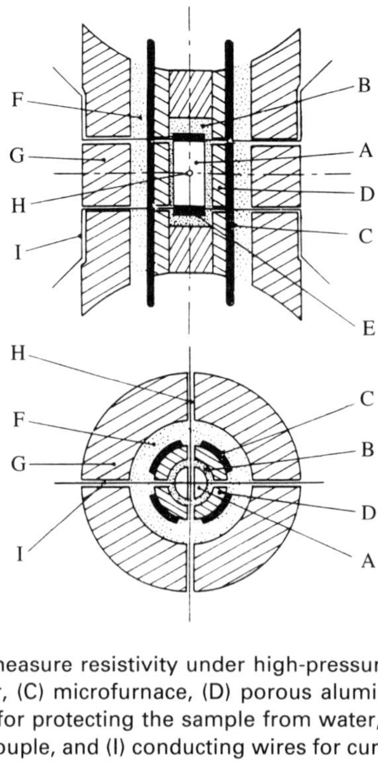

Fig. 7.18 A cell used to measure resistivity under high-pressure conditions: (A) cylindrical sample, (B) AlF_3 layer, (C) microfurnace, (D) porous alumina tube, (E) copper wires (voltage), (F) La_2O_3 layer for protecting the sample from water, (G) cell body made from pyrophyllite, (H) thermocouple, and (I) conducting wires for current measurements.[37,38]

Fig. 7.19 A general view of a hexahedral press.

experiments may be carried out in the liquid state. Reaction cells with sapphire windows can be used up to 1 GPa at room temperature.[42]

2.5.3 Crystal growth under very-high-pressure conditions

The synthesis of diamond[43] and cubic boron nitride[44] has strongly motivated improvements in the development of high-pressure equipment and increased the interest in these materials, which have exceptional properties.[45] Single crystals are required for optical and electronic applications. Consequently, specific crystal-growth processes have been set up under very high-pressure conditions. The principle is similar to that described, at lower pressures, for the preparation of single crystals of α-SiO_2.

2.5.4 Shaping of materials under very-high-pressure conditions

When the stabilization of the initial material requires high-pressure conditions, specific cells have been developed, in particular for the elaboration of diamond and cubic-boron-nitride ceramics.[46]

2.6 Conclusions

Pressure is an important physical variable in materials science. The development of high-pressure techniques has led, in particular, during the last fifty years, to advances in fundamental scientific research and to industrial applications.

Because the compressibilities of the gas, liquid, and solid phases differ, the structure of the equipment used with each of these phases is very different. In addition, specific modifications have been developed according to: (i) the objective of the high-pressure studies (synthesis, crystal growth, or physical measurements), and (ii) the chemical nature of the medium.

References

1. Jayaraman A.; *Sci. French Edn Sci. Am.* **1984**, *80*, pp. 36–42.
2. Goodenough, J. B.; Kafalas, T. A.; Longo, J. M. In *Preparative Methods in Solid State Chemistry*; Hagenmuller, P., ed.; Academic Press: New York, **1972**.
3. Spain, I. L.; Bolsaitis, P. *High Pressure Technology*, Spain, I. L.; Paauwe, J., eds.; Marcel Dekker: New York, **1977**; Vol. II, p. 109.
4. Papen, E. L. J. *High Pressure Technology*; Spain I. L.; Paauwe, J. eds.; Marcel Dekker: New York, **1977**; Vol. II, p. 339.
5. Pugh, H. D.; Donaldson, C. J. H. *High Pressure Technology*; Spain, I. L.; Paauwe, J., ed.; Marcel Dekker: New York, **1977**; Vol. II, p. 423.
6. Chevalier, B.; Demazeau, G.; Etourneau, J.; Hagenmuller, P. *Phys. Status Sol.* (b), **1979**, *93*, K63.
7. Demazeau, G. thesis, number 419, University Bordeaux I, **1973**.
8. Demazeau, G. *High Pressure Chemical Synthesis*, Jurczak, J. and Baranowski, B. eds.; Elsevier: Amsterdam, **1989**; Chapter 5, p. 101.
9. Demazeau, G. *High Pressure and Biotechnology*, Balny, C.; Hayashi, R; Heremans, R.; Masson, R., eds; Libbey Eurotext: Paris, **1992**; p. 481.
10. Demazeau, G.; Maestro, P.; Plante, Th.; Pouchard, M.; Hagenmuller, P. *Ann. Chim. Sci. Matériaux*, **1978**, *3*, 353.
11. Demazeau, G.; Maestro, P.; Plante, Th., Pouchard, M.; Hagenmuller, P. *Mat. Res. Bull.* **1979**, *14*, 121.
12. Plante, Th.; Demazeau, G.; Pouchard, M.; Hagenmuller, P. *Bull. Soc. Chim. Fr.* **1973**, *12*, 3301.
13. Putilin, S. N.; Antipov, E. V.; Marezio, M. *Physica C*, **1993**, *212*, 266.
14. Wang, A.; Capitain, F; Monnier, V.; Matar, S.; Demazeau, G. *J. Mater Synth. Processing*, in press.
15. Wilson, W. D. *High Pressure Technology*, Spain, I. L.; Paauwe, J., ed.; Marcel Dekker, New York, **1977**; Vol. II, p. 293.
16. Taki, S. *Prog. Crystal Growth Charact.* **1991**, *23*, 313.
17. Balny, C.; Heremans, K.; Masson, P. *Biofutur*, **1992**, *112*, 37.
18. Cheftel, J. C. *IAA*, **1991**, *141*.
19. Hagenmuller, P.; Demazeau, G. *Advances in Solid State Chemistry*; Rao, C. N. R.; *Proc, Indian Natn. Sci. Acad.*, **1986**; *52A*, no 7, p. 102.
20. *High Pressure and Materials*, Proceedings of the XXVIIIth EHPRG Conference, Bordeaux, 1990, Demazeau, G.; Cros, C.; Matar, S.; Rahm, A. Gordon Breach: Philadelphia, **1991**; pp. 361–390.
21. Newton, R. C. In *Advances in High Pressure Research*; Bradley, R. S., ed.; Academic Press: New York, **1966**, Vol I, p. 195.
22. Leger, J. M.; Yacoubi, N.; Loriers, J. *Mat. Res. Bull.* **1979**, *14*, 1431.
23. Bundy, F. P. *High Pressure Technology*; Spain, I. L.; Paauwe, I.; eds.; Marcel Dekker: New York, **1977**, Vol I, p. 321.
24. Hall, H. T. *Rev. Sci. Instrum.* **1960**, *31*, 125.
25. Contré, M. *French Patent* number 14 57 690, 23 September 1965, 26 September 1966.
26. Demazeau, G.; Pouchard, M.; Buffat, B.; Hagenmuller, P. *J. Physique* **1984**, *45*, 345.

27. Vel, L.; Demazeau, G.; Etourneau, J. *Mat. Sci. Eng. Sol. State Mat. Adv. Technol.*, B, **1991**, *10*, 149.
28. Singh, A. K., ed.; *Recent Trends in High Pressure Research*, Proceedings of the XIII AIRAPT, International Conference on High Pressure Science and Technology; Oxford, IBM: **1991**; pp. 3–103.
29. Putilin, S. N.; Antipov, E. V.; Chmaissen, E. V.; Marezio, M. *Nature*, **1993**, *362*, 226.
30. Demazeau G.; Plante, Th. *High Pres. Res.* **1991**, *7*, 328.
31. Kanamaru, F.; Miyamoto, H.; Mimura, Y.; Koizumi, M.; Shimada, M.; Kima, S. *Mat. Res. Bull.* **1970**, *5*, 257.
32. Demazeau, G.; Buffat, B.; Menil, F.; Fournès, L.; Pouchard, M.; Dance, J. M.; Fabritchnyi, P.; Hagenmuller, P. *Mat. Res. Bull.* **1981**, *16*, 1465.
33. Demazeau, G.; Parent, C.; Pouchard, M.; Hagenmuller, P. *Mat. Res. Bull.* **1972**, *7*, 913.
34. Hayashi, K.; Demazeau, G.; Pouchard, M.; Hagenmuller, P. *Mat. Res. Bull.* **1980**, *15*, 461.
35. Demazeau, G.; Jung, D. Y.; Sanchez, J. P.; Colineau, E.; Blaise, A.; Fournès, L. *Solid State Comm.* **1993**, *85*, 479.
36. Delmas, C.; Demazeau, G.; Devalette, M.; Fouassier, C.; Hagenmuller, P. *J. Sol. State Chem.* **1976**, *19*, 87.
37. Chevalier, B.; Demazeau, G.; Etourneau, J.; Hagenmuller, P. *Phys. Status Sol.* (b), **1979**, *93*, K63.
38. Langlais, F.; Demazeau, G.; Portier, J.; Tressaud, A.; Hagenmuller, P. *Sol. State Commun.*, **1979**, *29*, 473.
39. Bassett, W. A.; Takahashi, T. In *Advances in High Pressure Research*; Bradley, R. S., ed.; Academic Press: New York, **1974**; Vol IV, p. 165.
40. Arrouy, F.; Cros, C., Demazeau, G. *Physica* C, **1991**, *184*, 74.
41. Contré, M. *High Temp.–High Press.* **1969**, *1*, 339.
42. Whatley, L. S.; Van Valkenburg, A. In *Advances in High Pressure Research*; Bradley, R. S.; ed.; Academic Press: New York, **1966**; Vol I, p. 237.
43. Wentorf, Ir. R. H. In *Advances in High Pressure Research*; Bradley, R. S., ed.; Academic Press: New York; **1974**; Vol IV, p. 249.
44. Wentorf, Jr. R. H. *J. Chem. Phys.*, **1963**, *28(5)*, 1144.
45. Demazeau, G. *Diamond and Related Materials*, **1995**, *4*, 284.
46. Moore, A. W. In *High Pressure Technology*, Spain, I. L., Paauwe, J., eds.; Marcel Dekker: New York, **1977**; Vol II, p. 409.

3. Electrochemical measurements at elevated pressures

Thomas W. Swaddle

3.1 Introduction

Several laboratories[1–17] have developed high-pressure electrochemical techniques (here construed as excluding conductimetric techniques), with motivations ranging from the probing of supercritical aqueous electrolyte solutions[9] to the analysis of the volume of activation, $\Delta V^{\ddagger}$, for redox (reduction–oxidation) processes in solution in terms of Marcus–Hush theory.[18] Typically, one wishes to know the molar volume change ΔV accompanying a redox reaction. In principle, ΔV can be obtained from the algebraic sum of the molar volumes of the products and reactants,[19] but it is rare that all of these volumes are known, and it is better to obtain ΔV from the pressure dependence of the cell electrode potentials E, which gives a rather direct measure of the corresponding reaction volume, ΔV_{cell}, for a half-cell *relative to the particular reference electrode chosen*:

$$\Delta V_{\text{cell}} = -nF\left(\frac{\partial E}{\partial P}\right)_T, \tag{7.3.1}$$

where n is the number of moles of electrons transferred (usually equal to one mole) and F is the Faraday constant (96 487 A s mol^{-1}). Consequently, ΔV_{cell} contains an unknown contribution from the pressure dependence of the response of the reference electrode. Tregloan and co-workers[20] have proposed an ingenious method of separating out the contribution of the reference electrode to ΔV_{cell}, while others have attempted a separation of the half-cell contributions by assigning a molar volume to the electrons,[17,21] but in practice it is sufficient to record all values of ΔV_{cell} relative to a common reference electrode, since, in general, one wishes to obtain ΔV for a complete reaction from the difference in ΔV_{cell} for its constituent half-reactions.

3.2 Pressurizable electrochemical cell designs

A simple but effective system for the routine measurement of the pressure dependence of E up to 300 MPa is described below.[14] In practice, a working pressure range 0.1–200 MPa is adequate for most purposes; as the pressure range is increased, the experimental difficulties and uncertainties worsen. Furthermore, experience has shown that ΔV is generally more or less independent of pressure (that is, that plots of E against the pressure are linear within the experimental uncertainty) over this range. On the other hand, extension of the maximum pressure to 1 GPa can give information on the anticipated[14] pressure dependence of ΔV.[13,16,17]

The working electrode can be a simple cylindrical wire of platinum, gold, or some other suitable material, and this is entirely satisfactory for making cyclic voltammetric (CV), differential pulse voltammetric (DPV), or direct Nernstian potentiometric measurements under pressure. There are, however, important advantages (including reduced double-layer capacitance, fast voltammetric response, and small iR losses) to the use of (ultra)microelectrodes in high-pressure systems.[13] In our current work we are successfully using home-made disc or cylindrical platinum, gold, or carbon-fibre microelectrodes with diameters of 7–25 μm in place of the conventional 0.5 mm wire working electrode at pressures up to 250 MPa—the use of a Faraday cage, however, is essential to the accurate measurement of the low currents (nA to μA) obtained with disc microelectrodes.

The most convenient reference electrodes for high-pressure work are Ag/Ag^+ and $Ag/AgCl/Cl^-$ electrodes.[14,15] For aqueous systems, these electrodes happen to have a similar pressure response, so that the ΔV_{cell} values obtained with either of the reference electrode are interchangeable, with only a small correction.[22] For the Ag/AgCl/KCl(aq) electrode described in *Protocol 7*, 4.0 mol L^{-1} KCl is chosen over the traditional saturated KCl solution with KCl crystals present, because the solubility of KCl is pressure dependent,

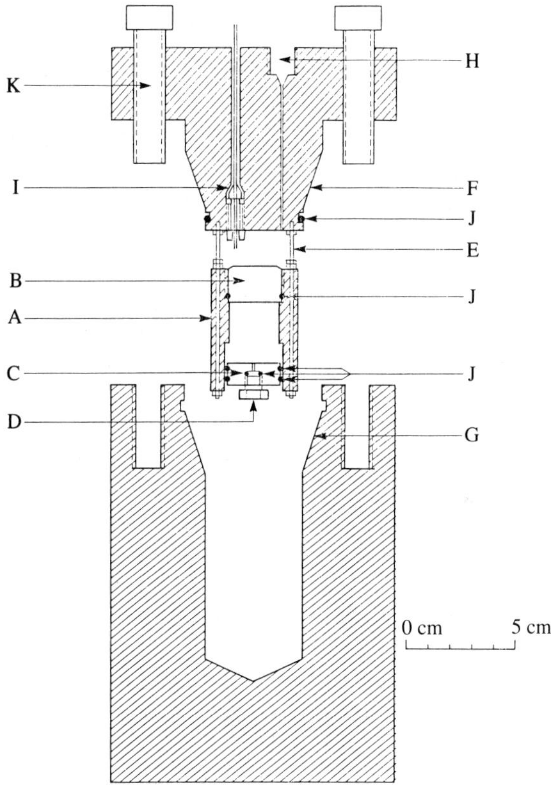

Fig. 7.20 A pressure vessel: (A) Teflon sample cell, (B) threaded Teflon plug in which the electrodes are mounted (see Fig. 7.22), (C) free piston (Teflon), (D) a Teflon-plug sealing–filling port, (E) threaded stainless-steel cell support, (F) main sealing surface of the pressure-vessel cap (39° cone), (G) main sealing surface of the pressure vessel body (40° cone), (H) standard $\frac{1}{4}$ inch pressure connection (in the plane in front of the cell supports, E), (I) high-pressure electrical leadthrough (in the plane in front of the cell supports, E; there are two more leadthroughs in the plane behind E, and the detailed structure of a leadthrough is shown in Fig. 7.21), (J) neoprene O-rings, and (K) grade 8 hexhead bolts (0.625 in NC, 2.25 in long; eight of these bolts are placed symmetrically around the pressure-vessel cap).[14] (Reprinted with permission from NRC Research Press.)

so that the electrode response may drift on changing the pressure as KCl dissolves or precipitates. An Ag/$AgNO_3$ (0.01 mol L^{-1}) reference electrode,[15] however, is easier to prepare, and it gives better reproducibility for ΔV_{cell}, and is therefore preferred unless halide or other ions that may form a precipitate with Ag^+ inside the junction frit are present in the sample. A ferrocene/ferricenium reference couple is a viable alternative for nonaqueous systems.

Figure 7.20 shows a pressure vessel (with a 130 mm outer diameter and a 44 mm inner diameter) which was machined from 17-4PH stainless steel and then heat treated (for 1 h at 480 °C, followed by air cooling) to Rockwell hardness C44. The designed operational pressure ceiling was 515 MPa, but in practice the pressure may be limited by the Vespel insulation in the electrical leadthroughs (Fig. 7.21), and a 300 MPa ceiling is recommended. The sample cell shown in Fig. 7.20 and its top plug B (Fig. 7.22) were machined from virgin Teflon to a maximum outer diameter of 38.5 mm and a minimum inner diameter of 24 mm (capacity ca. 14 mL); the dimensions may be reduced if necessary (a cell with a capacity of 4.5 mL was made for use with solutions of biological materials),[22] but better results have generally been obtained with the stated dimensions. The bottom plug C in Fig. 7.20 is a free piston that

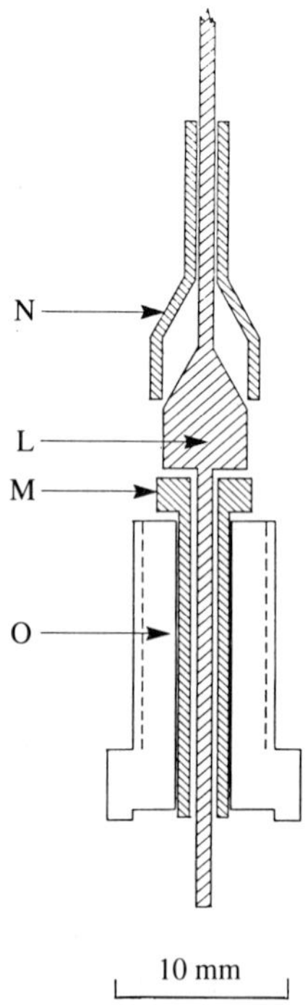

Fig. 7.21 A high-pressure electrical leadthrough: (L) a brass electrical connector, (M) a vespel (Du Pont polyimide resin) insulating sleeve, (N) a Vespel pressure-sealing sleeve, and (O) a bolt feeding into the bottom of the pressure-vessel cap.

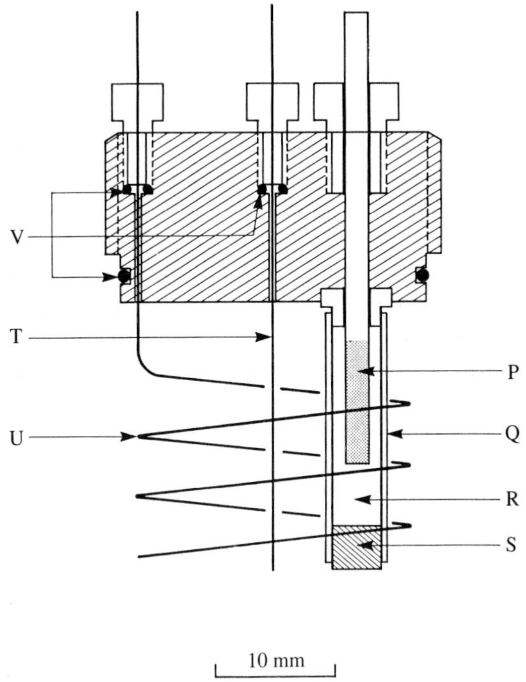

Fig. 7.22 The electrode assembly (plug B in Fig. 7.20): (P) a reference electrode (this is a 2.0 mm diameter Ag rod for an Ag/Ag^+ electrode, or an Ag rod with an AgCl coating over the lowest 2 cm for an Ag/AgCl electrode, (Q) heat-shrunk Teflon tubing, (R) aqueous $AgNO_3$ (0.01 mol L^- for an Ag/Ag^+ electrode) or KCl (4.0 mol L^{-1}, for an Ag/AgCl electrode), (S) a Vycor porous disc, (T) the working electrode (platinum wire), (U) the counter electrode (Platinum wire), and (V) neoprene O-rings.

allows for the compression of the cell contents; thus, the pressure differential between the pressurizing fluid and the solution in B is always small. Electrical connections between the protruding parts of the electrodes outside B and the brass high-pressure leadthroughs are made by soldering minimal lengths of copper wire to them.

The working electrode may be made of any suitable material; we have used 0.5 mm platinum or gold wire (any pretreatment procedures which are established in conventional electrochemical practice should be employed; for example, gold wire pretreated with 4,4′-dipyridyldisulphide was used for work on cytochrome *c* solutions).[22] The counter electrode, shown as a long helix in Fig. 7.22, need not be significantly larger than the working electrode. The shapes and sizes of these electrodes were found to be unimportant. In general, a reference electrode of the Ag/Ag^+ type is recommended, and this may be readily assembled from a polished silver rod (2.0 mm in diameter) immersed in 0.01 mol L−1 $AgNO_3$ solution in a sleeve of Teflon, previously heatshrunk at the bottom on to a Vycor porous plug (4 mm in diameter; obtained from E. G. and G. Instruments Division, Markham, Ontario). This sleeve is slipped on to a close-fitting adapter in B (cf. Fig. 7.22). The Teflon sleeve collapses slightly to accommodate the increased pressure, and, while this works satisfactorily, the design of Shalders *et al.*,[15] in which the porous plug is mounted in a free piston in a glass sleeve, offers a more controlled compression.

If, however, a reference electrode of the Ag/AgCl type is required (as in *Protocol 7*), the lower part of the Ag rod is first sanded, polished, rinsed with acetone and dilute acid, and then coated electrolytically with AgCl in a KCl solution. The electrolysis is performed by increasing the voltage slowly until a uniform AgCl film just appears; the electrolysis is then continued at this voltage for half an hour. The AgCl-coated rod should be aged by suspension in distilled water or in a KCl solution for 24 h or more before use, and contamination of the AgCl surface must be carefully avoided. The electrode is then assembled as described above, but with 4.0 mol L^{-1} KCl in place of the $AgNO_3$ solution (Fig. 7.22).

The pressure vessel is thermostated with a close-fitting outer jacket (not shown), sealed with O-rings, and fed with water from a circulating temperature controller. Alternatively, and less conveniently, the vessel may be immersed in a thermostated oil bath.

Protocol 7.
Experimental procedure

Caution! For safety procedures see Appendix 2.

1. Assemble the sample cell first (Figs 7.20 and 7.22) and complete the electrical connections to the leadthroughs (I).
2. Fill pressure vessel with an electrically insulating, electrochemically clean, pressurizing fluid—pure hexane is preferred—and bring it to the required temperature.
3. Deoxygenate the test solution—in this example, 1.5 mmol L^{-1} $K_3[Fe(CN)_6]$ in 0.5 mol L^{-1} aqueous KCl—by purging with N_2 for 1 h, and bring the test solution to near the required temperature (here, 25°C), and then inject it into the sample cell through the filling port D in the free piston C, leaving no bubbles.
4. Seal port D, and lower the head assembly into the body of the pressure vessel, and effect the pressure seal by progressively tightening opposite pairs of the eight bolts K.
5. Make pressure connections through H, either directly to a pump and a pressure gauge filled with hexane, or, if it is undesirable to expose the pump to hexane, via a free-piston separator to the oil-filled pump (the pressure gauge should be on the pressure-vessel side of the separator if possible).

6. Make electrical connections to, for example, a Princeton Applied Research Corp. model 174A polarographic analyser with an RE0074 *x–y* recorder (Shalders *et al.*[15,23] describe a microcomputer-based alternative system), and run a test cyclic voltammetric measurement to ensure that the system is fully functional. (For $Fe(CN)_6^{3-}$ versus Ag/AgCl, the voltage scan in direct-current mode may be from 0.6 V (or 0.8 V) to 0 V at 20 or 50 mV s^{-1}; the current will depend on the characteristics of the electrodes.)

7. Increase the pressure (even for zero pressure measurements, a small pressure (1–5 MPa above atmospheric) should be applied to firm up the seals and to collapse any remaining bubbles).

8. Collect cyclic voltammetric,[14] differential pulse voltammetric,[22] or potentiometric[15] data when thermal equilibration is complete (after up to 1.5 h for the first pressure, and after less than 50 min for successive pressures—see below). Repeat this routine at various increasing and then decreasing pressures. (Only data sets whose measurements are made on the upward and downward legs of the pressure cycle agree within the experimental uncertainty (±2 mV) should be accepted.)

Typical cyclic voltammetric measurements for $Fe(CN)_6^{3-}$ at two pressures are shown in Fig. 7.23. The mean of the positions of the current extrema *at a given pressure* gives *E*, which is clearly pressure dependent. The *spread* of the extrema may also be seen to change with pressure, from which it is possible

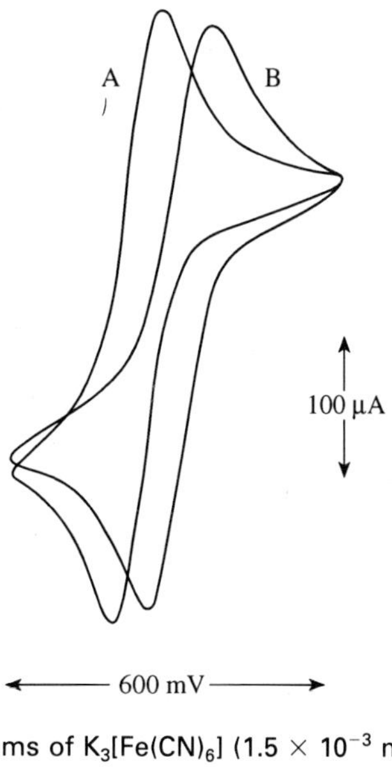

Fig. 7.23 Cyclic voltammograms of $K_3[Fe(CN)_6]$ (1.5×10^{-3} mol L^{-1} in 0.5 mol L^{-1} aqueous KCl), 0–600 mV versus Ag/AgCl, at 25 °C: (A) 5 MPa, and (B) 204 MPa.

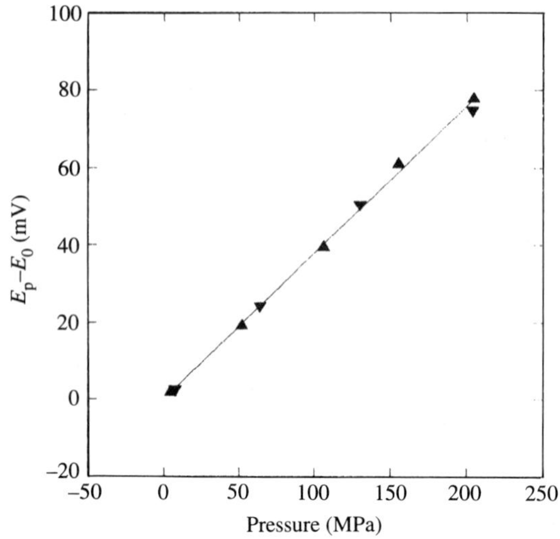

Fig. 7.24 The pressure dependence of *E* from cyclic voltammograms for the system of Fig. 7.23. (▲) increasing pressure, and (▼) decreasing pressure.

to estimate the *volume of activation*, $\Delta V^{\ddagger}$, for the electrode reaction.[24] The E versus pressure plot (Fig. 7.24) is linear, with a slope corresponding to $\Delta V_{\text{cell}} = -36.6 \pm 0.7$ cm^3 mol^{-1}

It is difficult to devise a pressure-proof, electrochemically inert temperature sensor with good thermal contact that could monitor the internal temperature of the pressurized sample solution internally during or between electrochemical measurements (such an arrangement could also introduce electrical noise). Instead, the equilibrium temperature of the interior of the cell may be measured before and/or after a series of pressure experiments with an ordinary thermistor, thermocouple, or platinum resistance thermometer. We have shown[14] that, in the apparatus described above, temperature perturbations caused by (de)pressurization decay to within $\pm 0.05\,^{\circ}$C of the temperature of an equilibrium sample near $25\,^{\circ}$C in less than 50 minutes; equilibration of the cell contents is relatively fast (5–10 min), despite the poor thermal conductivity of the Teflon cell walls, but the much larger volume of the pressurizing fluid determines the rate. During the initial set-up, a longer period (up to 1.5 h) may be allowed for the pressure vessel itself, (which has a large thermal inertia) to equilibrate thoroughly.

Acknowledgements

I thank Drs Y. Fu, R. D. Shalders, H. Takagi, and T. W. Whitcombe for their contributions to our electrochemical work, and the Natural Sciences and Engineering Research Council of Canada for financial support.

References

1. Cohen, E.; Piepenbroek, K. *Z. Phys. Chem. Abt.* A, **1934**, *167*, 365; **1934**, *170*, 145.
2. Ewald, A. H.; Lim, S. C. *J. Phys. Chem.* **1957**, *61*, 1443.
3. Hills, G. J.; Ovenden, P. J. *Adv. Electrochem. Electrochem. Eng.* **1966**, *4*, 185.
4. Hills, G. J. In *Advances in High Pressure Research*; Bradley, R. S., ed.; Academic Press: New York, **1969**; Vol. 2, Chapter 4.
5. Mesmer, R. E.; Baes Jr, C. F.; Sweeton, F. H. *J. Phys. Chem.* **1970**, *74*, 1937.
6. Fleischmann, M.; Gara, W. B.; Hills, G. J. *Electroanal. Chem. Interfac. Electrochem.* **1975**, *60*, 313.
7. Conway, B. E.; Currie, J. C. *J. Electrochem. Soc.* **1978**, *125*, 252; **1978**, *125*, 257.
8. Sato, M.; Yamada, T. In *High Pressure Science and Technology*; Vodar, B.; Marteau, P., ed.; Pergamon: Oxford, **1980**; vol. 2, pp. 812–814.
9. McDonald, A. C.; Fan, F.-R. F.; Bard, A. J. *J. Phys. Chem.* **1986**, *90*, 196.
10. Franklin, T. C.; Mathew, S. A. *J. Electrochem. Soc.* **1987**, *134*, 760.
11. Jovancicevic, V.; Bockris, J. O'M. *Rev. Sci. Instrum.* **1987**, *58*, 1251.
12. Krasinski, P.; Tkacz, M.; Baranowski, B.; Galus, Z. J. *J. Electroanal. Chem.* **1991**, *308*, 189.
13. Golas, J.; Drickamer, H. G.; Faulkner, L. R. *J. Phys. Chem.* **1991**, *95*, 10191.
14. Doine, H.; Whitcombe, T. W.; Swaddle, T. W. *Can. J. Chem*, **1992**, *70*, 81.
15. Sachinidis, J.; Shalders, R. D.; Tregloan, P. A. *J. Electroanal. Chem*, **1992**, 327, 219.
16. Cruañes, M. T.; Rodger, K. K.; Sligar, S. G. *J. Am. Chem. Soc.* **1992**, *114*, 9660.
17. Cruañes, M. T.; Drickamer, H. G.; Faulkner, L. R. *J. Phys. Chem.* **1992**, *96*, 9888.
18. Grace, M. R.; Takagi, H.; Swaddle, T. W. *Inorg. Chem.* **1994**, *33*, 1915.
19. Bänsch, B.; Martinez, P.; van Eldik, R. *J. Phys. Chem.* **1992**, *96*, 234.
20. Sachinidis, J. I.; Shalders, R. D.; Tregloan, P. A. *Inorg. Chem.* **1994**, *33*, 6180.
21. Heusler, K. E.; Gaisler, L. *Ber. Bunsenges. Phys. Chem.* **1969**, *73*, 1059.
22. Sun, J.; Wishart, J. F.; van Eldik, R.; Shalders, R. D.; Swaddle, T. W. *J. Am. Chem. Soc.* **1995**, *117*, 2600.
23. Shalders, R. D. Ph.D. thesis, University of Melbourne, Australia, 1992.
24. Takagi, H.; Swaddle, T. W. *Inorg. Chem.* **1992**, *31*, 4669.

8

Electrophoresis of proteins under high hydrostatic pressures

PATRICK MASSON

1. Introduction and basic principles

Elevated pressures can induce functional and structural alterations of proteins. The effects of pressure are governed by Le Chatelier's principle. According to this principle, an increase in pressure favours processes which reduce the overall volume of the system, and conversely increases in pressure inhibit processes which increase the volume. The effects of pressure on proteins depend on the relative contribution of the intramolecular forces which determine their stability and functions. Ionic interactions and hydrophobic interactions are disrupted by pressure. On the other hand, stacking interactions between aromatic rings and charge-transfer interactions are reinforced by pressure. Hydrogen bonds are almost insensitive to pressure. Thus, pressure acts on the secondary, tertiary, and quaternary structure of proteins. The extent and the reversibility, or irreversibility, of pressure effects depend on the pressure range, the rate of compression, and the duration of exposure to increased pressures. These effects are also influenced by other environmental parameters, such as the temperature, the pH, the solvent, and the composition of the medium.

Moderate pressures ($P < 1$ kbar) alter the binding properties of proteins and the activity of enzymes (inhibition, activation, or specificity changes), but they have no significant structural effects. Pressures ($P > 1.5$ kbar) can induce conformational transitions (isomerization, unfolding), dissociation of oligomers, denaturation and aggregation, but significant unfolding only takes place at pressure higher than 5 kbar (for reviews on the effects of pressure upon structure and functions of proteins see references 1–6). Thus, like temperature and physicochemical perturbants, pressure is a useful external variable for exploring binding-site features, dissecting enzyme mechanisms, and investigating the conformational stability of proteins.[7–10] Moreover, a new field of protein engineering called *high-pressure enzyme engineering* is emerging; it aims to alter enzyme specificity by pressure, to modify proteins by chemical means under pressure, and to increase protein stability.[11,12]

Most biophysical techniques have been used at elevated pressures. Among many techniques adapted for high-pressure biochemistry, polyacrylamide-gel electrophoresis has proved to be a direct and reliable tool for obtaining qualitative and quantitative information on the effects of pressure upon the structure and activity of proteins.

Electrophoresis consists of the migration of charged particles in an electric field. It allows proteins to be separated and characterized in complex mixtures.[13] Since the electrophoretic mobility depends on the charge, the size (related to the molecular weight and the subunit number), and the conformation of proteins, electrophoresis is also a very powerful tool for analysis of the structure and stability of proteins. Electrophoresis under high hydrostatic pressures allows the study of protein equilibria, such as protein–ligand interactions, unfolding transitions, or quaternary-structure changes. This technique was first designed by Hawley[14,15] for investigation of the structural stability of a single-chain protein. Some years later, Paladini *et al.*[16] used high-pressure slab-gel electrophoresis for the study of pressure-induced dissociation of an oligomeric enzyme. At about the same time we demonstrated the feasibility of high-pressure affinity electrophoresis for the study of protein–ligand interactions.[17] By operating at different temperatures, by incorporating denaturing agents or stabilizers into the gels, or by changing the solvent, more information can be obtained on the factors which determine protein stability, on the mechanisms of denaturation, on the forces involved in ligand binding, and on the microenvironment of binding sites.

2. Apparatus for rod-gel electrophoresis under high hydrostatic pressures

Electrophoresis under high hydrostatic pressures requires the implementation of special devices, and numerous technical problems have to be resolved.

2.1 High-pressure vessels and high-pressure equipment

The high-pressure electrophoresis apparatus used in my laboratory (shown in Fig. 8.1) is composed of two massive, cylindrical, high-pressure stainless-steel vessels (C and G) housed in thermostated jackets. Vessel C was first used by Hawley;[14] the other parts were built in my laboratory. The two high-pressure vessels are interconnected. The first vessel (C) is designed for electrophoresis at pressures up to 200 MPa (0.987 atm = 1 bar = 0.1 MPa), over the temperature range −10 to 60 °C; its working volume is 1400 mL to accommodate nine cylindrical single-tube electrophoresis cells with volumes of 40 mL whose configuration is essentially the same as in Hawley's cell[15] (see Fig. 8.3(a)). The second vessel (G) can operate at pressures up to 400 MPa, over the temperature range 0–50 °C, and it can accommodate one six-capillary-tube electrophoresis cell (Figs 8.2 and 8.3(b)). The high-pressure bombs are sealed by screw cylinder heads with electrical connections (Fig. 8.2). The first unit is suited for electrophoresis under different conditions: for example, electro-

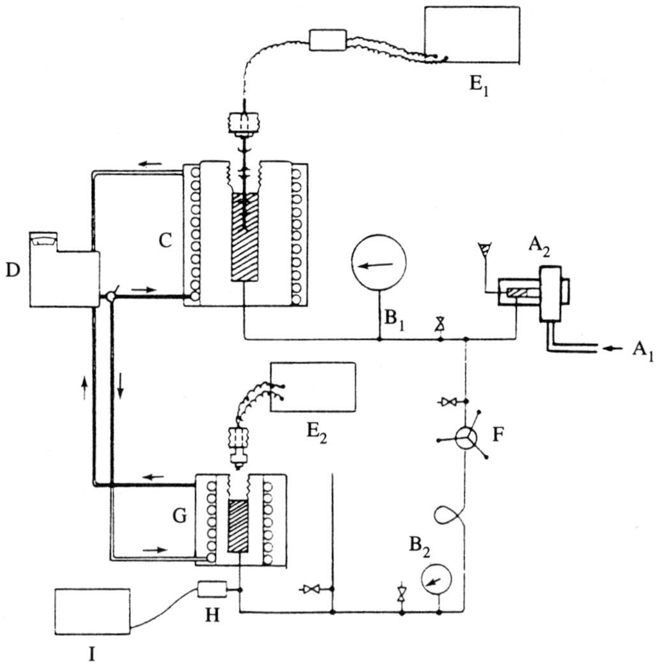

Fig. 8.1 A schematic representation of a high-pressure electrophoresis apparatus: (A_1) pneumatic compressor, (A_2) pneumatic pump with a hydraulic-fluid (paraffin oil. hexane, silicone oil, for example, Dow Corning DC 200 of 100 $mm^2 s^{-1}$ at 25°C or Rhodorsil 41V of 0.65 $mm^2 s^{-1}$) reservoir, (B_1, B_2) Bourdon gauges indicating the pressure in the first and second circuit, respectively, (C) Hawley's high-pressure vessel[14] with a thermostating jacket (this vessel operating up to 200 MPa can accommodate nine cylindrical electrophoresis cells similar to those used by Hawley[15]; the cells are fixed three-by-three along an axis screwed onto the interior of the sealing closure head), (D) a thermocryostat which allows the temperature of the cells to be maintained to within ±0.1°C of the desired temperature by circulating thermostated fluid through the jacket of the high-pressure vessels, (E_1, E_2) power supplies, (F) screw pump. (G) high-pressure vessel operating up to 400 MPa for microdisc electrophoresis in capillary tubes, (H) pressure sensor, and (I) digital pressure monitor.

phoresis in gels with different acrylamide concentrations, electrophoresis in the presence of specific ligands, electrophoresis in a hydro-organic environment, or in the presence of various protein-structure perturbants and denaturing agents. The second unit is suitable for these studies and for the study of the pressure-induced dissociation of oligomeric proteins and denaturation. The temperature of the electrophoresis cells is maintained to within ±0.1 °C of the desired temperature by circulating water or by a mixture of water and ethylene-glycol through the jacket of the high-pressure vessels. Hydrostatic pressure is generated by pumps and transmitted via an inert, noncompressible, and nonconducting hydraulic fluid.

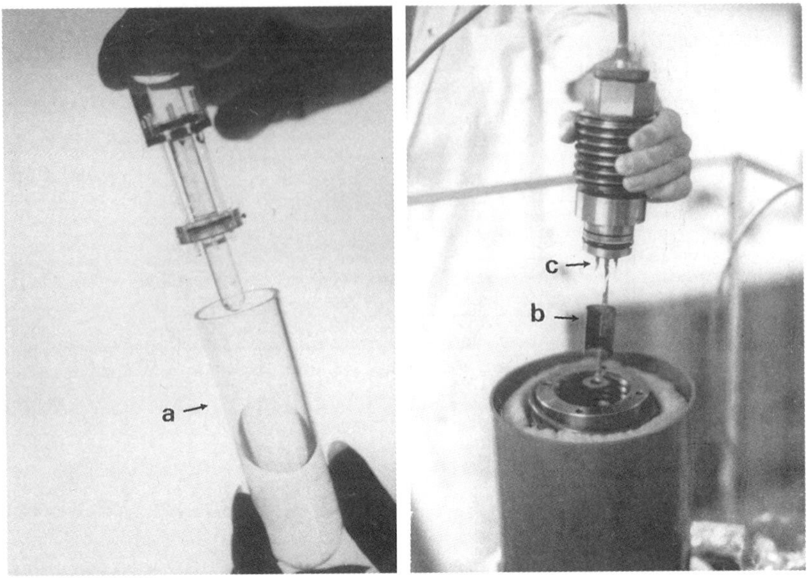

Fig. 8.2 A high-pressure vessel (G in Fig. 8.1) used for microdisc electrophoresis. The electrophoresis cell (a) is fitted into a stand (b) which is screwed onto the interior of the sealing-closure head. The electric leads (c) are soldered on inverted steel cones allowing connections through the plug to the power supply. The insulating shell of the cones is made of Teflon.

2.2 Electrophoresis cells

Electrophoresis can be carried out in different types of electrophoresis cells. For geometric reasons, these cells have to be miniaturized and they must be cylindrical. They consist of two concentric chambers. An outer glass or plastic centrifuge tube forms the anodal buffer chamber (cf. Fig. 8.2(a)). The volume of the buffer chambers must be large enough to minimize the increasing cell resistance.

Capillary tubes are filled to a height 60–80 mm by the capillarity of the acrylamide-gel-forming solution. The surface of the gel solution is carefully overlayed with water-saturated *n*-butanol by means of a 25 μL Hamilton microlitre syringe. Polymerization is achieved at room temperature in less than 60 min. A 10 mm high stacking gel can be added. The gel surface is then rinsed with the running buffer.

3. Apparatus for slab-gel electrophoresis under high-hydrostatic pressures

An apparatus for slab-gel electrophoresis under high hydrostatic pressures was constructed by Paladini *et al.*[16] to study the behaviour of oligomeric

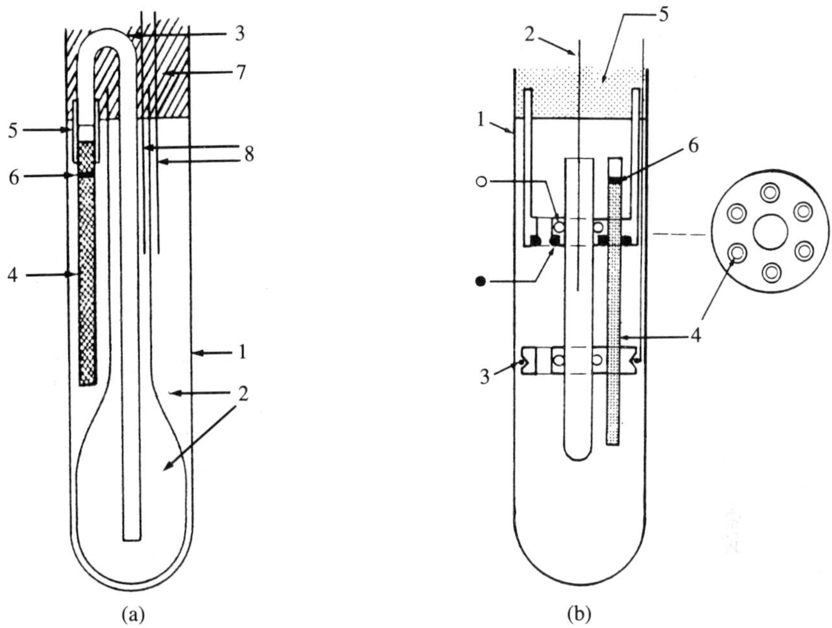

Fig. 8.3 Side views of two electrophoresis cells used in high-pressure rod-gel electrophoresis. (a) Hawley's cell: (1) a Pyrex centrifuge tube (100 mm long, with a diameter of 25 mm), (2) buffer chambers, (3) a U-tube containing a running buffer as a conductive bridge, (4) a gel tube (74 mm long, with an inner diameter of 2–3 mm), (5) a Tygon tubing joint, (6) the sample layer (5–10 μL), (7) hydraulic fluid, (8) platinum electrodes (reproduced from work by Hawley and Mitchel[15] with permission from the American Chemistry Society). (b) A Microdisc electrophoresis cell (M 4043) used for electrophoresis in capillary tubes: (1) a centrifuge tube (100 mm long, with a diameter of 28 mm), (2) the upper-buffer-chamber platinum electrode, (3) the bottom-chamber platinum electrode (this ring electrode is mounted on a plexiglass crown; an O-ring—Viton O-ring (●), with a diameter of 8 × 1.90 mm—allows the crown to slide on the inner tube which is a part of the upper buffer chamber), (4) a capillary gel tube (70–90 mm long, with an inner diameter of 1 mm); the six tubes (for example 100 μL Pyrex disposable microsampling pipettes) are held in a vertical position by the plexiglass crown and O-rings (silicone rubber O-rings (○), with a diameter of 1.7 × 1.78 mm) fitted into the upper chamber), (5) the pressure-transmitting fluid, (6) the sample layer (2 μL).

proteins under pressure (Fig. 8.4(a)). The gel is a thin hollow polyacrylamide cylinder (80 mm long, with a diameter of 0.5–3 mm) polymerized between two concentric glass tubes (Fig. 8.4(b)); the inner wall of the middle tube is siliconized prior to the gel polymerization. The gel has a capacity up to 12 wells. After the run, the middle glass tube is carefully removed and an incision along the axis of the gel allows the turning out of the slab.

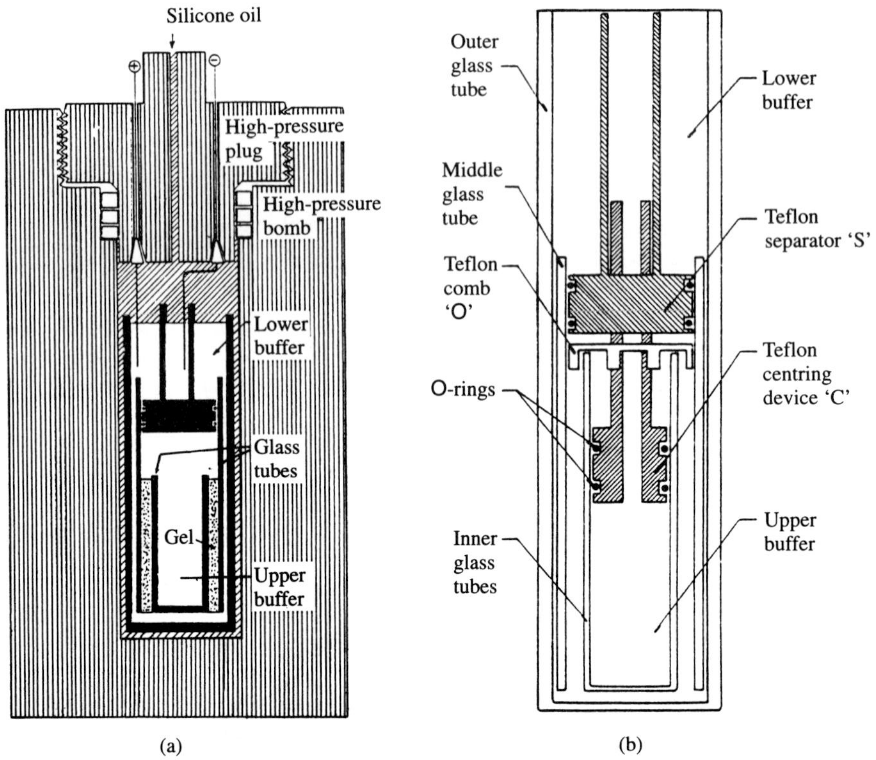

Fig. 8.4 A high-pressure apparatus for slab-gel electrophoresis under high pressures: (a) A side view of the high-pressure bomb with the electrophoresis cell in the running position, and (b) a Slab-gel electrophoresis cell. (Reproduced from a work by Paladini *et al.*[16] with permission from Academic Press.)

4. General conditions of electrophoresis under high hydrostatic pressures

4.1 Buffers for electrophoresis at elevated pressures

The ionization of weak acids varies with both the temperature and the pressure. This process is accompanied by electrostriction of water around the charges that are created. The pressure dependence of the pH is

$$\left(\frac{\partial(\mathrm{pH})}{\partial P}\right)_T \cong \frac{1}{2.303}\,\frac{\Delta V}{RT}. \tag{8.4.1}$$

For example, due to the large negative volume change of $H_2PO_4^-$ upon ionization, the phosphate buffer undergoes a pH shift of -0.44 for $\Delta P = 1$ kbar at 25 °C. On the other hand, the pH values of buffers made with conjugate acids of neutral bases such as Tris-HCl and imidazole–HCl are nearly pres-

Table 1. Some continuous buffer systems for polyacrylamide-gel electrophoresis[a]

pH range	Primary buffer component Usual concentration	ΔV ionization ($mL\ mol^{-1}$)	Reference	Chemical used to adjust the pH to the desired value
2.4–6.0	100 mM citric acid	−10.7, −12.3, and −22.3 at 25 °C	20	NaOH
2.8–3.8	50 mM formic acid	−60 at 25 °C	20	NaOH
3.5–4.5	50–350 mM β-alanine	−7.7/−6.7 at 25 °C	20	Acetic acid
4.0–5.5	50 mM acetic acid	−10.6/−11.5 at 25 °C	20, 21	NaOH or Tris
5.2–7.0	50 mM maleic acid	−5.1, and −24 at 25 °C	20	NaOH or Tris
6.0–8.0	50 mM K/NaH_2PO_4	−24.0 at 25 °C	21	NaOH
6.2–7.8	50 mM imidazole	−2.4 at 25 °C	22	HCl
7.0–8.5	50 mM veronal	−12.5 at 25 °C	20	HCl
7.2–9.0	50 mM Tris	+1 at 25 °C	21	HCl or glycine
8.5–10.0	15 mM $Na_2B_4O_7$	−32.1 at 22 °C	23	
	or boric acid	−30.2 at 25 °C	22	HCl or NaOH
9.0–10.5	50 mM glycine	+1.7 at 25 °C +1.9	20	NaOH
9.0–11.0	25 mM $NaHCO_3$	−25.4 at 22 °C	23	NaOH

[a] Adapted from work by Andrews.[24]

sure invariant (Table 8.1). However, a disadvantage with Tris buffers is that Tris-HCl dissociation is strongly dependent on the temperature. Indeed, the heat of ionization ΔH, of $Tris^+$ is large (10.9 kcal mol^{-1}) at 25 °C, and the pH of Tris buffers falls as the temperature rises. Thus, to design buffer systems for electrophoresis at elevated hydrostatic pressures, the effect of pressure and temperature on the ionization of weak acids must be taken into account. Indeed, changes in pH can affect the stability of the protein architecture and the affinity of the binding sites. In particular, pH changes shift the (P–T)-phase diagrams for the denaturation of proteins.[18,19] In addition, the electrophoretic mobility is strongly dependent on the pH. Thus, the pH during electrophoresis must be kept as constant as possible. It is attempting the impossible to try to keep the pH constant under *all* conditions; compromise solutions must be found for electrophoresis at different pressures and temperatures. In addition, (de)stabilization of protein structures may result from preferential interaction with buffer components. For example, glycine acts as a stabilizer; so its concentration in buffers must not be greater than 200 mM.

4.2 Preparation of polyacrylamide gels

Polyacrylamide gels with a total acrylamide content of 4–9 per cent (T (in per cent) = [acrylamide + bismethylene acrylamide]/100) and C = 3–4 per cent (C = [bismethylene acrylamide]/T) are prepared in the standard way using ammonium persulfate as the initiator and tetramethylethylenediamine (TEMED) as the catalyst for gel formation (for general preparation of polyacrylamide gels see references 13 and 24; for specific applications to electrophoresis under high pressures see references 15–17, 39, 40, 48, 49). Triton X-100 (0.1 per cent) may be added to the gel-forming solutions to facilitate the removal of the gels from capillary tubes, but osmolytes (such as glycerol or polyols) must be avoided because they may increase the protein stability.[5]

Protocol 1.
Standard discontinuous alkaline buffer system for polyacrylamide-gel electrophoresis of low-isoelectric-point proteins

Caution! For safety procedures see Appendix 2.

Materials

- Running buffer: 8.2 mM Tris/106 mM glycine, pH = 8.3 (20°C)
- Solutions for the separating gel: acrylamide solution (*T* = 24 per cent, *C* = 3.44 per cent), 23.172 g acrylamide, 0.827 g bismethylene–acrylamide, water to 100 mL, separating gel buffer (containing TEMED), 90 μL TEMED, 60 mM Tris/HCl, pH 8.7 (20°C) to 100 mL; 0.225 g ammonium persulfate per 100 mL of water
- Solutions for the stacking gel: (*T* = 7.5 per cent, *C* = 6.6 per cent), acrylamide solution: 7 g acrylamide, 0.5 g bismethylene–acrylamide, water to 100 mL; catalyst-buffer solution (store at 4°C in the dark): TEMED, 126 μL, 2 mg riboflavine, 10 mM Tris/HCl, pH = 5.8 (20°C) to 100 mL

1. Mix 300 μL of 24 per cent acrylamide solution, 300 μL of separating gel buffer, 300 μL of water and 300 μL of ammonium-persulfate solution for 1200 μL of separating gel (*T* = 6 per cent, *C* = 3.44 per cent). (These solutions must be deaerated by vacuum for 10 min before mixing.)
2. Fill each tube using a Hamilton syringe to a height of 60–80 mm and overlayer with 2–3 mm of water saturated *n*-butanol.
3. Allow polymerization to take place at room temperature for 30–60 min.
4. Prepare the stacking-gel-forming solution (*T*= 3.75 per cent, *C* = 6.66 per cent) as follows: mix one part of 7.5 per cent acrylamide solution with one part of catalyst–buffer solution.
5. Pour off the *n*-butanol layer and rinse the tops of the gels with the stacking-gel-forming solution.
6. Pour 10 mm of the stacking-gel-forming solution onto the tops of the separating gels and overlayer with *n*-butanol as above.
7. Illuminate the gels with a standard fluorescent light box (photopolymerization takes place in 15 min at room temperature).
8. Keep the gels at 4°C for 20 h prior to electrophoresis.

4.3 Electrophoresis conditions

Protein samples are diluted in the running buffer, which contains the tracking dye. In order to prevent diffusion of the samples, the samples may be diluted in a small amount of the stacking-gel-forming solution or in Sephadex G-200 (glycerol or sucrose must not be added to the samples to increase the viscosity because they can alter protein stability). Then the sample solutions are carefully loaded on the top of the gel tubes.

Certain pressure-induced unfolding transitions are known to occur slowly. The time required for conformational equilibration after the application of pressure has been found to depend on the protein itself, on the temperature, and on other environmental parameters. Therefore, as a general rule, samples should be allowed to equilibrate under pressure for 12 h prior to electrophoresis.

Electrophoresis is carried out at a constant intensity: 1 mA for Hawley's cell, and 0.1–0.3 mA per tube for microdisc electrophoresis cells.

4.4 Removal of gels from the tubes

Removal of gels from tubes with an internal diameter of 25 mm can be easily performed using extrusion needles. Removal of gels from capillary tubes is more difficult. Low-acrylamide-concentration gels are removed from the capillary tubes by applying water pressure. High-acrylamide-concentration gels ($T \geqslant 9$ per cent) can be removed by crushing the capillary glass tubes. Alternatively, the gels can be expelled by water pressure from siliconized capillary tubes. The tubes can be pretreated using dimethyldichlorosilane (2 per cent) in chloroform or in a water-soluble siliconizing fluid, such as AquaSil (0.1–1 per cent solution in water; Pierce Chemical Company Rockford, IL, USA) or Plot (Serva, Heidelberg, Germany).

4.5 Gel staining and densitometric analysis

Gels are stained for proteins with Coomassie brilliant blue R 250 by the usual procedure. Alternatively, gels can be stained for enzyme activity using specific staining methods.

Densitometric analysis of stained microgels can be easily performed using a scanning densitometer or a video densitometer. This is essential for accurate measurement of migration distances and estimation of the concentration distribution of the different protein states.

5. Study of protein–ligand interactions: Affinity electrophoresis

5.1 Theory and calculation of results

Affinity electrophoresis is an electrophoretic technique for the separation and characterization of proteins and nucleic acids on the basis of their affinity for the ligands that are immobilized within the separating matrix.[25] Proteins which interact with immobilized ligands are retarded in their migration. This decrease in mobility is a function of the concentration of the immobilized ligands, and it reflects the strength of the interaction (Fig. 8.5). Thus, affinity

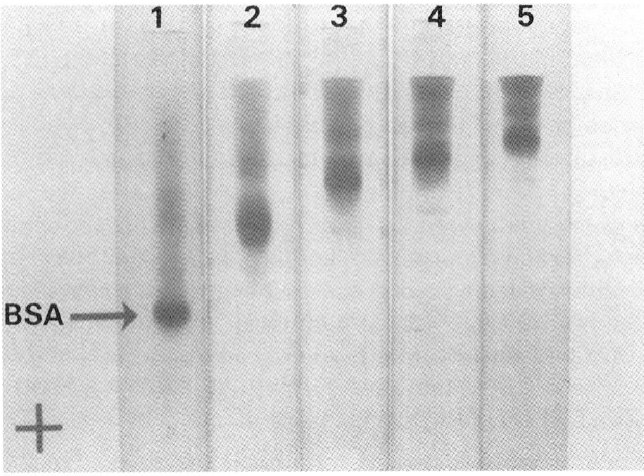

Fig. 8.5 Affinity-electrophoresis patterns for bovine serum albumin (BSA) on gels containing an immobilized derivative of stearic acid (*N*-acrylyl-6-amidocaproic acid *N*-octadecylamide). The electrophoreses were carried out at 25 °C under a pressure of 1.8 kbar at a constant current of 1 mA per tube. The total acrylamide concentration was T = 6 per cent (C = 3.44 per cent). The ligand concentration (in μM) in the gels was: (1) 0, (2) 0.90, (3) 1.81, (4) 3.62, and (5) 5.43 (adapted from work by Masson and Reybaud).[17]

electrophoresis is a convenient method of making direct qualitative and quantitative determinations of the affinity properties of proteins from minute amounts of complex protein mixtures and from highly purified proteins. It allows the estimation of the apparent dissociation constants of proteins with immobilized and nonimmobilized ligands; it is also a useful technique for probing binding sites and for identifying specific surface areas.

The theoretical corpus on affinity electrophoresis was essentially developed by Hořejši and co-workers in the nineteen seventies (reviews of affinity electrophoresis have been given by Hořejši and Tichá[26] and by Takeo[27]).

Let us consider the simplest case of a protein, P, which is assumed to have one binding site interacting with an immobilized ligand, L.

$$\mathrm{P} + \mathrm{L} \rightleftarrows \mathrm{PL}, \qquad K_{\mathrm{D}} = \frac{[\mathrm{P}]\,[\mathrm{L}]}{[\mathrm{PL}]}. \tag{8.5.1}$$

The estimation of the apparent dissociation constant, $K_{D_{app}}$, of the complex PL is based upon the assumptions: (i) the formation of the complex is very fast in comparison with the migration of the protein in the gel; (ii) the mobility of the complex PL is zero; and (iii) the concentration of the ligand is far higher than that of the protein, so that the total concentration of the ligand is approximately equal to that of the unbound ligands. Thus,

$$\frac{R_{m_i}}{R_{m_0}} = \frac{[\mathrm{P}]}{[\mathrm{P}] + [\mathrm{PL}]} \equiv \frac{[\mathrm{P}]_{\mathrm{Free}}}{[\mathrm{P}]_{\mathrm{Total}}}, \tag{8.5.2}$$

where R_{m_0} and R_{m_i} the relative mobility of protein (the migration distance in the separating gel related to the tracking dye band) in the absence and in the presence of immobilized ligand.

Equation (8.5.2) can be transformed into

$$\frac{R_{m_0}}{R_{m_i}} = \left(1 + \frac{[L]}{K_{DL}}\right). \tag{8.5.3}$$

A plot of $(R_{m_i})^{-1}$ against [L] gives a straight line that intercepts the abscissa axis at $-K_{DL}$ (Fig. 8.6).

Incorporation in the gel of a nonimmobilized ligand (I) acting as a competitor of L reverses the decrease in mobility. Assuming that the mobility of the complex PI is equal to that of the free protein, the apparent dissociation constants of the PL and PI complexes, K_{DL} and K_{DI}, respectively, can be determined by plotting $R_{m_i}/(R_{m_0} - R_{m_i})$ against [I] for various values of [L]:

$$\frac{R_{m_i}}{R_{m_0} - R_{m_i}} = \frac{K_{DL}}{[L]}\left(1 + \frac{[I]}{K_{DI}}\right). \tag{8.5.4}$$

The volume change, ΔV, accompanying the formation of protein–ligand complexes can be derived from the pressure dependence of the dissociation constants given by

$$\left(\frac{\partial (K_D)^{-1}}{\partial P}\right)_T = -\frac{\Delta V}{RT}, \tag{8.5.5}$$

where T is the absolute temperature and R is the gas constant (= 82 mL atm K^{-1} mol^{-1}; 1 atm = 1.013 bar).

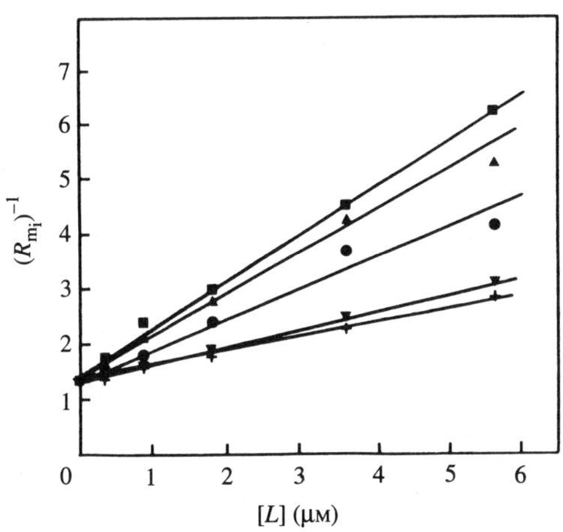

Fig. 8.6 Affinity plots of $(R_{m_i})^{-1}$ against [L] for the interaction between bovine serum albumin and an immobilized derivative of stearic acid (L) at 35 °C under high pressure: (■) atmospheric pressure, (▲) P = 0.35 kbar, (●) P = 0.7 kbar, (▼) P = 1.2 kbar, (+) P = 1.8 kbar). From eqn (8.5.3), the intercepts with the x-axis give the apparent dissociation constant of the complexes. (Reproduced from work by Masson and Reybaud.[17])

5.2 Preparation of an affinity-gel matrix

The most common way of preparing affinity-electrophoresis immobilized ligands is to synthesize a water-soluble macromolecular derivative of the ligand, called a macroligand. This is currently achieved by copolymerization of an acryloyl derivative of the ligand with acrylamide or a related monomer.

The copolymerization is achieved in water or in methanol under nitrogen. This method implies: (i) the correct ligand derivative is made without loss of its binding affinity; and (ii) a high-molecular-weight macroligand is synthesized which is fully entrapped within the matrix of the affinity gels (for reviews see references 26–28; specific examples are given in references 17, 29, and 30).

Another method consists of coupling the ligand to a polymeric backbone (meltable derivatives of agarose,[28] activated polyacrylamide, alterned copolymer of acrylamide and acrylic acid[29]). Block copolymerization of acryloyl ligand with acrylamide and bismethylene–acrylamide is also a possibility; but, due to the incomplete copolymerization of the ligand, any unbound ligand must be removed prior to electrophoresis. This step can be performed by soaking the gel in the electrophoresis buffer, or it can be achieved electrophoretically if the ligands are charged molecules.

5.3 Affinity-electrophoresis conditions

Polyacrylamide-gel-rod affinity electrophoresis is usually carried out in the alkaline discontinuous buffer system given by Ornstein and Davis.[31] Glass tubes 70–75 mm long and with an internal diameter of 2–3 mm are used. Typical separating gels with a total acrylamide concentration of T = 4–9 per cent and with C = 3.44 per cent contain various amounts of macroligand. The tubes are filled with the separating gel-forming mixture to a height of 55–60 mm. Gel polymerization is achieved at room temperature for 1 h (some macroligands have been found to slow the polymerization; for these macroligands the polymerization is continued for 3–5 h). Then, the standard stacking gel-forming solution is added (5–7 mm) and photopolymerized for 30 min. Alternatively, electrophoresis can be carried out in a continuous buffer system. Sample solutions (1–5 μL) are layered onto the top of the gels.

The runs are carried out at a constant current of 1 mA per tube and at a fixed temperature. The runs continue until the tracking dye (bromophenol blue) reaches the bottom of the separating gels. After removal, in order to mark the position of the migration front, the gels are cut with a razor blade at the level of the dye band.

The position of the internal markers (proteins with no affinity for the ligand) are revealed by staining the gels with Coomassie brilliant blue R 250. When serum albumin is used as a standard, its position is revealed during the migration by a blue band because it binds bromophenol blue.

To study the solvent–isotope effect of deuterium on binding and on the

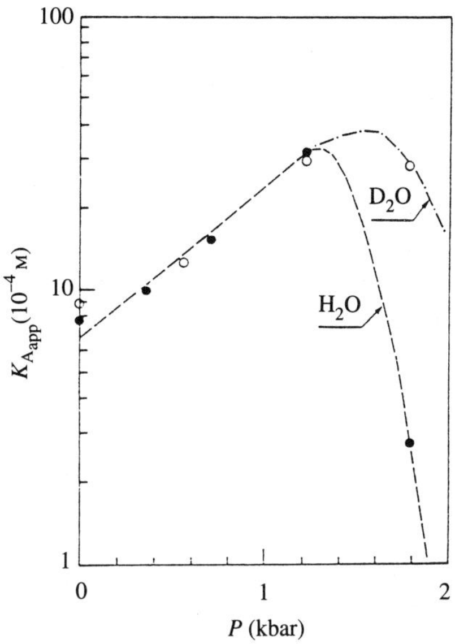

Fig. 8.7 The pressure dependence at 35°C of the apparent association constant ($K_{A_{app}} = (K_D)^{-1}$) of human butyrylcholinesterase with a derivative of *m*-aminophenyltrimethylammonium entrapped in the gel matrix for: (•) electrophoresis carried out in a discontinuous water-buffer system, and (○), electrophoresis carried out in a discontinuous heavy-water-buffer system. (Reproduced from work by Masson and Balny.[30])

protein structure, the buffers and gels can be prepared in heavy water.[30] For example, affinity electrophoresis of human butyrylcholinesterase on phenyltrimethylammonium-containing gels showed that there is no solvent–isotope effect on the affinity. In water, at pressures above 1.2 kbar, the affinity drops rapidly due to disruption of the active-site conformation. However, this pressure-induced loss of affinity is retarded when the electrophoresis is performed in a heavy-water-containing buffer (Fig. 8.7). This protective effect of heavy water may be a consequence of the enhancement of hydrophobic interactions by ^2H_2O.

Protocol 2.
Preparation of affinity gels (T = 6 per cent) with entrapped macroligand

Caution! For safety procedures see Appendix 2.

1. Use the alkaline discontinuous gel–buffer system as described in Protocol 1.
2. The macromolecular carriers of the ligand (macroligands: alternated linear copolymers of acrylamide and acryloyl-substituted ligand, ligand-substituted dextran) are in solution in water.

Protocol 2. *Continued*

3. Control gels (no ligand) are prepared as separating gels in *Protocol 1* (6 %T, 3.44 %C).
4. For 1200 μl of affinity gels, progressively increase the volume of macroligand solution up to 300 μl at the expense of water. Some macroligands may inhibit acrylamide polymerization; if this occurs, increase the concentration of polymerization catalysts.

5.4 Interpretation of data

The total volume change upon ligand binding is the sum of many elementary contributions, including the binding process, protein conformational change, and hydration change. Thus, care must be taken in interpreting ΔV in terms of volume changes associated with the formation or disruption of elementary interactions in the binding sites. However, qualitative information can be extracted from the magnitude and sign of the ΔV values. For example, affinity electrophoresis of human butyrylcholinesterase at high pressures on gels containing the immobilized derivative of phenyltrimethylammonium have shown that the binding strength is enhanced by pressures up to 1.2 kbar ($\Delta V_{app} = -33$ mL mol^{-1} at 35 °C), suggesting that aromatic ring stacking interactions occur in the binding rather than electrostatic interactions, which are weakened by the application of pressure.[30] The direct involvement of aromatic rings in the binding of aromatic ligands has recently been inferred from X-ray crystallography and from the molecular modelling of cholinesterases.[32] Similarly, the hydrophobic-interaction electrophoresis of serum albumin has produced positive volume changes (for example $\Delta V = 11.6$ mL mol^{-1} at 35 °C) as should be expected for the interaction of a long-chain aliphatic ligand into a hydrophobic pocket.[17]

6. Protein-conformation changes

Pressure-induced conformation changes in proteins can be studied using electrophoresis in multiple gel rods with different acrylamide concentrations and using Ferguson plots.

6.1 Theory

In polyacrylamide gel, the relative protein mobility, m, is related to the total acrylamide concentration, T (in per cent), by the empirical Ferguson relationship[33]

$$\log m = \log m_0 - K_R T, \qquad (8.6.1)$$

where m_o is the relative mobility at $T = 0$ per cent, and K_R is the retardation coefficient. The values of $m_0(Y_0)$ are expected to depend on the charge, the size, and the shape of the protein. The values of K_R have been found to depend on the size and protein conformation.[34] Assuming sphericity, there is a linear relationship between K_R and the protein size (the equivalent molecular radius, R) and the volume, V, given by

$$(K_R)^{1/2} = c(R + r) \quad \text{or} \quad K_R = c^2\left(\frac{3V}{4\pi}\right)^{2/3}, \tag{8.6.2}$$

where c is an experimental constant which depends on which electrophoresis system[33] is used (that is, on the degree of cross-linking, the pH, the ionic strength, and the temperature) and on the polyacrylamide fibre radius, r. Since K_R has been found to depend on the pressure, the sign and magnitude of ΔK_R give useful information about pressure-induced conformation transitions.[35] Indeed, since $R \gg r$, it follows from eqn (8.6.2) that

$$\Delta V = \frac{2\pi}{c^{3/2}} \Delta K_R \, (K_R)^{1/2}. \tag{8.6.3}$$

The apparent diffusion coefficients, D_{app} can also be estimated from the variance of the bandwidth, and it can then be plotted as a function of the gel concentration.[36,37] D_{app} is related to K_R and to R by the Stokes–Einstein relationship:

$$D_0 = \frac{k_B T}{6\pi\eta R}, \tag{8.6.4}$$

where D_0 is the value of D extrapolated to zero concentration, k_B is the Boltzmann constant, η is the viscosity of the medium, and T is the absolute temperature. Dynamic light-scattering studies of protein solutions at high pressures have shown that the pressure dependence of the diffusion coefficient reflects a pressure-induced conformational change in proteins.[38]

6.2 Experimental approach

Electrophoresis can be carried out either in microcapillary gels with different acrylamide concentrations[35,39] or on slab gels.[40] Continuous or discontinuous gel–buffer systems can be used. The gel-forming solutions must be carefully degassed under a vacuum prior to the polymerization, and the polymerization must be achieved at a constant temperature. Our capillary-gel-cell runs under pressure are currently performed with a constant intensity of 0.3 mA per gel cell at a temperature of 10–20 °C. Electrophoresis at subzero (Celsius) temperatures, can be performed down to −15 °C (polyacrylamide gels undergo a glass transition at around −15 °C; so electrophoresis at lower temperatures must be achieved in ternary copolymers of an acrylamide derivative[47]), to destabilize the protein structure.[39] After staining and densitometric analysis of the gels, Ferguson plots are constructed. As shown on Fig. 8.8(d), the value of K_R for butyrylcholinesterase decreases with increasing pressure, indicating that a small negative volume change occurs. However,

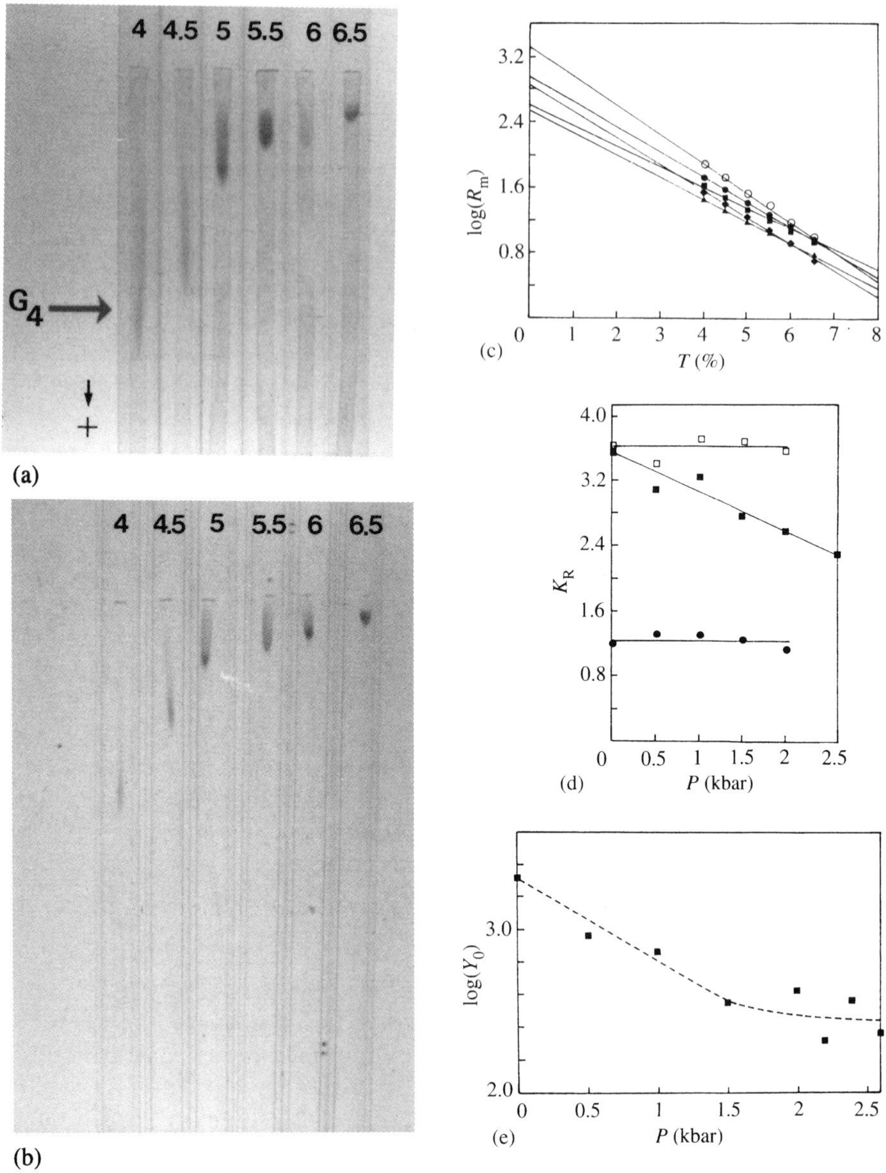

Fig. 8.8 Polyacrylamide-gel electrophoresis of the tetrameric form (G_4) of human butyrylcholinesterase in microcapillary gel tubes with different gel concentrations, *T*. Gel patterns for electrophoresis at 2 kbar and 20°C for: (a) butyrylcholinesterase, and (b) butyrylcholinesterase in the presence of 2 M sorbitol (the gels were stained for enzyme activity according to the procedure of Karnovsky and Roots[41]). The numbers indicate the acrylamide concentrations *T*. (c) Ferguson plots constructed for butyrylcholinesterase at different pressures: (○), atmospheric pressure, (●) 0.5 kbar, (◆) 1 kbar, (▲) 1.5 kbar, and (■) 2 kbar. (d) A secondary plot of K_R against pressure for: (●) bovine serum albumin, (■) butyrylcholinesterase, and (□) butyrylcholinesterase in the presence of 2 M sorbitol. (e) A secondary plot of log Y_0 against pressure for butyrylcholinesterase.

without further information it would be rash to interpret such a ΔV in terms of molecular structure. Indeed, this ΔV could reflect a change in the compressibility of the system, a change in the hydration of the enzyme, and/or a progressive transition towards denatured state of the enzyme. In order to examine the importance of hydration in stabilizing the native conformation of the enzyme, we have performed complementary experiments in the presence of 2 M sorbitol. Under these conditions, K_R became insensitive to pressure, at least up to 2 kbar, suggesting that the sorbitol increased the conformational stability of the enzyme. The effect of sorbitol may be interpreted in terms of preferential hydration, that is, in terms of the preferential exclusion of sorbitol from the enzyme surface: the hydration of sorbitol restrains the binding of water to the enzyme, which in turn counteracts the pressure-denaturing effect. Similar effects of osmolytes have already been observed in the pressure dissociation of oligomeric enzymes.[5,42,43]

Protocol 3. Procedure for electrophoresis in multiple gel rods with different acrylamide concentrations (continuous gel-buffer systems)

Caution! For safety procedures see Appendix 2.

Materials

- Acrylamide solution for separating gel (T = 30 per cent, C = 4.44 per cent): 28.67 g acrylamide, 1.33 g bismethylene acrylamide, and water to 100 mL
- TEMED solution: dilute TEMED ten times in water just before use
- Ammonium persulfate: 1.5 g per 100 mL water
- 5X gel and running buffers (store the solutions at 4°C): 250 mM imidazole/HCl, pH = 6.2 (20°C), 250 mM BisTris/HCl, pH = 7.0 (20°C), and 250 mM Tris/HCl, pH = 8.0 (20°C)

1. Deaerate the solutions by vacuum for 10 min at room temperature.
2. Mix the acrylamide solution, catalysts and 5X buffer according to the following table (volumes are in μL) for 1500 μL of separating gel:

	Final value of T (per cent)					
	4	5	6	7	8	9
5X buffer	300	300	300	300	300	300
T = 30 per cent acrylamide solution	200	250	300	350	400	450
Water	950	900	850	800	750	700
Persulfate solution	45	45	45	45	45	45
TEMED solution	5	5	5	5	5	5

3. Fill each tube as described in *Protocol 1*.
4. Allow polymerization to take place for 15 min at room temperature.

7. Dissociation of oligomeric proteins and protein denaturation

Since hydrophobic interactions are often involved in interactions between subunits, numerous oligomeric proteins dissociate into monomers under high pressure. So high-pressure techniques have long been used to follow dissociation of oligomeric proteins. The equilibrium dissociation constant, K_d, is

$$K_d = n^n \, [P]^{n-1} \, \alpha^n \, / \, (1 - \alpha), \qquad (8.7.1)$$

where α is the degree of dissociation of the oligomer, n is the number of subunits, and [P] is the oligomer concentration. The volume change on dissociation can be determined from the pressure dependence of K_d.

For example, an electrophoretic study by Paladini *et al.*[16] showed that at a pressure of 1.5 kbar the tryptophan synthase β_2-dimer is dissociated into non-denatured β-subunits and that the dimeric structure was fully recovered after the release of the pressure. On the other hand, an electrophoretic study of the multiheme hydroxylamine oxidoreductase under high pressures (< 3 kbar) at different temperatures and in different solvents[44] showed that pressure may induce irreversible structural changes. The changes occurred by a sequential process; their nature and extent were found to depend on the pressure range. No noticeable structural effects were observed below 1.5 kbar. Above 2 kbar, a partial dissociation of the enzyme was observed, and active aggregates and denatured aggregates formed above 3 kbar (Fig. 8.9). Similar results have been obtained with methanol dehydrogenase.[45] Electrophoresis of this enzyme at high pressures showed that pressure induces enzyme conformational heterogeneity: a slowly migrating inactive form appeared as the pressure increased. This form may be considered as a predenatured (partially unfolded) state. However, up to 2 kbar a significant fraction of the enzyme population remained active. Addition of urea to samples, and gels at concentrations up to 6 M, lowered the pressure at which oligomeric proteins dissociate.[40,45]

Hawley's pioneering works[14,15] demonstrated that the denaturation of single-chain proteins can be investigated by electrophoretic separation of the different conformational states. For reversible two-state transitions, if the electrophoretic mobilities of the conformational states N and D are different, and if the run times are lower than the relaxation times $\tau = (k_1 + k_{-1})^{-1}$, the areas under the respective peaks provide a rough measure of the denaturation constant, K_D.[15] Then, if we assume that the denaturation obeys the simple two-state model

$$N \underset{k_{-1}}{\overset{k_1}{\rightleftharpoons}} D, \qquad K_D = \frac{[D]}{[N]}. \qquad (8.7.2)$$

the apparent free-energy change, ΔG_D, and the volume change, ΔV_D, upon unfolding can be determined from

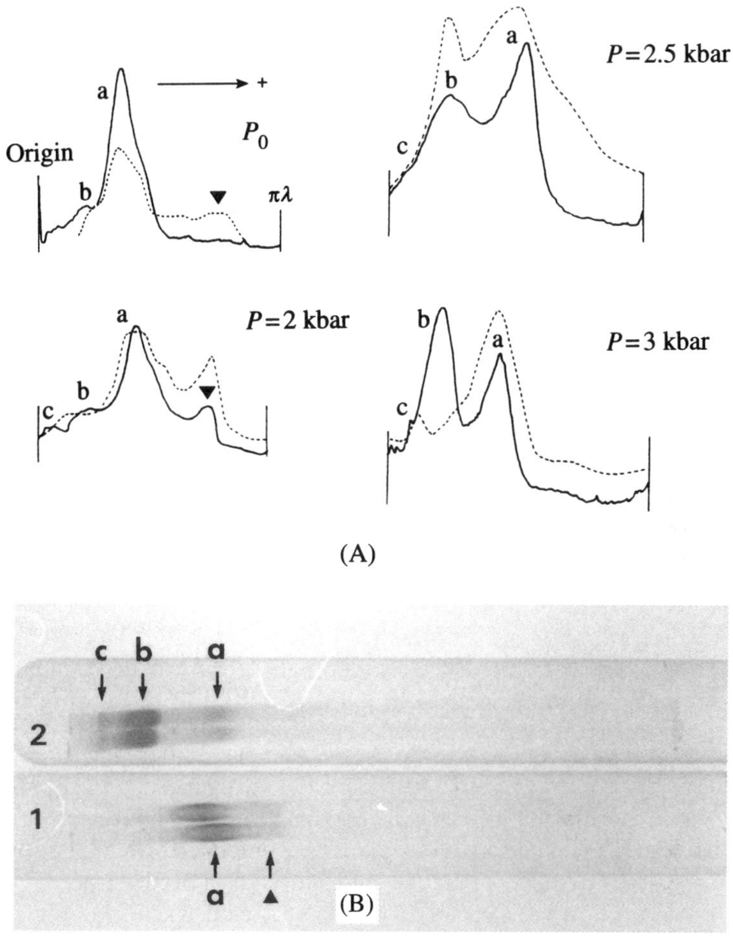

Fig. 8.9 Polyacrylamide-gel electrophoresis of hydroxylamine oxidoreductase in capillary gels (T = 4.5 per cent). (A) Densitometry patterns of the gels after electrophoresis at various pressures, P (in kbar): (——) protein-stained gels, (a) native enzyme (with a molecular mass of 150 kDa) and (b and c) aggregated forms of dissociated subunits; and (- - - -) Gels stained for enzyme activity; (▼) the position of a fast-migrating active form (dissociated subunit). (B) Gel patterns at 3 kbar: (1) gels stained for enzyme activity, and (2) gels stained for protein with Coomassie brilliant blue. (Adapted from work by Masson *et al.*[44])

$$\Delta G_D = -RT \ln K_D, \tag{8.7.3}$$

$$\Delta V_D = -RT\left(\frac{\partial(\ln K_D)}{\partial P}\right)_T. \tag{8.7.4}$$

More complex denaturation processes involving stable intermediates lead a to plurality of protein conformations, which can be probed by electrophoresis under pressure.[39]

8. Conclusions and future directions

So far there have been relatively few applications of electrophoresis at elevated hydrostatic pressures. However, this method has several advantages over other methods such as optical methods: it is a simple and direct means of studying dissociation and denaturation processes, and of describing the thermodynamics of protein–ligand interactions. These qualitative and quantitative studies can be performed using small amounts of pure proteins or complex protein mixtures. In addition, this technique permits separation and subsequent isolation of the different protein conformational states or subunits.

Future developments of high-pressure electrophoretic techniques should include isoelectrofocusing on immobilized pH gradients, and also capillary electrophoresis. Preliminary results have already demonstrated the feasibility of high-pressure and supercritical-fluid capillary electrophoresis.[46] Although this method was first used for the separation of small charged molecules, it could also be applied to biopolymers to study the effects of pressure and temperature on their zeta potentials and on their electrical double layers. A combination of slab-gel high-pressure electrophoresis with transverse-urea-gradient gel electrophoresis,[40] subzero (Celsius temperature) electrophoresis, and electrophoresis on new matrices compatible with hydroorganic media should produce a tenfold increase in the capabilities of this tool in the very near future.

Acknowledgement

This work was supported in part by La Direction des Recherches, Etudes et Techniques, grant number J. 91/1044 and 94/005.

References

1. Morild, E. *Adv. Prot. Chem.* **1981**, *34*, 93–166.
2. Heremans, K. *Ann. Rev. Biophys. Bioeng.* **1982**, *11*, 1–21.
3. Weber, G.; Drickamer, H. G. *Quart. Rev. Biophys.* **1983**, *16*, 89–112.
4. Gross, M., Jaenicke, R. *Eur. J. Biochem.* **1994**, *221*, 617–630.
5. Silva, J. L.; Weber, G. *Ann. Rev. Phys. Chem.* **1993**, *44*, 89–113.
6. Masson, P. In *High Pressure and Biotechnology*; Balny, C.; Hayashi, R.; Heremans, K.; Masson, P., ed.; Libbey Eurotext: Paris, **1992**; pp. 89–99.
7. Suzuki, K. *Meth. Enzymol.* C, **1972**, *26*, 424–452.
8. Hawley, S. A. *Meth. Enzymol.* **1978**, *49*, 14–24.
9. Douzou, P. *Quart. Rev. Biophys.* **1979**, *12*, 521–569.
10. Balny, C.; Masson, P.; Travers, F. *High Press. Res.* **1989**, *2*, 1–28.
11. Kunugi, S. *Ann. NY Acad. Sci.* **1992**, *672*, 293–304.
12. Kunugi, S. *Prog. Polym. Sci.* **1993**, *18*, 805–538.
13. Hames, B. D.; Rickwood, D. *Gel Electrophoresis of Proteins* 2nd edn; IRL Press: Oxford, **1990**.

14. Hawley, S. A. *Biochim. Biophys. Acta* **1973**, *317*, 236–239.
15. Hawley, S. A.; Mitchell, R. M. *Biochemistry* **1975**, *14*, 3257–3264.
16. Paladini, A. A.; Silva, J. L.; Weber, G. *Anal. Biochem.* **1987**, *161*, 358–364.
17. Masson, P.; Reybaud, J. *Electrophoresis* **1988**, *9*, 157–161.
18. Brandts, J. F.; Oliveira, R. J.; Westort, C. *Biochemistry* **1970**, *9*, 1038–1047.
19. Zipp, A.; Kauzmann, W. *Biochemistry* **1973**, *12*, 4217–4228.
20. Van Eldik, R.; Asano, T.; Le Noble, W. J. *Chem. Rev.* **1989**, *89*, 549–688.
21. Neuman, R. C.; Kauzmann, W.; Zipp, A. *J. Phys. Chem.* **1973**, *77*, 2687–2691.
22. Tsuda, M.; Shirotani, I.; Minomura, S.; Terayama, Y. *Bull. Chem. Soc. Jpn.* **1976**, *49*, 2952–2955.
23. Distèche, A.; Distèche, S. *J. Electrochem. Soc.* **1967**, *114*, 330–340.
24. Andrews, A. T. *Electrophoresis*; Clarendon Press: Oxford, **1981**; p. 17.
25. Boschetti, E. *J. Biochem. Biophys. Meth.* **1989**, *19*, 21–36.
26. Hořejši, V.; Tichá, M. *J. Chromatogr.* **1986**, *376*, 49–67.
27. Takeo, K. *Adv. Electroph*, **1987**, *1*, 229–279.
28. Masson, P. *Cell. Mol. Neurobiol.* **1991**, *11*, 173–189.
29. Masson, P.; Marnot, B.; Lombard, J. Y.; Morelis, P. *Biochimie* **1984**, *66*, 235–249.
30. Masson, P.; Balny. C. *Biochim. Biophys. Acta* **1990**, *1041*, 223–231.
31. Ornstein, L.; Davis B. J. *Ann. NY Acad. Sci.* **1964**, *121*, 321–427.
32. Harel, M.; Schalk, I.; Ehret-Sabatier, L.; Bouet, F.; Goeldner, M.; Hirth, C.; Axelsen, P. H.; Silman, I.; Sussman, J. L. *Proc. Natl. Acad. Sci. USA* **1993**, *90*, 9031–9035.
33. Rodbard, D.; Chrambach, A. *Anal. Biochem.* **1970**, *40*, 95–134.
34. Goldenberg, D. P.; Creighton, T. E. *Anal. Biochem.* **1984**, *138*, 1–18.
35. Masson, P., Cléry, C., Ghazarossian, F. *23rd Steenbock Symposium on High Pressure effects in Molecular Biophysics and Enzymology*; University of Wisconsin, Madison, 15–19 May, **1994**; p. 16.
36. Lunney, J.; Chrambach, A.; Rodbard, D. *Anal. Biochem.* **1971**, *40*, 158–173.
37. Chen, B.; Chrambach, A.; Rodbard, D. *Anal. Biochem.* **1979**, *96*, 120–130.
38. Nyström, B.; Roots, J. *J. Phys. Chem.* **1983**, *78*, 2833–2837.
39. Erijman, L.; Paladini, A. A.; Lorimer, G. H.; Weber, G. *J. Biol. Chem.* **1993**, *268*, 25914–15919.
40. Paladini, A. A.; Weber, G.; Erijman, L. *Anal. Biochem.* **1994**, *218*, 364–369.
41. Karnovsky, M. J.; Roots, L. *J. Histochem. Cytochem.* **1964**, *12*, 219–221.
42. Bonafe, C. F. S.; Villas-Boas, M.; Suarez, M. C. *J. Biol. Chem.* **1991**, *20*, 13210–13216.
43. Silva, J. L. *Brazil. J. Med. Biol. Res.* **1993**, *26*, 383–393.
44. Masson, P.; Arciero, D. M.; Hooper, A. B.; Balny, C. *Electrophoresis* **1990**, *11*, 128–133.
45. Heiber-Langer, I.; Cléry, C.; Frank, I.; Masson, P.; Balny, C. *Eur. Biophys. J.* **1992**, *21*, 241–250.
46. Yonker, C.; Smith, R. D. *J. Chromatogr.* **1990**, *517*, 573–581.
47. Curtil, C.; Channac, L.; Ebel, C.; Masson, P. *Biochim. Biophys. Acta* **1994**, *1208*, 1–7.
48. Cléry, C.; Renoult, F.; Masson, P. *FEBS Lett.* **1995**, *370*, 212–214.
49. Erijman, L.; Clegg, R. M. *Rev. Sci. Instrum.* **1996**, *67*, 813–817.

A1

List of suppliers of high-pressure equipment

Company	Product
Aad Hochdrucktechnik *Germany*: Postfach 1313, D-61283 Bad Homburg. Tel. 49-617290320	Autoclaves, pumps, and jet-cutting equipment
ABB Metallurgy AB *Sweden*: S-721 66 Västerås, Tel. 46-21327241	Isostatic pressure vessels, food processing equipment, and steels
ABB Industry Ltd, *UK*: Wilmslow Rd, Didsbury, Manchester M20 8RB. Tel. 44-161445 5555	
Abra Fluid A.G. *Switzerland*: Boschachstrasse 117, CH 9443 Widnau. Tel. 41-71724351	Isostatic presses to 5 kbar and 2000 °C
ACB Energie (Basset-Bretagne) *France*: Rue Léon du Rocher, 2X44040 Nantes, Cedex 01. Tel. 33-40411404. See also **GEC Alsthom ACB**	Hot and cold isostatic pressing equipment to 2 kbar and 2000 °C
American Instrument Co. (AMINCO) *USA*: Silver Spring, MD 20910.	Custom-built high-pressure reaction vessels (to about 1600 bar), optical cells (5 kbar), valves, tubing (6 kbar) and fittings, and liquid and gas pumps (to 4 kbar)
Andreas Hofer *Germany*: 4330 Mülheim an der Ruhr. Tel. 49-208460001	Pressure vessels for laboratory synthesis and ancilliary equipment

Aqua Hydraulics Ltd
UK: Star Trading Estate, Partridge Green, Horsham, West Sussex, RH13 8RA. Tel. 44-1403710835.

Autoclave Engineers Inc.
USA: PO Box 4007, Erie, PA 16512.

High-pressure vessels, autoclaves, valves, gauges, tubing, and a wide range of fittings

BASF Aktiengesellschaft
Germany: ZIF/AH HC, 6700 Ludwigshafen. Tel. 49-621603584

Betsa France
France: 2 rue de l'Industrie, F-77173 Chevry-Cossigny. Tel. 33-164053503

Membrane DACs for multiple purposes and wide ranges of temperature

Budenberg Gauge Co. Ltd
UK: PO Box 5 Altrincham, WA14 4ER. Tel. 44-1613535911/3

Bourdon gauges to 1600 bar

Burton Corblin
France: High Pressure Division, 78–80 Bd. Saint Marcel, 75005 Paris. Tel 33-15701111

Pressure vessels

Chas. S. Madan Co. Ltd.
UK: Vortex Works, Atlantic St, Broadheath, Altrincham, WA14 5DA. Tel. 44-1619282702

Diacell Products
Dr. D. M. Adams, 54 Ash Tree Rd, Leicester, LE2 5TD, UK. Tel. 44-1533712810

Manufacture DACs for most applications

Dilor GmbH
Germany: Wiesenstrasse 4, D-64625, Bensheim. Tel. 49-62514042

Membrane DACs for wide *p-T* regions: 10 K <T> 650 K

DuPont
USA: Research and Development, Experimental Station, PO Box 80356, Wilmington. DE. Tel 1-302695 9752 (Stanley W. Tozer)

Designs for miniature DACs for high magnetic fields and low temperatures

F.Uhde GmbH. Werk Hagen *Germany*: Buschmuelenstrasse 20, Postfach 4260, D-5800 Hagen. Tel. 49-23316920	Pumps to 8 kbar
GEC Alsthom ACB Boul. Prairie au Duc, 44945 Nantes, Cedex 9. Tel. 33-40411404	Laboratory and large-scale high pressure for food processing: e.g. 3 L, 7 kbar, −20 to 80 °C
George Meller Ltd *UK*: Orion Park, Northfield Avenue, Ealing, London, W13 9SJ. Tel. 44-1815792111 Also agent for **Autoclave Engineers** Units 12/13, Murcar Commercial Park, Denmore Rd, Bridge of Don, Aberdeen, Scotland, AB23 8JW. Tel. 44-1224822990	High-pressure fittings, pumps, gauges, valves, and tubing to 10 kbar
Ernst Haage *Germany*: Hauskampstrasse 58, Postfach 010740, 4330 Mülheim (Ruhr). 11. Tel. 49-208402051	Autoclaves and pressure-vessel systems, valves, tubing and fittings to 1 kbar
Harwood Engineering Co. Inc. *USA*: South St, Walpole, MA 1 2081-2799	Pressure vessels
Haskel Energy Systems Ltd *UK*: North Hylton Rd, Sunderland, SR4 3JD. Tel. 44-1915491212 *USA*: Haskel Inc, Burbank, CA 91502. Tel. 1-818843400	High-pressure pumps and systems for liquids and gases, and valves to 10 kbar
High Pressure Equipment Co *USA*: 1222 Linden ve, Erie, PA 16505. Tel. 1-8148382028	
Hi-Tech Scientific *UK*: Brunel Rd, Salisbury, SP2 7PU. Tel. 44-172223643	High-pressure stopped-flow spectrometer and spectrophotometer units to 2 kbar and 10 ms
Hydratron Ltd *UK*: Unit 1F Stag Industrial Estate, Atlantic St, Broadheath WA14 5DD. *USA*: Fluitron Inc, 30 Industrial Drive, Ivyland, Pa. 18974	

Hydro-Pac Inc *USA*: 7470 Market Rd, PO Box 921, Fairview, PA 16415. Tel. 1-8144741511	Gas compressors, liquid pumps to 13 kbar, and water-jet cutting equipment
Interros *Russia*: Moscow. Tel./Fax 7-952462103	Membrane DACs for wide p-T regions: 10 K $\leqslant$ T $\geqslant$ 650 K, p to 200 GPa
Intersonde Ltd. *UK*: Queen Mary's Avenue, Watford, Herts, WD1 7JR. Tel. 44-1923227761	High-pressure transducers (strain gauges) and instrumentation to 2 kbar
ISS *USA*: 2604 North Mattis Avenue, Campaign, Illinois 61821. Tel. 1-2173597879	High-pressure cells for absorption, fluorescence, and Raman spectroscopy to 3 kbar
Klaus Dunze Hochdrucktechnik *Germany*: Postfach 1313, Kaiser-Friedrich-Promenade 7, D-6380 Bad Homberg. Tel. 49-6172 2511	Pumps to 14 kbar, valves, gauges and fittings
Kulite Sensors Ltd *UK*: Stroudley Road, Kingsland Business Park, Basingstoke, Hants, RG24 8UK. Tel. 44-1256461646	
L. B. Couplings *UK*: Grifin Lane, Aylesbury, Bucks, HP19 3BP.	UK agents for SITEC, TOP INDUSTRIE
Lot (Oriel) *Germany*: Im Tiefen See 58, D-64293 Darmstadt. Tel. 49-24996600 (Jutta Riekenbrauck), 49-6151880613 (Manfred Berger)	DACs for optical spectroscopy and X-ray diffraction
Metallwerke Plansee GmbH *Austria*: A 6600 Reutte.	
National Forge Europe *Belgium*: Pressure Systems Division, Industriepark, Noord 7, B9100-Sint Niklaas. Tel. 32-37773511 *USA*: National Forge Co., PO Box 565, 155 River Street, Andover, MA 01810. Tel. 1-5084701620	Isostatic pressing systems, food processing equipment, and laboratory units

Nova Werke *Switzerland*: Vogelsangstrasse 24, CH-Effretikon. Tel. 41-52511111 *Germany*: Nova Werke Deutschland GmbH Sölderstrasses 152, D-44289 Dortmund. Tel. 49-23140427	A wide range of laboratory high-pressure equipment, such as vessels, pumps, optical cells, valves, and tubing, mostly to 4 kbar
Orion Diamond Manufacturing Co. *USA*: 576 Fifth Avenue, Suite 203, New York, NY 10036, SA *Israel*: PO Box 3053 Ramot Gan.	Diamonds for DACs
Pacson High Pressure Technology *UK*: Panmure Industrial Estate, Carnoustie, DD7 7NP, Scotland. Tel. 44-24154762	Liquid and gas pumps, and valves
PSIKA Pressure Systems Ltd. *UK*: K. Ashcroft, 1A Jackson's Buildings, Victoria Street, Glossop, Derbyshire, SK13 8AQ. Tel. 44-145743358	Piston–cylinder apparatus for the preparative scale to 20 kbar, and intensifiers, manganin gauges, valves, and fittings to 13 kbar
Schmidt Kranz and Co. GmbH *Germany*: Postfach 110348, D-5620 Verbert-Langenberg 11. Tel. 49-20523011	Hydraulic pumps and gas boosters to 5.5 kbar
Simac Ltd *UK*: Forum Drive, Rugby, Warwickshire, CV21 1PD. Tel. 44-178860766	
Sitec *Switzerland*: Sieber Engineering AG, Aschbach 621, CH-8124, Maur/Zurich	Autoclaves, high-pressure optical cells, valves, pumps, gauges, fittings, and intensifiers to 7 kbar
Staffordshire Hydraulics Services Ltd. *UK*: PO Box 453, Stoke on Trent. Tel. 44-1782771225	
Stanstead Fluid Power Ltd *UK*: 70 Bentfield Road, Stanstead, Essex, CM24 8HT.	Isostatic pressing systems, food processing equipment, and laboratory-scale presses
Swagelok Companies More than 80 outlets in the USA and worldwide	A wide variety of relatively low-pressure fittings of all types

USA: Norwalk Valve and Fitting Co., PO Box 393, 120 Long Hill Cross Road, Shelton, CT 06484. Tel. 1-2039296341
UK: South London Valve and Fitting Co., Burgess Hill, West Sussex, RH15 9TL. Tel. 44-1444248048.

Top Industrie
France: ZAC 'Le Plateau de Bière', Rue Ampère, BP 95, 77193 Dammarie-les-Lys, Cedex. Tel. 33-164396969
UK: see **L. B. Couplings, Ltd.**

A wide variety of high-pressure fittings, pumps, valves, and gauges systems, and special cells to 6 kbar and 400 °C

Unipress
Poland: High Pressure Research Centre, Sokotowska 29/37, 01–142 Warsaw. Telex 817618 zwc pl

Laboratory-scale pressure vessels, optical cells, pressure gauges to 2.5 GPa, X-ray cells, high- and low-temperature cells, and high-pressure microwave and NMR cells

A2

Safety procedures for high-pressure systems

NEIL S. ISAACS AND ALAN SHARP

There is inevitably a hazard risk when using high pressures of which the operator must be aware. One main hazard arises from the potential energy stored in the system by virtue of the compression of fluid contents and the possibility of its being converted to kinetic energy and transferred to components of the system which could cause injury to nearby personnel. The other matter of safety is awareness of the limitations of materials which, if exceeded, could result in damage to or failure of the often expensive equipment.

Clearly the amount of energy stored is a crucial factor. For a fluid such as water or a hydraulic oil, pressurization to 3–4 kbar causes compression of about 10 per cent and 25 per cent by about 10 kbar. The corresponding stored energies are of the order 0.5 kJ mol^{-1} $kbar^{-1}$. By comparison, a gas such as helium would store around 2000 kJ mol^{-1} $kbar^{-1}$ and so it follows that leakage of a very small proportion of pressurized liquid through a fracture relieves the pressure and removes the hazard whereas for a gas, almost all must be vented in the event of failure before the system is safe. Greater precautions against failure must be taken when working with high pressure gases than with liquids and with large volume systems than with small.

Meeting the exacting requirements of safety means using good engineering and operating techniques which will lead to reliable systems. The High pressure Safety Code of the British High Pressure Technology Association[1] is an excellent guide to this. Safe working requires forward planning since post provision for safety and failure of equipment are both expensive.

It is essential that the main groups involved in a high pressure system, designers, manufacturers, and users, should appreciate and identify the aspects of their particular contribution to the situation; no person can be isolated from the lack of care for others.

It is difficult to be general in prescribing safety practises since they will depend upon the specific application. Diamond anvil cells pose hardly any hazard since the volumes of pressurized materials are so small and, moreover, they are usually solid. Small-scale equipment for 1–2 kbar use may be made in a good workshop, but inexperienced operators should take the advice of laboratories who have built and run such equipment as to materials, design, and practice.

Large-scale equipment is necessarily manufactured by a specialist engineering firm whose responsibility it is to ensure the safety of those using their products. However, the installation may involve matters which are in the control of the user, as discussed below. When accepting such equipment, it is usual, mandatory in some cases, for the designer to provide the user and any inspection authority with all the relevant information for the safe operation of the system and to determine its physical condition at regular intervals. Systems should be designed such that inspection may be carried out in accordance with the requirements. Acceptance of the equipment should be made only after being satisfied that all design criteria are met.

The manufacturer of a pressure system should be made aware of the user's requirements; the maximum pressure and temperature at which it is to be used, the materials with which the various parts will come into contact, and, importantly, the lifetime required and the number of pressurization cycles to be made. On the last matter, the number of cycles which a vessel will withstand depends very much on how close to the pressure maximum it will be used. If the vessel is subjected to pressures close to its rated strength, the lifetime will be uncertain and dependent on how good the actual sample of steel is which has been chosen. If a large number of cycles are required, the specifications should be such that the vessel is run at pressures considerably below the maximum rating.

Close cooperation between designer, manufacturer, and user is essential. One should none the less accept that even with the greatest care, flaws in steels used may not be revealed until failure under pressure occurs.

Installation

Large-scale equipment will need to be housed in a suitable location, but properly designed and built equipment such as 10 kbar vessels for synthesis or for food processing of several litres capacity are commonly sited in ordinary laboratories. These vessels are usually double-walled for strength so that even if failure of the inner vessel occurs, no flying fragments of metal are generated. It is our experience that a sudden rupture of an inner vessel was accompanied by nothing dramatic beyond immediate loss of pressure. Duplex tubing used to transmit fluid from intensifiers at up to 15 kbar has similar inbuilt safety. On the other hand, the installation may require flexible pressure hose to transmit hydraulic pressure to an intensifier at pressure, for example up to 500 bar. This should be kept as short as possible, should preferably be wire reinforced, and inspected regularly for aging. A screen of 1/8″ mild steel or 1/2″ polycarbonate may be placed between the hose and the operator so that, if a pinhole were to develop, a jet of high pressure oil would not cause injury. Valves should be sited with stems vertical so that, in the event of a valve-stem becoming detached it will not be projected towards an operator. When pumping through high pressure valves, it is not necessary to loosen beyond about one turn; leave the maximum amount of thread retaining the valve stem in place. Other safety considerations are;

(a) Allow adequate access to the area for movement of operators and equipment for tools and maintenance.

(b) High pressure equipment may be extremely heavy. Ensure that the floor loading is not exceeded. If necessary spread the load by placing the equipment on a steel sheet.

(c) Ensure that adequate ventilation exists in the event of runaway reaction producing noxious fumes or failure causing leakage of toxic materials.

(d) It makes good sense to place shields of 1/4″ steel sheet around vessels going to 10 kbar and beyond. Pistons which may be transmitting 10 tonne force are made of brittle tool steel and, should a fracture occur, could eject fragments at high velocity.

(e) Operation of equipment containing large volumes of high pressure gases should be performed beyond blast-proof walls, the specifications of which should be laid down by specialists.

(f) Windows of sapphire or other transparent material are likely to be weaker than the steel vessels in which they are housed. One should never, therefore, look directly into a window which is under pressure. The use of a mirror is also not without hazard since rupture of the window could create fragments which reflect off the mirror to the observer. The best arrangement by far, and one which has the facility both for observation and for storage of images, is the use of a video camera. Failing this, and for relatively low pressures (1–2 kbar), a 1/4″ sheet of polycarbonate between window and observer should give adequate protection. Polymethacrylate is not recommended since it is brittle.

(g) The operator should be aware of the properties of the materials being subjected to high pressure. Liquids may solidify under pressure and cause binding of moving parts. Undue force may then damage the equipment. Tables of freezing pressures should be consulted.[2] Chemical hazards may be experienced such as reactions which undergo pressure-induced explosion leading to volumes of high pressure gas.[2]

Newcomers to the high pressure field should make early contact with experienced operators since experience is an important factor in the safe and satisfactory operation of all manner of such equipment.[3]

References

1. 'High Pressure Safety Code'. High Pressure Technology Association. Dr G. Saville, Dept. of Chemical Engineering, Imperial College of Science and Technology, London SW7 2BY, UK.
2. Isaacs, N. S. *Liquid Phase High Pressure Chemistry*, Wiley, Chichester, **1981**.
3. Munday, G. *The Basis of Design for Safe High Pressure Equipment*, H. L. D. Pugh, Ed., Institute of Mechanical Engineers, **1971**.

Index

Index